MASTER HANDBOOK OF
ELECTRONIC
TABLES & FORMULAS
4TH EDITION

BY MARTIN CLIFFORD

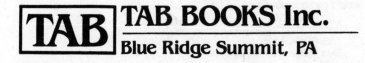

TAB BOOKS Inc.
Blue Ridge Summit, PA

FIRST EDITION

SIXTH PRINTING

Printed in the United States of America

Reproduction or publication of the content in any manner, without express permission of the publisher, is prohibited. The publisher takes no responsibility for the use of any of the materials or methods described in this book, or for the products thereof.

Library of Congress Cataloging in Publication Data

Clifford, Martin, 1910-
    Master handbook of electronic tables & formulas.

    Includes index.
    1. Electronics—Handbooks, manuals, etc.
2. Electronics—Tables. I. Title.
TK7825.C56 984   621.381′0212   84-8529
ISBN 0-8306-0625-4
ISBN 0-8306-1625-X (pbk.)

TAB BOOKS Inc. offers software for sale. For information and a catalog, please contact TAB Software Department, Blue Ridge Summit, PA 17294-0850.

Questions regarding the content of this book should be addressed to:

    Reader Inquiry Branch
    TAB BOOKS Inc.
    Blue Ridge Summit, PA 17294-0214

Contents

Other TAB Books by the Author

Introduction

Problems in electronics can be solved in a number of ways. Possibly the most common method is to use a formula and to plug in or substitute numerical values. This technique calls for some arithmetic dexterity, and, quite often, a good working knowledge of algebraic and trigonometric functions, and sometimes a bit of calculus. Aside from the work involved, the use of a formula has the disadvantage in that it supplies a single solution.

This disadvantage is overcome by the use of nomographs. A nomograph not only gives the desired solution to a problem, but also supplies the user with a fairly good number of alternate possibilities. Thus, nomographs are well-suited for the designer who is not only interested in the solution to a problem, but is confronted with the need for specifying practical and easily-obtained components. The ideal solution to a problem is not always the most practical one.

This book represents still another way of solving electronics problems. It consists of electronics data arranged in tabular form. In a few instances some arithmetic may be needed, but for the most part, if the elements of a problem are known, the answer is supplied immediately by a table.

The tables in this book are based on formulas commonly used in electronics. Many of the tables supply answers with a much higher order of accuracy than is generally needed in the solution of problems in electronics. Also, as in the case of nomographs, the tables supply a number of possible solutions, allowing the user a choice of practical component values that may be needed for a circuit.

There is a limit to the number of electronics tables that can be prepared. Tables are easily developed when only two variables are involved. Thus, it is simple enough to set up a table for capacitors in series or for resistance vs. conductance. For involved formulas it is better to use the formulas directly, or to use nomographs if these are available.

What is the purpose of having a book of tables? Its function is to save time and work. Actually, there is no single best method for problem solving. Those who must solve problems in electronics as part of their educational training or work will find it helpful to be able to have a variety of techniques at their command — solving problems by formulas, solving problems by nomographs, by using a calculator and, with the help of this book, solving problems by tables.

I wish to thank the Digital Equipment Corporation, Maynard, Massachusetts for their kind permission to use their "Powers of Two" table. My thanks also go to my friend Marcus G. Scroggie and to Iliffe Books Ltd. for granting permission to use the "Decibel" table that originally appeared in his *Radio Laboratory Handbook*, and to Dr. Bernhard Fischer and the Macmillan Company for the use of their "Vector Conversion" table.

1

Resistance and Conductance

EQUIVALENT RESISTANCE—RESISTORS IN PARALLEL

Whenever two resistors are connected in parallel, the total value of the shunt combination must always be less than that of the value of the smaller unit. From a practical viewpoint, if one of the two parallel resistors has a value that is 10 or more times that of the other resistor, the equivalent value can be taken as being approximately equal to that of the smaller resistor. Use Table 1-1 to find the equivalent resistance of two resistors in parallel.

The tables shown on the following pages can also be used to find the equivalent resistance of three or more resistors in parallel (Fig. 1-1) if the problem is handled on a step-by-step basis. First, take any two of the resistors, and, using the Tables, find the equivalent resistance. Consider this equivalent resistance just as though it were a physical unit and combine its value with the remaining resistor, again using the Tables.

Sometimes a design problem involving resistors will yield a value that is not practical—not practical in the sense that a resistor having such a value will be unavailable. In this instance the tables can again be used to advantage. Simply locate the nearest value in the Tables and then move left to get the value of R2 and upward to get the value of R1. R1 and R2 will be standard, available resistors, which can be connected in parallel to supply the required resistance.

1

☐ **Example:**
What is the equivalent resistance of two resistors in parallel, one having a value of 5.6 ohms and the other a value of 9.1 ohms?

Locate 5.6 at the left, in the column marked R2 in Table 1-1. Move across and locate the column headed by 9.1. The equivalent resistance is shown to be 3.467 ohms.

☐ **Example:**
What two resistors in parallel will give an equivalent value of 3 ohms?

The Table shows possible combinations. You could use 3.3 and 33 ohms, or 3.6 and 18 ohms, or 3.9 and 13 ohms.

☐ **Example:**
What is the equivalent shunt resistance of a 68-ohm resistor and a 27-ohm resistor in parallel?

Table 1-1. Equivalent Resistance of Two Resistors in Parallel.

R1

R2	2.7	3.0	3.3	3.6	3.9	4.3	4.7	5.1
2.7	1.350	1.421	1.485	1.543	1.595	1.659	1.715	1.766
3.0	1.421	1.500	1.571	1.636	1.695	1.767	1.831	1.889
3.3	1.485	1.571	1.650	1.722	1.788	1.867	1.939	2.004
3.6	1.543	1.636	1.722	1.800	1.872	1.959	2.039	2.110
3.9	1.595	1.695	1.788	1.872	1.950	2.045	2.131	2.210
4.3	1.659	1.767	1.867	1.959	2.045	2.150	2.246	2.333
4.7	1.715	1.831	1.939	2.039	2.131	2.246	2.350	2.446
5.1	1.766	1.889	2.004	2.110	2.210	2.333	2.446	2.550
5.6	1.822	1.953	2.076	2.191	2.299	2.432	2.555	2.669
6.2	1.881	2.022	2.154	2.278	2.394	2.539	2.673	2.798
6.8	1.933	2.082	2.222	2.354	2.479	2.634	2.779	2.914
7.5	1.986	2.143	2.292	2.432	2.566	2.733	2.889	3.036
8.2	2.031	2.196	2.353	2.502	2.643	2.821	2.988	3.144
9.1	2.082	2.256	2.422	2.580	2.730	2.920	3.099	3.268
10	2.126	2.308	2.481	2.647	2.806	3.007	3.197	3.377
11	2.168	2.351	2.538	2.712	2.879	3.092	3.293	3.484
12	2.204	2.400	2.588	2.769	2.943	3.166	3.377	3.579
13	2.236	2.438	2.632	2.819	3.000	3.231	3.452	3.663
15	2.288	2.500	2.705	2.903	3.095	3.342	3.579	3.806
16	2.310	2.526	2.736	2.939	3.136	3.389	3.633	3.867

				R1				
R2	2.7	3.0	3.3	3.6	3.9	4.3	4.7	5.1
18	2.348	2.571	2.789	3.000	3.205	3.471	3.727	3.974
20	2.379	2.609	2.833	3.051	3.264	3.539	3.806	4.064
22	2.405	2.640	2.870	3.094	3.313	3.597	3.873	4.140
24	2.427	2.667	2.901	3.130	3.355	3.647	3.930	4.206
27	2.455	2.700	2.941	3.176	3.408	3.709	4.003	4.290
30	2.477	2.737	2.973	3.214	3.451	3.761	4.063	4.359
33	2.496	2.750	3.000	3.246	3.488	3.804	4.114	4.417
36	2.512	2.769	3.023	3.273	3.519	3.841	4.157	4.467
39	2.525	2.786	3.043	3.296	3.545	3.873	4.195	4.510
43	2.540	2.804	3.065	3.322	3.576	3.909	4.237	4.559
47	2.553	2.820	3.083	3.344	3.601	3.939	4.273	4.601
51	2.564	2.833	3.099	3.363	3.623	3.966	4.303	4.636
56	2.576	2.847	3.116	3.383	3.646	3.993	4.336	4.674
62	2.587	2.862	3.133	3.402	3.669	4.021	4.369	4.712
68	2.597	2.873	3.147	3.419	3.688	4.044	4.396	4.744
75	2.606	2.885	3.161	3.435	3.707	4.067	4.423	4.775
82	2.614	2.894	3.172	3.449	3.723	4.086	4.445	4.801
91	2.622	2.904	3.185	3.463	3.740	4.106	4.469	4.829
100	2.629	2.913	3.195	3.475	3.754	4.123	4.489	4.853

R2	5.6	6.2	6.8	7.5	8.2	9.1	10.
2.7	1.822	1.881	1.933	1.986	2.031	2.082	2.126
3.0	1.953	2.022	2.082	2.143	2.196	2.256	2.308
3.3	2.076	2.154	2.222	2.292	2.353	2.422	2.481
3.6	2.191	2.278	2.354	2.432	2.502	2.580	2.647
3.9	2.299	2.394	2.479	2.566	2.643	2.730	2.806
4.3	2.432	2.539	2.634	2.733	2.821	2.920	3.007
4.7	2.555	2.673	2.779	2.889	2.988	3.099	3.197
5.1	2.669	2.798	2.914	3.036	3.144	3.268	3.377
5.6	2.800	2.942	3.071	3.206	3.328	3.467	3.590
6.2	2.942	3.100	3.243	3.394	3.531	3.688	3.827
6.8	3.071	3.243	3.400	3.566	3.717	3.892	4.048
7.5	3.206	3.394	3.566	3.750	3.917	4.111	4.286
8.2	3.328	3.531	3.717	3.917	4.100	4.313	4.505
9.1	3.467	3.688	3.892	4.111	4.313	4.550	4.764
10	3.590	3.827	4.048	4.286	4.505	4.764	5.000

	R1						
R2	5.6	6.2	6.8	7.5	8.2	9.1	10.
11	3.711	3.965	4.202	4.459	4.698	4.980	5.238
12	3.818	4.088	4.340	4.615	4.871	5.175	5.455
13	3.914	4.198	4.465	4.756	5.028	5.353	5.652
15	4.078	4.387	4.679	5.000	5.302	5.664	6.000
16	4.148	4.468	4.772	5.106	5.421	5.801	6.154
18	4.271	4.612	4.935	5.294	5.634	6.044	6.429
20	4.375	4.733	5.075	5.455	5.816	6.254	6.667
22	4.464	4.837	5.194	5.593	5.974	6.437	6.875
24	4.541	4.927	5.299	5.714	6.112	6.598	7.059
27	4.638	5.042	5.432	5.870	6.290	6.806	7.297
30	4.719	5.138	5.543	6.000	6.440	6.982	7.500
33	4.788	5.219	5.638	6.111	6.568	7.133	7.674
36	4.846	5.289	5.720	6.207	6.679	7.264	7.826
39	4.897	5.350	5.790	6.290	6.775	7.378	7.959
43	4.955	5.419	5.871	6.386	6.887	7.511	8.113
47	5.004	5.477	5.941	6.468	6.982	7.624	8.246
51	5.046	5.528	6.000	6.538	7.064	7.722	8.361
56	5.091	5.582	6.064	6.614	7.153	7.828	8.485
62	5.136	5.636	6.128	6.691	7.242	7.935	8.611
68	5.174	5.682	6.182	6.755	7.318	8.026	8.718
75	5.211	5.727	6.235	6.818	7.392	8.115	8.824
82	5.242	5.764	6.279	6.872	7.455	8.191	8.913
91	5.275	5.805	6.327	6.929	7.522	8.273	9.010
100	5.303	5.838	6.367	6.977	7.579	8.341	9.091
R2	11	12	13	15	16	18	20
2.7	2.168	2.204	2.236	2.288	2.310	2.348	2.379
3.0	2.357	2.400	2.438	2.500	2.526	2.571	2.609
3.3	2.538	2.588	2.632	2.705	2.736	2.789	2.833
3.6	2.712	2.769	2.819	2.903	2.939	3.000	3.051
3.9	2.879	2.943	3.000	3.095	3.136	3.205	3.264
4.3	3.092	3.166	3.231	3.342	3.389	3.471	3.539
4.7	3.293	3.377	3.452	3.579	3.633	3.727	3.806
5.1	3.484	3.579	3.663	3.806	3.867	3.974	4.064
5.6	3.711	3.818	3.914	4.078	4.148	4.271	4.375
6.2	3.965	4.088	4.198	4.387	4.468	4.612	4.733

Table 1-1. Equivalent Resistance of Two Resistors in Parallel (cont'd).

R1

R2	11	12	13	15	16	18	20
6.8	4.202	4.340	4.465	4.679	4.772	4.935	5.075
7.5	4.459	4.615	4.756	5.000	5.106	5.294	5.455
8.2	4.698	4.871	5.028	5.302	5.421	5.634	5.816
9.1	4.980	5.175	5.353	5.664	5.801	6.044	6.254
10	5.238	5.455	5.652	6.000	6.154	6.429	6.667
11	5.500	5.739	5.958	6.346	6.519	6.828	7.097
12	5.739	6.000	6.240	6.667	6.857	7.200	7.500
13	5.958	6.240	6.500	6.964	7.172	7.548	7.879
15	6.346	6.667	6.964	7.500	7.742	8.182	8.571
16	6.519	6.857	7.172	7.742	8.000	8.471	8.889
18	6.828	7.200	7.548	8.182	8.471	9.000	9.474
20	7.097	7.500	7.879	8.571	8.889	9.474	10.000
22	7.333	7.765	8.171	8.919	9.263	9.900	10.476
24	7.543	8.000	8.432	9.231	9.600	10.286	10.909
27	7.816	8.308	8.775	9.643	10.047	10.800	11.489
30	8.049	8.571	9.070	10.000	10.435	11.250	12.000
33	8.250	8.800	9.326	10.313	10.776	11.647	12.543
36	8.426	9.000	9.551	10.588	11.077	12.000	12.857
39	8.580	9.176	9.750	10.833	11.345	12.316	13.220
43	8.759	9.382	9.982	11.121	11.661	12.689	13.651
47	8.914	9.559	10.183	11.371	11.937	13.015	14.030
51	9.048	9.714	10.359	11.591	12.179	13.304	14.366
56	9.194	9.882	10.551	11.831	12.444	13.622	14.737
62	9.342	10.054	10.747	12.078	12.718	13.950	15.122
68	9.468	10.200	10.914	12.289	12.952	14.233	15.455
75	9.593	10.345	11.080	12.500	13.187	14.516	15.789
82	9.699	10.468	11.221	12.680	13.388	14.760	16.078
91	9.814	10.602	11.375	12.877	13.607	15.028	16.396
100	9.910	10.714	11.504	13.043	13.793	15.254	16.667

Table 1-1. Equivalent Resistance of Two Resistors in Parallel (cont'd).

R2	R1							
	22	24	27	30	33	36	39	43
2.7	2.405	2.427	2.455	2.477	2.496	2.512	2.525	2.540
3.0	2.640	2.667	2.700	2.727	2.750	2.769	2.786	2.804
3.3	2.870	2.901	2.941	2.973	3.000	3.023	3.043	3.065
3.6	3.094	3.130	3.176	3.214	3.246	3.273	3.296	3.322
3.9	3.313	3.355	3.408	3.451	3.488	3.519	3.545	3.576
4.3	3.597	3.647	3.709	3.761	3.804	3.841	3.873	3.909
4.7	3.873	3.930	4.003	4.063	4.114	4.157	4.195	4.237
5.1	4.140	4.206	4.290	4.359	4.417	4.467	4.510	4.559
5.6	4.464	4.541	4.638	4.719	4.788	4.846	4.897	4.955
6.2	4.837	4.927	5.042	5.138	5.219	5.289	5.350	5.419
6.8	5.194	5.299	5.432	5.543	5.638	5.720	5.790	5.871
7.5	5.593	5.714	5.870	6.000	6.111	6.207	6.290	6.386
8.2	5.974	6.112	6.290	6.440	6.568	6.679	6.775	6.887
9.1	6.437	6.598	6.806	6.982	7.133	7.264	7.378	7.511
10	6.875	7.059	7.297	7.500	7.674	7.826	7.959	8.113
11	7.333	7.543	7.816	8.049	8.250	8.426	8.580	8.759
12	7.765	8.000	8.308	8.571	8.800	9.000	9.176	9.382
13	8.171	8.432	8.775	9.070	9.326	9.551	9.750	9.982
15	8.919	9.231	9.643	10.000	10.313	10.588	10.833	11.121
16	9.263	9.600	10.047	10.435	10.776	11.077	11.345	11.661
18	9.900	10.286	10.800	11.250	11.647	12.000	12.316	12.689
20	10.476	10.909	11.489	12.000	12.543	12.857	13.220	13.651
22	11.000	11.478	12.122	12.692	13.200	13.655	14.066	14.554
24	11.478	12.000	12.706	13.333	13.895	14.400	14.857	15.404
27	12.122	12.706	13.500	14.211	14.850	15.429	15.955	16.586
30	12.692	13.333	14.211	15.000	15.714	16.364	16.957	17.671
33	13.200	13.895	14.850	15.714	16.500	17.217	17.875	18.671
36	13.655	14.400	15.429	16.364	17.217	18.000	18.200	19.595
39	14.066	14.857	15.955	16.957	17.875	18.720	19.500	20.451
43	14.554	15.403	16.586	17.671	18.671	19.595	20.451	21.500

Table 1-1. Equivalent Resistance of Two Resistors in Parallel (cont'd).

				R1				
R2	22	24	27	30	33	36	39	43
47	14.986	15.887	17.149	18.312	19.388	20.386	21.314	22.456
51	15.370	16.320	17.654	18.889	20.036	21.103	22.100	23.330
56	15.795	16.800	18.217	19.535	20.764	21.913	22.989	24.323
62	16.238	17.302	18.809	20.217	21.537	22.776	23.941	25.390
68	16.622	17.739	19.326	20.816	22.218	23.538	24.785	26.342
75	17.010	18.182	19.853	21.423	22.917	24.324	25.658	27.331
82	17.346	18.566	20.312	21.964	23.530	25.017	26.430	28.208
91	17.717	18.991	20.822	22.562	24.218	25.795	27.300	29.201
100	18.033	19.355	21.260	23.077	24.812	26.471	28.058	30.070

R2	47	51	56	62	68	75	82	91	100
2.7	2.553	2.564	2.576	2.587	2.597	2.606	2.614	2.622	2.629
3.0	2.820	2.833	2.847	2.862	2.873	2.885	2.894	2.904	2.913
3.3	3.083	3.099	3.116	3.133	3.147	3.161	3.172	3.185	3.195
3.6	3.344	3.363	3.383	3.402	3.419	3.435	3.449	3.463	3.475
3.9	3.601	3.623	3.646	3.669	3.688	3.707	3.723	3.740	3.754
4.3	3.939	3.966	3.993	4.021	4.044	4.067	4.086	4.106	4.123
4.7	4.273	4.303	4.336	4.369	4.396	4.423	4.445	4.469	4.489
5.1	4.601	4.636	4.674	4.712	4.744	4.775	4.801	4.829	4.853
5.6	5.004	5.046	5.091	5.136	5.174	5.211	5.242	5.275	5.303
6.2	5.477	5.528	5.582	5.636	5.682	5.727	5.764	5.805	5.838
6.8	5.941	6.000	6.064	6.128	6.182	6.235	6.279	6.327	6.367
7.5	6.468	6.538	6.614	6.691	6.755	6.818	6.872	6.929	6.977
8.2	6.982	7.064	7.153	7.242	7.318	7.392	7.455	7.522	7.579
9.1	7.624	7.722	7.828	7.935	8.026	8.115	8.191	8.273	8.341
10.0	8.246	8.361	8.485	8.611	8.718	8.824	8.913	9.010	9.091
11	8.914	9.048	9.194	9.342	9.468	9.593	9.699	9.814	9.910
12	9.559	9.714	9.882	10.054	10.200	10.345	10.468	10.602	10.714
13	10.183	10.359	10.551	10.747	10.914	11.080	11.221	11.375	11.504
15	11.371	11.591	11.831	12.078	12.289	12.500	12.680	12.877	13.043
16	11.937	12.179	12.444	12.718	12.952	13.187	13.388	13.607	13.793

Table 1-1. Equivalent Resistance of Two Resistors in Parallel (cont'd).

	R1								
R2	47	51	56	62	68	75	82	91	100
18	13.015	13.304	13.622	13.950	14.233	14.516	14.760	15.028	15.254
20	14.030	14.366	14.737	15.122	15.455	15.789	16.078	16.396	16.667
22	14.986	15.370	15.795	16.238	16.622	17.010	17.346	17.717	18.033
24	15.887	16.320	16.800	17.302	17.739	18.182	18.566	18.991	19.355
27	17.149	17.654	18.217	18.809	19.326	19.853	20.312	20.822	21.260
30	18.312	18.889	19.535	20.217	20.816	21.423	21.964	22.562	23.077
33	19.388	20.036	20.764	21.537	22.218	22.917	23.530	24.218	24.812
36	20.836	21.103	21.913	22.776	23.538	24.324	25.017	25.795	26.471
39	21.314	22.100	22.989	23.941	24.785	25.658	26.430	27.300	28.058
43	22.456	23.330	24.323	25.390	26.342	27.331	28.208	29.201	30.070
47	23.500	24.459	25.553	26.734	27.791	28.893	29.876	30.993	31.973
51	24.459	25.500	26.692	27.982	29.143	30.357	31.444	32.683	33.775
56	25.553	26.692	28.000	29.424	30.710	32.061	33.275	34.667	35.897
62	26.734	27.982	29.424	31.000	32.431	33.942	35.306	36.876	38.272
68	27.791	29.143	30.710	32.431	34.000	35.664	37.173	38.918	40.476
75	28.893	30.357	32.061	33.942	35.664	37.500	39.172	41.114	42.857
82	29.876	31.444	33.275	35.306	37.173	39.172	41.000	43.133	45.055
91	30.993	32.683	34.667	36.876	38.918	41.114	43.133	45.500	47.644
100	31.973	33.775	35.897	38.272	40.476	42.857	45.055	47.644	50.000

R2	10	11	12	13	15	16	18	20	22
100	9.091	9.910	10.714	11.504	13.043	13.793	15.254	16.667	18.033
110	9.167	10.000	10.820	11.626	13.200	13.968	15.469	16.923	18.333
120	9.231	10.076	10.909	11.729	13.333	14.118	15.652	17.143	18.592
130	9.286	10.142	10.986	11.818	13.448	14.247	15.811	17.333	18.816
150	9.375	10.248	11.111	11.963	13.636	14.458	16.071	17.647	19.186
160	9.412	10.292	11.163	12.023	13.714	14.545	16.180	17.778	19.341
180	9.474	10.366	11.250	12.124	13.846	14.694	16.364	18.000	19.604
200	9.524	10.427	11.321	12.207	13.953	14.815	16.514	18.182	19.820
220	9.565	10.476	11.380	12.275	14.043	14.915	16.639	18.333	20.000
240	9.600	10.518	11.423	12.332	14.118	15.000	16.744	18.462	20.153

Table 1-1. Equivalent Resistance of Two Resistors in Parallel (cont'd).

	R1								
R2	10	11	12	13	15	16	18	20	22
270	9.643	10.569	11.489	12.403	14.211	15.105	16.875	18.621	20.342
300	9.677	10.611	11.538	12.460	14.286	15.190	16.981	18.750	20.497
330	9.706	10.645	11.579	12.507	14.348	15.260	17.069	18.857	20.625
360	9.730	10.674	11.613	12.547	14.400	15.319	17.143	18.947	20.733
390	9.750	10.698	11.642	12.581	14.444	15.369	17.206	19.024	20.825
430	9.773	10.726	11.674	12.619	14.494	15.426	17.277	19.111	20.929
470	9.792	10.748	11.701	12.650	14.536	15.473	17.336	19.184	21.016
510	9.808	10.768	11.724	12.677	14.571	15.513	17.386	19.245	21.090
560	9.825	10.788	11.748	12.705	14.609	15.556	17.439	19.310	21.168
620	9.841	10.808	11.772	12.733	14.646	15.597	17.492	19.375	21.246
680	9.855	10.825	11.792	12.756	14.676	15.632	17.536	19.429	21.311
750	9.868	10.841	11.811	12.779	14.706	15.666	17.578	19.481	21.373
820	9.880	10.854	11.827	12.797	14.731	15.694	17.613	19.524	21.425
910	9.891	10.869	11.844	12.817	14.757	15.724	17.651	19.570	21.481
1000	9.901	10.880	11.858	12.833	14.778	15.748	17.682	19.608	21.526

R2	24	27	30	33	36	39	43	47	51
100	19.355	21.260	23.077	24.812	26.471	28.058	30.070	31.973	33.775
110	19.701	21.679	23.571	25.385	27.123	28.792	30.915	32.930	34.845
120	20.000	22.041	24.000	25.882	27.692	29.434	31.656	33.772	35.789
130	20.260	22.357	24.375	26.319	28.193	30.000	32.312	34.520	36.630
150	20.690	22.881	25.000	27.049	29.032	30.952	33.420	35.787	38.060
160	20.870	23.102	25.263	27.358	29.388	31.357	33.892	36.329	38.673
180	21.176	23.478	25.714	27.887	30.000	32.055	34.709	37.269	39.740
200	21.429	23.789	26.087	28.326	30.508	32.636	35.391	38.057	40.637
220	21.639	24.049	26.400	28.696	30.938	33.127	35.970	38.727	41.402
240	21.818	24.270	26.667	29.011	31.304	33.548	36.466	39.303	42.062
270	22.041	24.545	27.000	29.406	31.765	34.078	37.093	40.032	42.897
300	22.222	24.771	27.273	29.730	32.143	34.513	37.609	40.634	43.590
330	22.373	24.958	27.500	30.000	32.459	34.878	38.043	41.141	44.173
360	22.500	25.116	27.692	30.229	32.727	35.188	38.412	41.572	44.672
390	22.609	25.252	27.857	30.426	32.958	35.455	38.730	41.945	45.102

	R1								
R2	24	27	30	33	36	39	43	47	51
430	22.731	25.405	28.043	30.648	33.219	35.757	39.091	42.369	45.593
470	22.834	25.533	28.200	30.835	33.439	36.012	39.396	42.727	46.008
510	22.921	25.642	28.333	30.994	33.626	36.230	39.656	43.034	46.364
560	23.014	25.758	28.475	31 164	33.826	36.461	39.934	43.361	46.743
620	23.106	25.873	28.615	31.332	34.024	36.692	40.211	43.688	47.124
680	23.182	25.969	28.732	31.473	34.190	36.885	40.443	43.961	47.442
750	23.256	26.062	28.846	31.609	34.351	37.072	40.668	44.228	47.753
820	23.318	26.139	28.941	31.723	34.486	37.229	40.857	44.452	48.014
910	23.383	26.222	29.043	31.845	34.630	37.397	41.060	44.692	48.293
1000	23.438	26.290	29.126	31.946	34.749	37.536	41.227	44.890	48.525

R2	56	62	68	75	82	91	100
100	35.897	38.272	40.476	42.857	45.055	47.644	50.000
110	37.108	39.651	42.022	44.595	46.979	49.801	52.381
120	38.182	40.879	43.404	46.154	48.713	51.754	55.545
130	39.140	41.979	44.646	47.561	50.283	53.529	56.522
150	40.777	43.868	46.789	50.000	53.017	56.639	60.000
160	41.481	44.685	47.719	51.064	54.215	58.008	61.538
180	42.712	46.116	49.355	52.941	56.336	60.443	64.286
200	43.750	47.328	50.746	54.545	58.156	62.543	66.667
220	44.638	48.369	51.944	55.932	59.735	64.373	68.750
240	45.405	49.272	52.987	57.143	61.118	65.982	70.588
270	46.380	50.422	54.320	58.696	62.898	68.061	72.973
300	47.191	51.381	55.435	60.000	64.398	69.821	75.000
330	47.876	52.194	56.382	61.111	65.680	71.330	76.744
360	48.462	52.891	57.196	62.069	66.767	72.639	78.261
390	48.969	53.496	57.904	62.903	67.754	73.784	79.592
430	49.547	54.187	58.715	63.861	68.867	75.106	81.132
470	50.038	54.774	59.405	64.679	69.819	76.239	82.456
510	50.459	55.280	60.000	65.385	70.642	77.221	83.607
560	50.909	55.820	60.637	66.142	71.526	78.280	84.848
620	51.361	56.364	61.279	66.906	72.422	79.353	86.111

Table 1-1. Equivalent Resistance of Two Resistors in Parallel (cont'd).

R2	56	62	68	75	82	91	100
			R1				
680	51.739	56.819	61.818	67.550	73.176	80.259	87.179
750	52.109	57.266	62.347	68.182	73.918	81.153	88.235
820	52.420	57.642	62.793	68.715	74.545	81.910	89.130
910	52.754	58.045	63.272	69.289	75.222	82.727	90.099
1000	53.030	58.380	63.670	69.767	75.786	83.410	90.909

Locate 6.8 ohms in R1 column. Move the decimal point one place to the right so that 6.8 becomes 68. Locate 2.7 ohms in the R2 row and consider it now as 27 ohms. These two values meet at 1.933 in the Table. However, this is now 19.33 ohms, since you must again move the decimal point one place to the right.

This problem can also be done directly, in two ways. Locate 68 in the R2 row and then move to the right to find the answer in the 27 column under the general heading of R1. Or, locate 27 in the R2 row and move to the right to find the answer in the 68 column under the general heading of R1.

The values shown in Table 1-1 are median values and do not take tolerances into consideration. Resistor tolerances are usually 20 percent or less, and may be plus or minus. However, Table 1-1 does supply a practical guide for the quick determination of two resistors in parallel, or for finding parallel resistor combinations which will be equivalent to a desired resistance value.

The assumption is made in Table 1-1 that the resistors to be wired in parallel are those having similar units, that is, in ohms, kilohms, or megohms. If one resistor has its value specified in ohms, and another in kilohms, for example, it would be necessary to change kilohms to ohms or ohms to kilohms.

☐ **Example:**

What is the equivalent resistance of a 1 kilohm resistor shunted by a 100 ohm resistor?

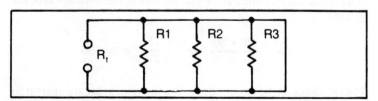

Fig. 1-1. Resistors in parallel.

1 kilohm is equal to 1,000 ohms. Locate 1,000 in the R2 column in Table 1-1 and then 100 in the last column to the right. The equivalent resistance is given as 90.909 ohms.

FORMULAS FOR RESISTORS IN PARALLEL

While Table 1-1 is suitable for finding the equivalent resistance of parallel resistors, there are components other than physical resistors that have resistance. A length of wire has resistance and so does a coil made of that wire. A metal bar may have resistance and so may a normally nonconductive material. Heating elements used for electric irons, ovens, and soldering irons have resistance. And, of course, there are also resistors that do not fit into the values shown in Table 1-1.

You can calculate the value of any pair of resistors using the following formula:

$$R_t = \frac{R1 \times R2}{R1 + R2}$$

R_t is the total resistance. R1 is either of the resistors; R2 is the other. R1 and R2 must be in similar units: ohms, kilohms, or megohms.

If you have three resistors in parallel, you can still use this formula, but in a two-step process. Find the equivalent value of any two of the resistors. Then, using the formula again, combine the answer with the remaining resistor. Alternatively, for three resistors you can use this formula:

$$R_t = \frac{1}{\dfrac{1}{R1} + \dfrac{1}{R2} + \dfrac{1}{R3}}$$

☐ **Example:**

What is the equivalent resistance of two resistors with values of 45 ohms and 95 ohms? Note that neither of these values appears in Table 1-1.

$$R_t = \frac{45 \times 95}{45 + 95} = \frac{4275}{140} = 30.536$$

You can check on the answer by finding the values that are nearest to those given in the problem and using Table 1-1. The closest to 95 ohms is 91 ohms. The closest to 45 ohms is 47 ohms. Table 1-1 shows the equivalent resistance to be 30.993 ohms.

□ **Example:**
    What is the effective resistance of three parallel resistors with values of 75 ohms, 95 ohms, and 120 ohms?

$$R_t = \cfrac{1}{\cfrac{1}{R1} + \cfrac{1}{R2} + \cfrac{1}{R3}} = \cfrac{1}{\cfrac{1}{75} + \cfrac{1}{95} + \cfrac{1}{120}}$$

$$1/75 = 0.0133 \qquad 1/95 = 0.0105 \qquad 1/120 = 0.0083$$

$$R_t = \cfrac{1}{0.0133 + 0.0105 + 0.0083} = \cfrac{1}{0.0321} = 31.15 \text{ ohms}$$

The formula for finding three resistors in parallel can be extended to find the equivalent resistance of four or more shunt (parallel) resistors.

    Still another formula for finding the equivalent resistance of a number of resistors in parallel is as follows:

$$\frac{1}{R_t} = \frac{1}{R1} + \frac{1}{R2} + \frac{1}{R3}$$

Although the formula indicates just three resistors, it can be expanded for any number of resistors.

□ **Example:**
    What is the equivalent resistance of 46 ohms, 53 ohms, and 75 ohms wired in parallel?

$$\frac{1}{R_t} = \frac{1}{R1} + \frac{1}{R2} + \frac{1}{R3}$$

$$= \frac{1}{46} + \frac{1}{53} + \frac{1}{75}$$

$$= 0.0217 + 0.019 + 0.013$$

$$= 0.537$$

This answer isn't the resistance, but is actually the conductance. To find the resistance, divide the answer into 1.

$$\frac{1}{R_t} = 0.537$$

$$R_t = \frac{1}{0.0537} = 19.23 \text{ ohms}$$

This answer is an approximation since the answers for the division of 46 into 1, 53 into 1 and 75 into 1 were carried out to only two or three decimal places. Thus, 1/46 is actually 0.021739; 1/53 is 0.018867 and 1/75 is 0.013333. If these numbers are used the shunt resistance is 18.539 ohms. This degree of accuracy may or may not be needed, depending on the work that is being done.

RESISTORS IN SERIES

For resistors in series, the formula is as follows

$$R_t = R1 + R2 + R3$$

All that is needed is to add the values of the resistors in the series circuit (Fig. 1-2). The resistance value must be in the same units: ohms, kilohms, or megohms.

RESISTANCE VS. CONDUCTANCE

The opposition to the movement of an electrical current can be expressed in terms of resistance (measured in ohms) or in terms of conductance (expressed in siemens). It is often very convenient to be able to move back and forth quickly and easily between resis-

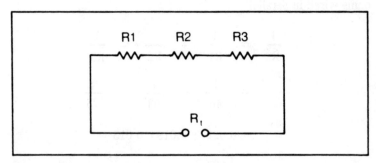

Fig. 1-2. Resistors in series.

tance and conductance in the solution of electronics problems. This can be readily done since resistance and conductance are reciprocals.

Sometimes, in working with resistances you will find the values are such that they are not covered by the tables and that using a formula to solve the problem will involve some laborious arithmetic. In that case it may be easier and quicker to work with conductances. To find the total conductance of resistors in parallel simply consider them as conductors and add the values of the individual units. Thus, if you have a number of resistors in parallel, use Table 1-2 to find the equivalent conductance of each resistor. Add the conductances and then use the table once again to find the equivalent resistance.

The symbol for resistance is R; that used for conductance is G. The relationship between the two is expressed as R equals 1/G or G equals 1/R. If you are considering a complete circuit, that is, a circuit consisting of a number of resistors in parallel, then the total resistance of the circuit is the reciprocal of the total conductance.

Conductance can be substituted into the different forms of Ohm's law. Thus, for resistance, we would have R equals E/I. For conductance we would have G equals I/E.

☐ **Example:**

What is the conductance of a resistor whose value, as measured, is 64 ohms?

Locate 64 in the column marked ohms. The value of conductance, as shown in the column (siemens) to the right, is 0.0156 siemen.

Table 1-2. Resistance (Ohms) vs. Conductance (Siemens).

Ohms	Siemens	Ohms	Siemens	Ohms	Siemens
0.1	10.0000	1.0	1.0000	10	0.1000
0.2	5.0000	2	0.5000	11	0.0909
0.3	3.3333	3	0.3333	12	0.0833
0.4	2.5000	4	0.2500	13	0.0769
0.5	2.0000	5	0.2000	14	0.0714
0.6	1.6667	6	0.1667	15	0.0667
0.7	1.4286	7	0.1429	16	0.0625
0.8	1.2500	8	0.1250	17	0.0588
0.9	1.1111	9	0.1111	18	0.0556

Table 1-2. Resistance (Ohms) vs. Conductance (Siemens) (cont'd).

Ohms	Siemens	Ohms	Siemens	Ohms	Siemens
19	0.0526	48	0.0208	74	0.0135
20	0.0500	49	0.0204	75	0.0133
21	0.0476	50	0.0200	76	0.0132
22	0.0455	51	0.0196	77	0.0130
23	0.0435	52	0.0192	78	0.0128
24	0.0417				
25	0.0400	53	0.0189	79	0.0127
26	0.0385	54	0.0185	80	0.0125
27	0.0370	55	0.0182	81	0.0123
28	0.0357	56	0.0179	82	0.0122
29	0.0345	57	0.0175	83	0.0120
30	0.0333				
31	0.0323	58	0.0172	84	0.0119
32	0.0313	59	0.0169	85	0.0118
33	0.0303	60	0.0167	86	0.0116
34	0.0294	61	0.0164	87	0.0115
35	0.0286	62	0.0161	88	0.0114
36	0.0278	63	0.0159		
37	0.0270			89	0.0112
		64	0.0156	90	0.0111
38	0.0263	65	0.0154	91	0.0110
39	0.0256	66	0.0152	92	0.0109
40	0.0250	67	0.0149	93	0.0108
41	0.0244	68	0.0147	94	0.0106
42	0.0238			95	0.0105
43	0.0233	69	0.0145	96	0.0104
44	0.0227	70	0.0143	97	0.0103
45	0.0222	71	0.0141	98	0.0102
46	0.0217	72	0.0139	99	0.0101
47	0.0213	73	0.0137	100	0.0100

☐ **Example:**

What is the resistance of a component whose conductance is 0.0556 siemen.

The value of 0.0556, in the siemens column, corresponds to 18 ohms, as indicated in the ohms column.

☐ **Example:**

The values of four resistors, measured on a bridge, are 90 ohms, 83, 79, and 71 ohms, respectively. What is the equivalent resistance when these units are connected in parallel?

Using Table 1-2 you will find that the corresponding conductance values are 0.0111, 0.0120, 0.0127, and 0.0141 siemen, respectively. Adding these results in a total conductance of 0.0499 siemen. Using Table 1-2 once again, the closest conductance value is 0.0500 siemen, and, as shown by the table, corresponds to 20 ohms.

STANDARD EIA VALUES FOR COMPOSITION RESISTORS

Composition resistors are available in values based on the recommendations of the Electronics Industries Association (EIA). These values are shown in Table 1-3.

Table 1-3. Standard EIA Values for Composition Resistors.

Ohms	Ohms	Ohms	Ohms	Ohms	Ohms
2.7	39	560	8.2K	120K	1.8 Meg
3.0	43	620	9.1K	130K	2.0 Meg
3.3	47	680	10K	150K	2.2 Meg
3.6	51	750	11K	160K	2.4 Meg
3.9	56	820	12K	180K	2.7 Meg
4.3	62	910	13K	200K	3.0 Meg
4.7	68	1K	15K	220K	3.3 Meg
5.1	75	1.1K	16K	240K	3.6 Meg
5.6	82	1.2K	18K	270K	3.9 Meg
6.2	91	1.3K	20K	300K	4.3 Meg
6.8	100	1.5K	22K	330K	4.7 Meg
7.5	110	1.6K	24K	360K	5.1 Meg
8.2	120	1.8K	27K	390K	5.6 Meg
9.1	130	2K	30K	430K	6.2 Meg
10	150	2.2K	33K	470K	6.8 Meg
11	160	2.4K	36K	510K	7.5 Meg
12	180	2.7K	39K	560K	8.2 Meg
13	200	3K	43K	620K	9.1 Meg
15	220	3.3K	47K	680K	10 Meg
16	240	3.6K	51K	750K	11 Meg
18	270	3.9K	56K	820K	12 Meg
20	300	4.3K	62K	910K	13 Meg
22	330	4.7K	68K	1 Meg	15 Meg
24	360	5.1K	75K	1.1 Meg	16 Meg
27	390	5.6K	82K	1.2 Meg	18 Meg
30	430	6.2K	91K	1.3 Meg	20 Meg
33	470	6.8K	100K	1.5 Meg	22 Meg
36	510	7.5K	110K	1.6 Meg	. . . . . .

17

Table 1-4. Conductance of Standard EIA Values for Composition Resistors (Resistance, R, in Ohms; Conductance, G, in Siemens).

R	G	R	G	R	G	R	G
2.7	0.37037	47	0.02128	820	0.001220	15K	0.000066
3.0	0.33333	51	0.01961	910	0.001100	16K	0.000062
3.3	0.30303	56	0.01786	1K	0.001000	18K	0.000055
3.6	0.27778	62	0.01613	1.1K	0.000909	20K	0.000050
3.9	0 25641	68	0.01471	1.2K	0.000833	22K	0.000045
4.3	0.23256	75	0.01333	1.3K	0.000769	24K	0.000041
4.7	0.21277	82	0.01220	1.5K	0.000666	27K	0.000037
5.1	0.19608	91	0.01099	1.6K	0.000625	30K	0.000033
5.6	0.17857	100	0.01000	1.8K	0.000555	33K	0.000030
6.2	0.16129	110	0.00909	2K	0.000500	36K	0.000027
6.8	0.14706	120	0.00833	2.2K	0.000454	39K	0.000025
7.5	0.13333	130	0.00769	2.4K	0.000416	43K	0.000023
8.2	0.12195	150	0.00667	2.7K	0.000370	47K	0.000021
9.1	0.10989	160	0.00625	3K	0.000333	51K	0.000019
10	0.10000	180	0.00556	3.3K	0.000303	56K	0.000017
11	0.09091	200	0.00500	3.6K	0.000277	62K	0.000016
12	0.08333	220	0.00455	3.9K	0.000256	68K	0.000014
13	0.07692	240	0.00417	4.3K	0.000232	75K	0.000013
15	0.06667	270	0.00370	4.7K	0.000212	82K	0.000012
16	0.06250	300	0.00333	5.1K	0.000196	91K	0.000011
18	0.05556	330	0.00303	5.6K	0.000178	100K	0.000010
20	0.05000	360	0.00278	6.2K	0.000161		
22	0.04545	390	0.00256	6.8K	0.000147		
24	0.04167	430	0.00233	7.5K	0.000133		
27	0.03704	470	0.00213	8.2K	0.000121		
30	0.03333	510	0.00196	9.1K	0.000109		
33	0.03030	560	0.00179	10K	0.000100		
36	0.02778	620	0.00161	11K	0.000090		
39	0.02564	680	0.00147	12K	0.000083		
43	0.02326	750	0.00133	13K	0.000077		

K means kilo, or multiply by 1000. 120 K equals 120 × 1000, or 120,000 ohms. Meg means multiply by 1,000,000. 1.8 meg equals 1.8 × 1,000,000, or 1,800,000 ohms.

Table 1-4 shows conductance values for standard composition resistors. Column R indicates the resistance values in ohms; column G the corresponding conductance values in siemens. To find the conductance of any number of parallel resistors, add the conductances and then locate the nearest equivalent resistance value. Thus, 10-ohm, 18-ohm, and 30-ohm resistors in parallel have conductances of 0.1 siemen, 0.05556 siemen, and 0.03333 siemen. 0.1 + 0.05556 + 0.03333 equals 0.1889 siemen. The nearest value in the G column corresponding to 0.1889 siemen is 0.19608 siemen, equivalent to a standard resistance value of 5.1 ohms.

To find a more accurate value of resistance, divide the sum of the conductances into 1. Thus, 1 divided by 0.1889 equals 5.294 ohms. However, this is not a standard value. The closest standard value is 5.1 ohms.

METER MULTIPLIERS

Not only actual resistors, but components or components plus resistors can be put in parallel. Thus, a resistor can be mounted in shunt with a current meter, often an ammeter. The purpose of the shunt is to extend the range of the meter.

A formula for finding the value of shunt resistance is as follows:

$$R = \frac{R_i \times A}{I - A}$$

In this formula, R_i is the internal resistance of the meter. This value can be obtained from the manufacturer of the meter. A is the maximum reading on the existing scale of the meter and I is the current to be measured.

☐ **Example:**

A meter scale indicates a maximum possible reading of 1 ampere on a meter whose internal resistance is 0.01 ohm. What value of shunt resistance should be used to extend the scale to 10 amperes? See Fig. 1-3.

Based on the information supplied R_i is 0.01 ohm, A is 1 ampere and I is 10 amperes.

$$R = \frac{R_i \times A}{I - A} = \frac{0.01 \times 1}{10 - 1} = 0.0011 \text{ ohm}$$

Unlike ordinary resistors, such as those used in radio and television receivers, shunts must have a high order of accuracy and are often supplied by the manufacturers of the meters.

To determine the multiplying effect of the shunt, use this formula:

$$MP = \frac{R_i}{R} + 1$$

MP is the multiplying value, R_i is the internal resistance, and R is the value of the shunt.

In the example just given, R_i is 0.01 ohm, R is 0.0011 ohm.

$$MP = \frac{0.01}{0.0011} + 1 = 9.0 + 1 = 10$$

The formula need not have been used since it is fairly obvious that a meter whose full scale reading is 1 ampere and then becomes 10 amperes with the use of a shunt, has had its current reading ability increased by a factor of 10.

Sometimes more than one formula can be used to solve a problem in electronics. Although such formulas may look different, they are generally the same. They just use other symbols or have a different arrangement.

When two resistive elements are in parallel, the same voltage appears across both. Thus:

$$I_m R_m = I_s R_s$$

In this formula, I_m is the current through the meter, R_m is the meter resistance, I_s is the shunt current, and R_s is the value of shunt resistance. The formula can be transposed to solve for any one of the unknowns.

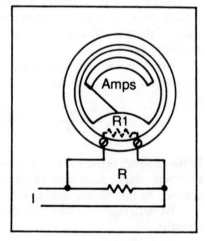

Fig. 1-3. Shunt used to extend the range of a current-reading meter.

$$R_s = \frac{I_m R_m}{I_s}$$

□ Example:

Assume you have a milliameter having a full scale deflection of 1 milliampere and an internal resistance of 50 ohms. Suppose you want to extend the full scale reading by a factor of 20.

$$R_s = \frac{I_m R_m}{I_s} = \frac{(0.001)\,(50)}{0.020} = \frac{0.05}{0.02} = 2.5 \text{ ohms}$$

Since the shunt must carry 20 times as much current as the meter, it must have 1/20th of the resistance of the meter. The internal resistance of the meter is 50 ohms; 1/20th of 50 ohms is 2.5 ohms.

A resistor wired in series with a voltmeter will extend its range based on the following formula:

$$R = \frac{E}{I} - R_i$$

R_i is the internal resistance of the meter, I is the current needed for full-scale deflection and E is the voltage to be measured.

□ Example:

You have a meter whose internal resistance is 1,000 ohms. It requires 1 milliampere for full-scale deflection. How can this instrument be made to measure 200 volts?

The maximum voltage the instrument can measure, based on its electrical characteristics, is 1 volt. One milliampere flowing through the internal resistance of 1,000 ohms is an electrical pressure of 1 volt. (E = I × R. 0.001 × 1,000 = 1 volt.).

$$R = \frac{E}{I} - R_i$$

$$= \frac{200}{0.001} - 1,000 = 200,000 - 1,000 = 199,000 \text{ ohms}$$

By using a resistor in series with the meter, as shown in Fig. 1-4, the meter can be used to measure up to 200 volts. This is dc voltage,

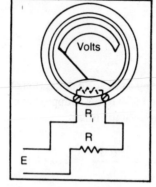

Fig. 1-4. Series resistor used to extend voltmeter range.

not ac. In this example, the multiplying power is obviously 200. It can also be calculated by a formula.

$$MP = \frac{R_i + R}{R_i}$$

$$= \frac{1,000 + 199,000}{1,000} = \frac{200,000}{1,000} = 200$$

Table 1-5. Multiplier Resistor Values.

range full scale (volts)	meter sensitivity microamperes				milliamperes	
	50	100	200	500	1	5
1	20,000	10,000	5,000*	2,000*	1,000*	200*
1.5	30,000	15,000	7,500	3,000*	1,500*	300*
2.5	50,000	25,000	12,500	5,000*	2,500*	500*
3	60,000	30,000	15,000	6,000	3,000*	600*
5	100,000	50,000	25,000	10,000	5,000*	1,000*
6	120,000	60,000	30,000	12,000	6,000	1,200
10	200,000	100,000	50,000	20,000	10,000	2,000
12	240,000	120,000	60,000	24,000	12,000	2,400
15	300,000	150,000	75,000	30,000	15,000	3,000
25	500,000	250,000	125,000	50,000	25,000	5,000
30	600,000	300,000	150,000	60,000	30,000	6,000
50	1,000,000	500,000	250,000	100,000	50,000	10,000
60	1,200,000	600,000	300,000	120,000	60,000	12,000
100	2,000,000	1,000,000	500,000	200,000	100,000	20,000
150	3,000,000	1,500,000	750,000	300,000	150,000	30,000
250	5,000,000	2,500,000	1,250,000	500,000	250,000	50,000
300	6,000,000	3,000,000	1,500,000	600,000	300,000	60,000
500	10,000,000	5,000,000	2,500,000	1,000,000	500,000	100,000
1,000	20,000,000	10,000,000	5,000,000	2,000,000	1,000,000	200,000
1,500	30,000,000	15,000,000	7,500,000	3,000,000	1,500,000	300,000

* Meter internal resistance should be subtracted.

R_i is the internal resistance and R is the value of the multiplier resistance.

Table 1-5 supplies resistor values for commonly used voltmeters. Thus, a meter having a full-scale deflection of 1 volt (first column) and requiring a current of 50 microamperes (second column) has a resistance of 20,000 ohms per volt. In the left-hand column, move down until you reach the number 100. The column at the immediate right shows a resistance value of 2,000,000 ohms. To extend the range to 100 volts would require a multiplier resistance of 2 megohms.

CONDUCTANCE OF METALS

One of the distinguishing characteristics of metals is their ability to conduct electricity. As metals differ in hardness, ductility, density, tensile strength, malleability and melting point, they also differ in inherent conductance of electrical flow. See Table 1-6.

RESISTIVITIES OF CONDUCTORS

Yes, metals do conduct electricity, but some do it better than others. Table 1-7 shows how much resistance various conductors exhibit. Silver is the best electrical conductor and presents the least resistance to electrical flow. Although gold is the third best conductor, its cost makes gold impractical to use as such. Consequently, gold has not been included in Table 1-7.

OHM'S LAW FOR DC

Of all electronic formulas, Ohm's law is probably the best

Table 1-6. Relative Conductance of Various Metals.

Substance	Relative conductance (Silver= 100%)
Silver	100
Copper	98
Gold	78
Aluminum	61
Tungsten	32
Zinc	30
Platinum	17
Iron	16
Lead	15
Tin	9
Nickel	7
Mercury	1
Carbon	0.05

Table 1-7. Resistivities of Conductors.

Resistivities of Conductors at 0° C

Substance	Microhms Centimeter cube	Inch cube	Ohms-Circular mils per foot Round Wires
Aluminum	3.21	1.26	19.3
Carbon	4000 to 10,000	1600 to 2800	24,00 to 42,000
Constantan (Cu 60%, Ni 40%)	49	19.3	295
Copper	1.72	0.68	10.4
Iron	12 to 14	4.7 to 5.5	72 to 84
Lead	20.8	8.2	125
Manganin (Cu 84%, Ni 4%, Mn 12%)	43	16.9	258
Mercury	95.76	37.6	575
Nichrome (Ni 60%, Cr 12%, Fe 26%, Mn 2%)	110	43	660
Platinum	11.0	4.3	66
Silver	1.65	0.65	9.9
Tungsten	5.5	2.15	33
Zinc	6.1	2.4	36.7

known and the most widely used. Basically set up as $E = IR$, it can be used to find an unknown value when the other two are known. It can also be used to determine the power being dissipated. The various arrangements of this law are shown in Table 1-8.

Basic Units

The basic units in Ohm's law are the volt, ampere and ohm. Multiples and submultiples of these units are also used in the solution of problems. See Table 1-9.

Table 1-8. Ohm's Law Formulas for Dc.

Formula for Finding Unknown Values of . . .

Known Values	I =	R =	E =	P =
I & R			IR	I^2R
I & E		$\dfrac{E}{I}$		EI
I & P		$\dfrac{P}{I^2}$	$\dfrac{P}{I}$	
R & E	$\dfrac{E}{R}$			$\dfrac{E^2}{R}$
R & P	$\sqrt{\dfrac{P}{R}}$		$\sqrt{PR}$	
E & P	$\dfrac{P}{E}$	$\dfrac{E^2}{P}$		

Table 1-9. Units and Symbols.

Unit	Symbol	Multiple	Value
volt	E	kilovolt (kv)	1000 volts
volt	E	millivolt (mv)	1/1000 volt
volt	E	microvolt (μv)	1/1,000,000 volt
ohm	R	kilohm	1000 ohms
ohm	R	megohm	1,000,000 ohms
ampere	I	milliampere (ma)	1/1000 ampere
ampere	I	microampere (μa)	1/1,000,000 ampere

When large numbers are used in the solution of Ohm's law problems it is more convenient to use exponents. See Table 1-10.

Table 1-10. Units in Exponential Notation.

1 volt	= 10^3 millivolts	= 10^6 microvolts
1 millivolt	= 10^{-3} volt	= 10^3 microvolts
1 microvolt	= 10^{-6} volt	= 10^{-3} millivolt
1 ohm	= 10^{-3} kilohm	= 10^{-6} megohm
1 kilohm	= 10^3 ohms	= 10^{-3} megohm
1 megohm	= 10^6 ohms	= 10^3 kilohms
1 ampere	= 10^3 milliampere	= 10^6 microamperes
1 milliampere	= 10^{-3} ampere	= 10^3 microamperes
1 microampere	= 10^{-6} ampere	= 10^{-3} milliamperes

Power

The basic unit of power is the watt. Large whole numbers or decimals may be involved in the solution of power problems. Using exponents can make the work easier. Power formulas are valid only for linear resistors; that is, resistors which follow Ohm's law. See Table 1-11.

Table 1-11. Symbols, Multiple and Exponential Values of Dc Power.

Unit	Symbol	Multiple	Value
watt	P	microwatt	1/1,000,000 watt
watt	P	milliwatt	1/1000 watt
watt	P	megawatt	1,000,000 watts

In Terms of Exponents		
1 watt	= 10^3 milliwatts	= 10^6 microwatts
1 milliwatt	= 10^{-3} watt	= 10^3 microwatts
1 microwatt	= 10^{-6} watt	= 10^{-3} milliwatt

FIXED ATTENUATORS

The insertion loss of a fixed attenuator network, or pad, is the ratio of the power input to power output, given in dB, and assuming equal impedances for the source and the load. Table 1-12 is for use where these impedances are the same. The values in the table are based on input and output impedances of 600 ohms. Figure 1-5 shows the types of pads for which Table 1-12 may be used.

☐ **Example**:

A simple pad is required to supply an insertion loss of 40 dB. How many resistors are needed and what is their value?

A T-pad can be used to solve this problem. The circuit is a four-terminal network as shown in the corresponding circuit diagram. Locate 40 dB in the left-hand column of Table 1-12. To the right of this are the values for R1 and R2. R1 represents two resistors, each having a value of 588.1 ohms. R2 is shown as 12 ohms. Note that a π pad could also be used to supply the same insertion loss, except that 30,000 ohms is needed for R1 while R2 consists of two resistors, each of which is 612.1 ohms. The nearest standard values can be selected from Table 1-3, or series and parallel combinations can be used to obtain the required resistance.

☐ **Example**:

It is necessary to drop the output voltage of a 600-ohm source by 3 dB. What type of pad can do this, and what are the values of the resistors in the attenuator?

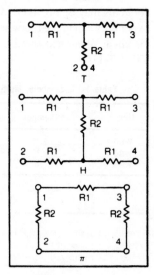

Fig. 1-5. Attenuator pads. Design values for these networks are given in Table 1-12.

Find the number 3 in the column headed Loss, dB. As shown in Table 1-12, a number of different pads can be used to get the same result. For an H pad four resistors (R1) will be needed, each having a value of 51.3 ohms. A single resistor (R2) of 1703 ohms will complete the network.

L ATTENUATOR

An L attenuator (Fig. 1-6) uses a pair of resistors operated by a single control, consequently their resistance values are changed at the same time. R1 and R_s, shown in the drawing, are in series with each other with respect to the source voltage. As the value of R1 is increased, that of R2 is decreased by the common control. Consequently, the load imposed by the attenuator remains constant for all settings of its control. The total resistance of the L pad is made equal to the impedance of the source voltage to obtain maximum transfer of electrical energy to the load.

In Fig. 1-6, R2 is in shunt with the load, and R1 is in series with it. The total resistance of the attenuator and the load, as measured across points 1 and 2 is found with the following equation:

$$R_t = R1 + \frac{(R2)\,(R_{load})}{R2 + R_{load}}$$

Because this combination is equal to the resistance of the source, then the source resistance is:

$$R_s = R1 + \frac{R2 \times R_{load}}{R2 + R_{load}}$$

OHM'S LAW FOR AC

When Ohm's law is used for ac circuits containing reactive

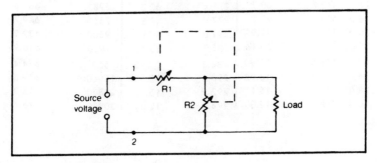

Fig. 1-6. L attenuator.

Table 1-12. Design Values for Attenuator Networks.

Loss, dB	T-PAD R1	T-PAD R2	H-PAD R1	H-PAD R2	πPAD R1	πPAD R2
0.1	3.58	50204	1.79	50204	7.20	100500
0.2	6.82	26280	3.41	26280	13.70	57380
0.3	10.32	17460	5.16	17460	20.55	34900
0.4	13.79	13068	6.90	13068	27.50	26100
0.5	17.20	10464	8.60	10464	34.40	20920
0.6	20.9	8640	10.45	8640	41.7	17230
0.7	24.2	7428	12.1	7428	48.5	14880
0.8	27.5	6540	13.75	6540	55.05	13100
0.9	31.02	5787	15.51	5787	62.3	11600
1.0	34.5	5208	17.25	5208	68.6	10440
2.0	68.8	2582	34.4	2582	139.4	5232
3.0	102.7	1703	51.3	1703	212.5	3505
4.0	135.8	1249	67.9	1249	287.5	2651
5.0	168.1	987.6	84.1	987.6	364.5	2141
6.0	199.3	803.4	99.7	803.4	447.5	1807
7.0	229.7	685.2	114.8	685.2	537.0	1569
8.0	258.4	567.6	129.2	567.6	634.2	1393
9.0	285.8	487.2	142.9	487.2	738.9	1260
10.0	312.0	421.6	156.0	421.6	854.1	1154
11.0	336.1	367.4	168.1	367.4	979.8	1071
12.0	359.1	321.7	179.5	321.7	1119	1002
13.0	380.5	282.8	190.3	282.8	1273	946.1
14.0	400.4	249.4	200.2	249.4	1443	899.1
15.0	418.8	220.4	209.4	220.4	1632	859.6
16.0	435.8	195.1	217.9	195.1	1847	826.0
17.0	451.5	172.9	225.7	172.9	2083	797.3
18.0	465.8	152.5	232.9	152.5	2344	772.8
19.0	479.0	136.4	239.5	136.4	2670	751.7
20.0	490.4	121.2	245.2	121.2	2970	733.3
22.0	511.7	95.9	255.9	95.9	3753	703.6
24.0	528.8	76.0	264.4	76.0	4737	680.8
26.0	542.7	60.3	271.4	60.3	5985	663.4
28.0	554.1	47.8	277.0	47.8	7550	649.7
30.0	563.0	37.99	281.6	37.99	9500	639.2
32.0	570.6	30.16	285.3	30.16	11930	630.9
34.0	576.5	23.95	288.3	23.95	15000	624.4
36.0	581.1	18.98	290.6	18.98	18960	619.3
38.0	585.1	15.11	292.5	15.11	23820	615.3
40.0	588.1	12.00	294.1	12.00	30000	612.1

Table 1-12. Dsign Values for Attenuator Networks (cont'd).

Loss, dB	R1	R2	R1	R2
0.1	3.60	100500	3.58	100500
0.2	6.85	57380	6.82	57380
0.3	10.28	34900	10.32	34900
0.4	13.80	26100	13.79	26100
0.5	17.20	20920	17.20	20920
0.6	20.85	17230	20.9	17230
0.7	24.25	14880	24.2	14880
0.8	27.53	13100	27.5	13100
0.9	31.2	11600	31.02	11600
1.0	34.3	10440	34.5	10440
2.0	69.7	5232	68.8	5232
3.0	106.2	3505	102.7	3505
4.0	143.8	2651	135.8	2651
5.0	182.3	2141	168.1	2141
6.0	223.8	1807	199.3	1807
7.0	268.5	1569	229.7	1569
8.0	317.1	1393	258.4	1393
9.0	369.4	1260	285.8	1260
10.0	427.0	1154	312.0	1154
11.0	489.9	1071	336.1	1071
12.0	559.5	1002	359.1	1002
13.0	636.3	946.1	380.5	946.1
14.0	721.5	899.1	400.4	899.1
15.0	816.0	859.6	418.8	859.6
16.0	923.2	826.0	435.8	826.0
17.0	1042	797.3	451.5	797.3
18.0	1172	772.8	465.8	772.8
19.0	1335	751.7	479.0	751.7
20.0	1485	733.3	490.4	733.3
22.0	1877	703.6	511.7	703.6
24.0	2369	680.8	528.8	680.8
26.0	2992	663.4	542.7	663.4
28.0	3775	649.7	554.1	649.7
30.0	4750	639.2	563.2	639.2
32.0	5967	630.9	570.6	630.9
34.0	7500	624.4	576.5	624.4
36.0	9480	619.3	581.1	619.3
38.0	11910	615.3	585.1	615.3
40.0	15000	612.1	588.1	612.1

Table 1-12. Design Values for Attenuator Networks (cont'd).

Loss, dB	T-PAD R1	H-PAD R2	πPAD R1	R2
0.1	7.2	50000	3.6	50000
0.2	13.8	26086	6.9	26086
0.3	21.0	17143	10.5	17143
0.4	28.2	12766	14.1	12766
0.5	35.4	10169	17.7	10169
0.6	43.2	8333	21.6	8333
0.7	50.4	7143	25.2	7143
0.8	57.6	6250	28.8	6250
0.9	65.4	5504	32.7	5504
1.0	73.2	4918	36.6	4918
2.0	155.4	2316	77.7	2316
3.0	247.8	1452	123.9	1452
4.0	351.0	1025	175.5	1025
5.0	466.8	771.2	233.4	771.2
6.0	597.0	603.0	298.5	603.0
7.0	743.4	484.3	371.7	484.3
8.0	907.2	396.8	453.6	396.8
9.0	1091	329.9	545.5	329.9
10.0	1297	277.5	648.5	277.5
11.0	1529	235.5	764.5	235.5
12.0	1788	201.3	894	201.3
13.0	2080	173.1	1040	173.1
14.0	2407	149.6	1204	149.6
15.0	2773	129.8	1387	129.8
16.0	3186	113.0	1598	113.0
17.0	3648	98.68	1824	98.68
18.0	4166	86.4	2083	86.4
19.0	4748	75.8	2374	75.8
20.0	5400	66.66	2700	66.66
22.0	6954	51.72	3477	51.72
24.0	8910	40.4	4455	40.4
26.0	11370	31.66	5685	31.66
28.0	14472	24.87	7236	24.87
30.0	18372	19.58	9186	19.58
32.0	23286	15.46	11643	15.46
34.0	29472	12.21	14736	12.21
36.0	37260	9.66	18630	9.66
38.0	47058	7.65	23529	7.65
40.0	59400	6.06	29700	6.06

components the phase angle (θ) between voltage and current becomes part of the calculations. Table 1-13 is a summary of Ohm's law formulas for ac.

Table 1-13. Power and Ohm's Law Formulas for Ac.

Watts	Amperes	Volts	Impedance
$P =$	$I =$	$E =$	$Z =$
I^2R	E/Z	IZ	E/I
$EI \cos \theta$	$\dfrac{P}{E \cos \theta}$	$\dfrac{P}{I \cos \theta}$	$\dfrac{E^2 \cos \theta}{P}$
$\dfrac{E^2 \cos \theta}{Z}$	$\sqrt{\dfrac{P}{Z \cos \theta}}$	$\sqrt{\dfrac{PZ}{\cos \theta}}$	$\dfrac{P}{I^2 \cos \theta}$
$I^2Z \cos \theta$	$\sqrt{\dfrac{P}{R}}$	$\dfrac{\sqrt{PR}}{\cos \theta}$	$\dfrac{R}{\cos \theta}$

CONDUCTANCE

Conductance, the reciprocal of resistance can be considered as resistive siemens while susceptance is reactive siemens.

A wire (or any other component) can be either resistive or conductive. While the terms are reciprocals they are two different ways of describing the same thing. See Table 1-14.

INTERNAL VOLTAGE DROP OF A BATTERY

Any load put across a battery is effectively in series with the internal resistance of that battery. Unlike the load which is generally a fixed value, the internal resistance of a battery can vary depending on the state of charge. A battery has minimum internal resistance when fully charged, with that resistance increasing when the battery discharges.

The current flowing through the internal resistance produces a voltage drop, thus supplying less voltage to the load. If the terminal voltage of a battery is 12 volts and the internal voltage drop is 0.3

Table 1-14. Resistance and Related Symbols.

	Symbol		Symbol
Resistance	R	Conductance	G
Reactance	X	Susceptance	B
Impedance	Z	Admittance	Y

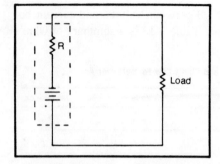

Fig. 1-7. Internal resistance of battery, R_i, increases as battery is used or gets older and decreases when battery is charged.

volt, then the actual voltage that can be delivered to the load is 12.0 − 0.3 = 11.7 volts. Measuring the voltage across a battery without its load is of no significance, since under such circumstances the high internal resistance of the voltmeter means that the voltage drop inside the battery will be very small. Consequently, the measured terminal voltage of the battery will be maximum or close to it. The true voltage of the battery is obtained only when it is connected to its load. (See Fig. 1-7.)

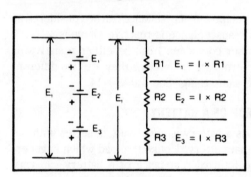

Fig. 1-8. Batteries (left) in series aiding and voltage drops in series aiding (right).

VOLTAGES IN SERIES AIDING

When voltages are wired in series aiding, either through the use of batteries or voltage drops as shown in Fig. 1-8, the total voltage is the sum of the individual voltages. For batteries, the total voltage is:

$$E_t = E_1 + E_2 + E_3$$

For voltage drops, the total voltage is:

$$E_t = (I \times R1) + (I \times R2) + (I \times R3)$$
$$= I (R1 + R2 + R3)$$
$$E_t = E1 + E2 + E3$$

Voltage and Current

SINE WAVES

A single sine wave consists of a complete cycle, as shown in Fig. 2-1. Each sine wave has two peaks; a maximum positive and a maximum negative. One peak is at 90°; the other at 270° with 360° required for a complete waveform. 180° of a cycle is one alternation or a half wave.

PEAK, PEAK-TO-PEAK, AVERAGE, AND RMS
(EFFECTIVE) VALUES OF CURRENTS OR VOLTAGES OF SINE WAVES

The peak value of a sine wave of voltage or current is measured at either 90 degrees or 270 degrees. For this reason peak (or peak-to-peak) values can be considered as instantaneous values. The average of all the instantaneous values over a complete cycle is zero; hence, average is generally understood to be the average of the instantaneous values over a half cycle. The average value is also equal to 2 divided by π. Taking the value of π as 3.14159265, then the average value of a sine wave of voltage or current is 0.636619, generally rounded off to 0.637, the value used in this book. In some texts, however, you will find the average value given as 0.636. Average value in Table 2-1 is 0.637 times the peak value.

The effective or root-mean-square (rms) value is also a form of instantaneous value averaging. Arithmetically, the effective value is obtained by dividing 1 by the square root of 2. Taking the square

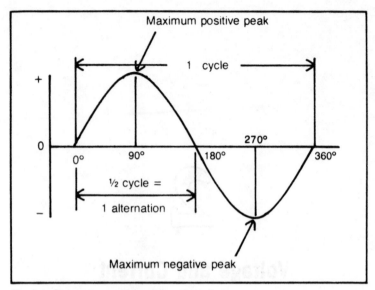

Fig. 2-1. Sine wave consists of two alternations.

root of 2 as equal to 1.414213, then the effective or rms value of a sine wave of voltage or current is equal to 1/1.414213, or 0.707107. In this text, as in other books on electricity and electronics, the rms value is rounded off to 0.707 times the peak value.

The data in Table 2-1 allows rapid movement among peak, peak-to-peak, average, and rms values of currents or voltages of sine waves. Also see Fig. 2-2.

□ **Example:**

What is the peak value of a sine wave current whose effective

**Table 2-1. Peak, Peak-to-Peak,
Average and rms (Effective) Values of Sine Wave Currents or Voltages.**

Peak	Peak-to-Peak	Average	rms
1	2	0.637	0.707
2	4	1.274	1.414
3	6	1.911	2.121
4	8	2.548	2.828
5	10	3.185	3.535
6	12	3.822	4.242
7	14	4.459	4.949
8	16	5.096	5.656
9	18	5.733	6.363
10	20	6.370	7.070

34

Table 2-1. Peak, Peak-to-Peak,
Average and rms (Effective) Values of Sine Wave Currents or Voltage (cont'd).

Peak	Peak-to-Peak	Average	rms
11	22	7.007	7.777
12	24	7.644	8.484
13	26	8.281	9.191
14	28	8.918	9.898
15	30	9.555	10.605
16	32	10.192	11.312
17	34	10.829	12.019
18	36	11.466	12.726
19	38	12.103	13.433
20	40	12.740	14.140
21	42	13.377	14.847
22	44	14.014	15.554
23	46	14.651	16.261
24	48	15.288	16.968
25	50	15.925	17.675
26	52	16.562	18.382
27	54	17.199	19.089
28	56	17.836	19.796
29	58	18.473	20.503
30	60	19.110	21.210
31	62	19.747	21.917
32	64	20.384	22.624
33	66	21.021	23.331
34	68	21.658	24.038
35	70	22.295	24.745
36	72	22.932	25.452
37	74	23.569	26.159
38	76	24.206	26.866
39	78	24.843	27.573
40	80	25.480	28.280
41	82	26.117	28.987
42	84	26.754	29.694
43	86	27.391	30.401
44	88	28.028	31.108
45	90	28.665	31.815

Peak	Peak-to-Peak	Average	rms
46	92	29.302	32.522
47	94	29.939	33.229
48	96	30.576	33.936
49	98	31.213	34.643
50	100	31.850	35.350
51	102	32.487	36.057
52	104	33.124	36.764
53	106	33.761	37.471
54	108	34.398	38.178
55	110	35.035	38.885
56	112	35.672	39.592
57	114	36.309	40.299
58	116	36.946	41.006
59	118	37.583	41.713
60	120	38.220	42.420
61	122	38.857	43.127
62	124	39.494	43.834
63	126	40.131	44.541
64	128	40.768	45.248
65	130	41.405	45.955
66	132	42.042	46.662
67	134	42.679	47.369
68	136	43.316	48.076
69	138	43.953	48.783
70	140	44.590	49.490
71	142	45.227	50.197
72	144	45.864	50.904
73	146	46.501	51.611
74	148	47.138	52.318
75	150	47.775	53.025
76	152	48.412	53.732
77	154	49.049	54.439
78	156	49.686	55.146
79	158	50.323	55.853
80	160	50.960	56.560
81	162	51.597	57.267
82	164	52.234	57.974

Table 2-1. Peak, Peak-to-Peak,
Average and rms (Effective) Values of Sine Wave Currents or Voltage (cont'd).

Peak	Peak-to-Peak	Average	rms
83	166	52.871	58.681
84	168	53.508	59.388
85	170	54.145	60.095
86	172	54.782	60.802
87	174	55.419	61.509
88	176	56.056	62.216
89	178	56.693	62.923
90	180	57.330	63.630
91	182	57.967	64.337
92	184	58.604	65.044
93	186	59.241	65.751
94	188	59.878	66.458
95	190	60.515	67.165
96	192	61.152	67.872
97	194	61.789	68.579
98	196	62.426	69.286
99	198	63.063	69.993
100	200	63.700	70.700
101	202	64.337	71.407
102	204	64.974	72.114
103	206	65.611	72.821
104	208	66.248	73.528
105	210	66.885	74.235
106	212	67.522	74.942
107	214	68.159	75.649
108	216	68 796	76.356
109	218	69.433	77.063
110	220	70.070	77.770
111	222	70.707	78.477
112	224	71.344	79.184
113	226	71.981	79.891
114	228	72.618	80.598
115	230	73.255	81.305
116	232	73.892	82.012
117	234	74.529	82.719
118	236	75.166	83.426
119	238	75.803	84.133
120	240	76.440	84.840

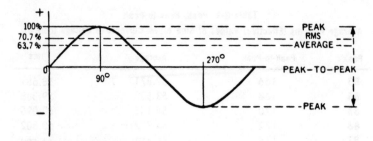

Fig. 2-2. Relationships between peak, peak-to-peak, average, and rms values of sine waves of voltage or current.

(rms) value is measured as 17½ volts?

Locate the nearest value in the rms (effective) column. This is 17.675. Move to the left along the same line and locate 25 as the answer in the column marked peak.

☐ **Example:**

What is the average value of a voltage sine wave whose peak value is 160 volts?

The maximum peak value shown in Table 2-1 is 120. You can extend the table, however, by multiplying each value by 10. Do this by adding a zero to the right of each whole number. Thus, in the peak column, 16 becomes 160. Move to the right and locate 10.192 in the average column. Multiply this value by 10 by moving the decimal point one place to the right. The average value is 101.92 volts.

VALUES OF VOLTAGE OR CURRENT OF SINE WAVES

Peak-to-peak, peak, rms and average values of voltage or current can be calculated from data in Table 2-2.

☐ **Example:**

The rms value of a sine wave is 3.14 volts. What is its peak value?

Locate rms in the left column of Table 2-2. Move across to the peak column. The multiplication factor is 1.414, 1.414 × 3.14 = 4.4399 volts.

INSTANTANEOUS VALUES OF VOLTAGE OR CURRENT OF SINE WAVES

The instantaneous value of a wave is a function of the phase angle. At 0 degrees, 180 degrees, and 360 degrees the instantane-

Given This Value	Multiply by this value to get			
	Average	rms (Effective)	Peak	p-p
Average	-	1.11	1.57	1.274
rms (Effective)	0.9	-	1.414	2.828
Peak	0.637	0.707	-	2.0
p-p	0.3185	0.3535	0.50	-

ous value of a sine wave is zero. It is a peak at 90 degrees and 270 degrees. See Fig. 2-3. These are the only values which may be known without the use of a table or formula. The instantaneous value of a sine voltage is E equals E max sin ωt (ω equals $2\pi f$) and for a sine current is I equals I max sin ω t. Table 2-3 gives the instantaneous values of either voltage or current through a complete sine wave cycle of 360 degrees for values of voltage ranging from 1 to 10. Other ranges may be obtained by moving the decimal point.

The values given in Table 2-3 under the heading of Peak Voltage or Current are those obtained from a table of natural trigonometric functions, and represent sine values. Thus, the sine of 10 degrees equals 0.1736 as shown by locating 10 in the extreme left-hand column and moving to the right and stopping in column 1. Greater accuracy can be obtained by consulting tables of natural trigonometric functions that supply a larger number of decimal places. Thus, a 5-place table would show the sine of 10 degrees as 0.17365. For example, if a sine wave of voltage or current has a peak

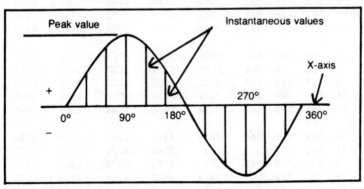

Fig. 2-3. Instantaneous values of voltage or current of a sine wave.

value of 1 volt, its instantaneous value when the wave reaches 10 degrees is 0.1736 or 0.17365 volt depending on the accuracy you want. For a 2-volt peak, the value would be 2 × 0.1736 or 0.3472. For a 3-volt peak, the value would be 3 × 0.1736 or 0.5208 volt, as shown in the respective columns headed 2 and 3 in Table 2-3. Using this technique, the instantaneous value of any sine wave or voltage can be found.

□ **Example**:

What is the instantaneous value of a sine wave at 27 degrees if the peak value of the wave is 138.2 volts?

Locate 27 degrees in the table. Move to the right and in column 1 find 0.4540 volt. This is the instantaneous value at 27 degrees when the peak value is 1 volt. For 138.2 volts, multiply 0.4540 by 138.2. That is, 0.4540 × 138.2 equals 62.7428 volts. A 5-place table

Table 2-3. Instantaneous Values of Voltage or Current of Sine Waves.

Phase Angle (degrees)				Peak Voltage or Current				
				1	2	3	4	5
0	180	180	360	.0000	.0000	.0000	.0000	.0000
1	179	181	359	.0175	.0350	.0525	.0700	.0875
2	178	182	358	.0349	.0698	.1047	.1396	.1745
3	177	183	357	.0523	.1046	.1569	.2092	.2615
4	176	184	356	.0698	.1396	.2094	.2792	.3490
5	175	185	355	.0872	.1744	.2616	.3488	.4360
6	174	186	354	.1045	.2090	.3135	.4180	.5225
7	173	187	353	.1219	.2438	.3657	.4876	.6095
8	172	188	352	.1392	.2784	.4176	.5568	.6960
9	171	189	351	.1564	.3128	.4692	.6256	.7820
10	170	190	350	.1736	.3472	.5208	.6944	.8680
11	169	191	349	.1908	.3816	.5724	.7632	.9540
12	168	192	348	.2079	.4158	.6237	.8316	1.0395
13	167	193	347	.2250	.4500	.6750	.9000	1.1250
14	166	194	346	.2419	.4838	.7257	.9676	1.0095
15	165	195	345	.2588	.5176	.7764	1.0352	1.2940
16	164	196	344	.2756	.5512	.8268	1.1024	1.3780
17	163	197	343	.2924	.5848	.8772	1.1696	1.4620
18	162	198	342	.3090	.6180	.9270	1.2360	1.5450
19	161	199	341	.3256	.6512	.9768	1.3024	1.6280
20	160	200	340	.3420	.6840	1.0260	1.3680	1.7100

Phase Angle (degrees)				Peak Voltage or Current				
				1	2	3	4	5
21	159	201	339	.3584	.7168	1.0752	1.4336	1.7920
22	158	202	338	.3746	.7492	1.1238	1.4984	1.8730
23	157	203	337	.3907	.7814	1.1721	1.5628	1.9535
24	156	204	336	.4067	.8134	1.2201	1.6268	2.0335
25	155	205	335	.4226	.8452	1.2678	1.6904	2.1130
26	154	206	334	.4384	.8768	1.3152	1.7536	2.1920
27	153	207	333	.4540	.9080	1.3620	1.8160	2.2700
28	152	208	332	.4695	.9390	1.4085	1.8780	2.3475
29	151	209	331	.4848	.9696	1.4544	1.9392	2.4240
30	150	210	330	.5000	1.0000	1.5000	2.0000	2.5000
31	149	211	329	.5150	1.0300	1.5450	2.0600	2.5750
32	148	212	328	.5299	1.0598	1.5897	2.1196	2.6495
33	147	213	327	.5446	1.0892	1.6338	2.1784	2.7320
34	146	214	326	.5592	1.1184	1.6776	2.2368	2.7960
35	145	215	325	.5736	1.1472	1.7208	2.2944	2.8680
36	144	216	324	.5878	1.1756	1.7634	2.3512	2.9390
37	143	217	323	.6018	1.2036	1.8054	2.4072	3.0090
38	142	218	322	.6157	1.2314	1.8471	2.4628	3.0785
39	141	219	321	.6293	1.2586	1.8879	2.5172	3.1465
40	140	220	320	.6428	1.2856	1.9284	2.5712	3.2140
41	139	221	319	.6561	1.3122	1.9683	2.6244	3.2805
42	138	222	318	.6691	1.3382	2.0073	2.6764	3.3455
43	137	223	317	.6820	1.3640	2.0460	2.7280	3.4100
44	136	224	316	.6947	1.3894	2.0841	2.7788	3.4735
45	135	225	315	.7071	1.4142	2.1213	2.8284	3.5355
46	134	226	314	.7193	1.4386	2.1579	2.8772	3.5965
47	133	227	313	.7314	1.4628	2.1942	2.9256	3.6570
48	132	228	312	.7431	1.4862	2.2293	2.9724	3.7155
49	131	229	311	.7547	1.5094	2.2641	3.0188	3.7735
50	130	230	310	.7660	1.5320	2.2980	3.0640	3.8300
51	129	231	309	.7771	1.5542	2.3313	3.1084	3.8855
52	128	232	308	.7880	1.5760	2.3640	3.1520	3.9400
53	127	233	307	.7986	1.5972	2.3958	3.1954	3.9930
54	126	234	306	.8090	1.6180	2.4270	3.2360	4.0450
55	125	235	305	.8192	1.6384	2.4576	3.2768	4.0960

Table 2-3. Instantaneous Values of Voltage or Current of Sine Waves (cont'd).

Phase Angle (degrees)				Peak Voltage or Current				
				1	2	3	4	5
56	124	236	304	.8290	1.6580	2.4870	3.3160	4.1450
57	123	237	303	.8387	1.6774	2.5161	3.3548	4.1935
58	122	238	302	.8480	1.6960	2.5440	3.3920	4.2400
59	121	239	301	.8572	1.7144	2.5716	3.4288	4.2860
60	120	240	300	.8660	1.7320	2.5980	3.4640	4.3300
61	119	241	299	.8746	1.7492	2.6238	3.4984	4.3730
62	118	242	298	.8829	1.7658	2.6487	3.5316	4.4145
63	117	243	297	.8910	1.7820	2.6730	3.5640	4.4550
64	116	244	296	.8988	1.7976	2.6964	3.5952	4.4940
65	115	245	295	.9063	1.8126	2.7189	3.6252	4.5315
66	114	246	294	.9135	1.8270	2.7405	3.6540	4.5675
67	113	247	293	.9205	1.8410	2.7615	3.6820	4.6025
68	112	248	292	.9272	1.8544	2.7816	3.7088	4.6360
69	111	249	291	.9336	1.8672	2.8008	3.7344	4.6680
70	110	250	290	.9397	1.8794	2.8191	3.7588	4.6985
71	109	251	289	.9455	1.8910	2.8365	3.7820	4.7275
72	108	252	288	.9511	1.9022	2.8533	3.8044	4.7555
73	107	253	287	.9563	1.9126	2.8689	3.8252	4.7815
74	106	254	286	.9613	1.9226	2.8839	3.8452	4.8065
75	105	255	285	.9659	1.9318	2.8977	3.8636	4.8295
76	104	256	284	.9703	1.9406	2.9109	3.8812	4.8515
77	103	257	283	.9744	1.9488	2.9232	3.8976	4.8720
78	102	258	282	.9781	1.9562	2.9343	3.9124	4.8905
79	101	259	281	.9816	1.9632	2.9448	3.9264	4.9080
80	100	260	280	.9848	1.9696	2.9544	3.9392	4.9240
81	99	261	279	.9877	1.9754	2.9631	3.9508	4.9385
82	98	262	278	.9903	1.9806	2.9709	3.9612	4.9515
83	97	263	277	.9925	1.9850	2.9775	3.9700	4.9625
84	96	264	276	.9945	1.9890	2.9835	3.9780	4.9725
85	95	265	275	.9962	1.9924	2.9886	3.9848	4.9810
86	94	266	274	.9976	1.9952	2.9928	3.9904	4.9880
87	93	267	273	.9986	1.9972	2.9958	3.9944	4.9930
88	92	268	272	.9994	1.9988	2.9982	3.9976	4.9970
89	91	269	271	.9998	1.9996	2.9994	3.9992	4.9990
90	90	270	270	1.0000	2.0000	3.0000	4.0000	5.0000

Table 2-3. Instantaneous Values of Voltage or Current of Sine Waves (cont'd).

Phase Angle (degrees)				Peak Voltage or Current				
				6	7	8	9	10
0	180	180	360	.0000	.0000	.0000	.0000	.0000
1	179	181	359	.1050	.1225	.1400	.1575	.1750
2	178	182	358	.2094	.2443	.2792	.3141	.3490
3	177	183	357	.3138	.3661	.4184	.4707	.5230
4	176	184	356	.4188	.4886	.5584	.6282	.6980
5	175	185	355	.5232	.6104	.6976	.7848	.8720
6	174	186	354	.6270	.7315	.8360	.9405	1.0450
7	173	187	353	.7314	.8533	.9752	1.0971	1.2190
8	172	188	352	.8352	.9744	1.1136	1.2528	1.3920
9	171	189	351	.9384	1.0948	1.2512	1.4076	1.5640
10	170	190	350	1.0416	1.2152	1.3888	1.5624	1.7360
11	169	191	349	1.4448	1.3356	1.5264	1.7172	1.9080
12	168	192	348	1.2474	1.4553	1.6632	1.8711	2.0790
13	167	193	347	1.3500	1.5750	1.8000	2.0250	2.2500
14	166	194	346	1.4514	1.6933	1.9352	2.1771	2.4190
15	165	195	345	1.5528	1.8116	2.0704	2.3292	2.5880
16	164	196	344	1.6536	1.9292	2.2048	2.4804	2.7560
17	163	197	343	1.7544	2.0468	2.3392	2.6316	2.9240
18	162	198	342	1.8540	2.1630	2.4720	2.7810	3.0900
19	161	199	341	1.9536	2.2792	2.6048	2.9304	3.2560
20	160	200	340	2.0520	2.3940	2.7360	3.0780	3.4200
21	159	201	339	2.1504	2.5088	2.8672	3.2256	3.5840
22	158	202	338	2.2476	2.6222	2.9968	3.3714	3.7460
23	157	203	337	2.3442	2.7349	3.1256	3.5163	3.9070
24	156	204	336	2.4402	2.8469	3.2536	3.6603	4.0670
25	155	205	335	2.5356	2.9582	3.3808	3.8034	4.2260
26	154	206	334	2.6304	3.0688	3.5072	3.9456	4.3840
27	153	207	333	2.7240	3.1730	3.6320	4.0860	4.5400
28	152	208	332	2.8170	3.2865	3.7560	4.2255	4.6950
29	151	209	331	2.9088	3.3936	3.8784	4.3632	4.8480
30	150	210	330	3.0000	3.5000	4.0000	4.5000	5.0000
31	149	211	329	3.0900	3.6050	4.1200	4.6350	5.1500
32	148	212	328	3.1794	3.7093	4.2392	4.7691	5.2990
33	147	213	327	3.2676	3.8122	4.3568	4.9014	5.4460
34	146	214	326	3.3552	3.9144	4.4736	5.0328	5.5920
35	145	215	325	3.4416	4.0152	4.5888	5.1634	5.7360

43

Table 2-3. Instantaneous Values of Voltage or Current of Sine Waves (cont'd).

Phase Angle (degrees)				Peak Voltage or Current				
				6	7	8	9	10
36	144	216	324	3.5268	4.1146	4.7004	5.2902	5.8780
37	143	217	323	3.6108	4.2126	4.8144	5.4162	6.0180
38	142	218	322	3.6942	4.3099	4.9256	5.5413	6.1570
39	141	219	321	3.7758	4.4051	5.0344	5.6637	6.2930
40	140	220	320	3.8568	4.4996	5.1424	5.7852	6.4280
41	139	221	319	3.9366	4.5927	5.2488	5.9049	6.5610
42	138	222	318	4.0146	4.6837	5.3528	6.0219	6.6910
43	137	223	317	4.0920	4.7740	5.4560	6.1380	6.8200
44	136	224	316	4.1682	4.8629	5.5576	6.2523	6.9470
45	135	225	315	4.2426	4.9497	5.6568	6.3639	7.0710
46	134	226	314	4.3158	5.0351	5.7544	6.4737	7.1930
47	133	227	313	4.3884	5.1198	5.8512	6.5826	7.3140
48	132	228	312	4.4586	5.2017	5.9448	6.6879	7.4310
49	131	229	311	4.5282	5.2829	6.0376	6.7923	7.5470
50	130	230	310	4.5960	5.3620	6.1280	6.8940	7.6600
51	129	231	309	4.6626	5.4397	6.2168	6.9939	7.7710
52	128	232	308	4.7280	5.5160	6.3040	7.0920	7.8800
53	127	233	307	4.7916	5.5902	6.3888	7.1874	7.9860
54	126	234	306	4.8540	5.6630	6.4720	7.2810	8.0900
55	125	235	305	4.9152	5.7344	6.5536	7.3728	8.1920
56	124	236	304	4.9740	5.8030	6.6320	7.4610	8.2900
57	123	237	303	5.0322	5.8709	6.7096	7.5083	8.3870
58	122	238	302	5.0880	5.9360	6.7840	7.6320	8.4800
59	121	239	301	5.1432	6.0004	6.8576	7.7148	8.5720
60	120	240	300	5.1960	6.0620	6.9280	7.7940	8.6600
61	119	241	299	5.2476	6.1222	6.9968	7.8714	8.7460
62	118	242	298	5.2974	6.1803	7.0632	7.9461	8.8290
63	117	243	297	5.3460	6.2370	7.1280	8.0190	8.9100
64	116	244	296	5.3928	6.2916	7.1904	8.0892	8.9880
65	115	245	295	5.4378	6.3441	7.2504	8.1567	9.0630
66	114	246	294	5.4810	6.3945	7.3080	8.2215	9.1350
67	113	247	293	5.5230	6.4435	7.3640	8.2845	9.2050
68	112	248	292	5.5632	6.4904	7.4176	8.3448	9.2720
69	111	249	291	5.6016	6.5352	7.4688	8.4024	9.3360
70	110	250	290	5.6382	6.5779	7.5176	8.4573	9.3970

Table 2-3. Instantaneous Values of Voltage or Current of Sine Waves (cont'd).

Phase Angle (degrees)				Peak Voltage or Current				
				6	7	8	9	10
71	109	251	289	5.6730	6.6185	7.5640	8.5095	9.4550
72	108	252	288	5.7066	6.6577	7.6088	8.5599	9.5110
73	107	253	287	5.7378	6.6941	7.6504	8.6067	9.5630
74	106	254	286	5.7678	6.7291	7.6904	8.6517	9.6120
75	105	255	285	5.7954	6.7613	7.7272	8.6931	9.6590
76	104	256	284	5.8218	6.7921	7.7624	8.7327	9.7030
77	103	257	283	5.8464	6.8208	7.7952	8.7696	9.7440
78	102	258	282	5.8686	6.8467	7.8248	8.8029	9.7810
79	101	259	281	5.8896	6.8712	7.8528	8.8344	9.8160
80	100	260	280	5.9088	6.8936	7.8784	8.8632	9.8480
81	99	261	279	5.9262	6.9139	7.9016	8.8893	9.8770
82	98	262	278	5.9418	6.9321	7.9224	8.9127	9.9030
83	97	263	277	5.9550	6.9475	7.9400	8.9325	9.9250
84	96	264	276	5.9670	6.9615	7.9560	8.9505	9.9450
85	95	265	275	5.9772	6.9734	7.9696	8.9658	9.9620
86	94	266	274	5.9856	6.9832	7.9808	8.9784	9.9760
87	93	267	273	5.9916	6.9902	7.9888	8.9874	9.9860
88	92	268	272	5.9964	6.9958	7.9952	8.9946	9.9940
89	91	269	271	5.9988	6.9986	7.9984	8.9982	9.9980
90	90	270	270	6.0000	7.0000	8.0000	9.0000	10.0000

of natural trigonometric functions shows the value of 27 degrees as 0.45399. 0.45399 × 138.2 equals 62.741418 volts. Whether this greater accuracy is desirable depends on the work being done. The actual difference is 62.7428 − 62.741418 equals 0.001382 volt. Although the example mentions peak in terms of volts, peak can be volts, millivolts, or microvolts, amperes, milliamperes or microamperes.

☐ **Example:**

What is the instantaneous value of voltage at a phase angle of 37 degrees when the peak value is 3 volts?

Locate 37 degrees in the left-hand column of the table. Move horizontally until the 3-volt column is reached. The required voltage is 1.8054 volts.

☐ **Example:**

At what phase angles will the instantaneous voltage of a sine wave be 68 percent of its peak value?

Consider peak as 1 or 100 percent. Locate the nearest value to 68 in the column headed by the number 1. This value is 0.6820. Move to the left of this number and you will see that the phase angle is 43 degrees. Multiples of this value are also given. We have 137 degrees (180 degrees − 43 degrees); 223 degrees (43 degrees + 180 degrees) and 317 degrees (360 degrees − 43 degrees).

□ **Example:**

What is the instantaneous value of a sine wave of current at a phase angle of 77 degrees when its peak value is 30 mA?

Locate 77 degrees in the left-hand column of the table and move horizontally to the right to intercept 2.9232 in the column headed by the number 3. Multiply 3 by 10 to obtain the peak value specified in the question. However, since 3 was changed to 30 by multiplying it by 10 (or by moving the decimal point one place to the right) the answer must be similarly treated. The value is 29.232 mA.

PERIOD AND FREQUENCY

The time required for the completion of one complete cycle by a periodic function, such as a sine wave, is known as its period. The relationship between the period and the frequency of a wave is a reciprocal one and is shown in the formula T equals 1/f. In this formula, T, the period of the wave, is the time required for the completion of one full cycle; f is the frequency in hertz (cycles per second).

Table 2-4 permits the rapid conversion between the period of a wave and its frequency. Values not given in the table can be obtained by moving the decimal point. However, since the relationship is an inverse one, the decimal point for frequency and for time will move in opposite directions. Thus, for a frequency of 10 hertz, the time is 0.1 second. For a frequency of 100 hertz, move the decimal point one place to the right, changing 10 to 100. For the corresponding value of time, however, move the decimal point one place to the left. This would change 0.1 second to 0.01 second.

□ **Example:**

The sine wave input to a power supply is 60 Hz. What is the period of this wave?

Locate 60 in the frequency column. Immediately adjacent you will see it requires 0.0167 second to complete one single cycle of this waveform.

46

Table 2-4. Period vs. Frequency.

Frequency (Hz)	Time (sec.)	Frequency (Hz)	Time (sec.)	Frequency (Hz)	Time (sec.)
1	1.0000	34	.0294	67	.0149
2	.5000	35	.0286	68	.0147
3	.3333	36	.0278	69	.0145
4	.2500	37	.0270	70	.0143
5	.2000	38	.0263	71	.0141
6	.1667	39	.0256	72	.0139
7	.1429	40	.0250	73	.0137
8	.1250	41	.0244	74	.0135
9	.1111	42	.0238	75	.0133
10	.1000	43	.0233	76	.0132
11	.0909	44	.0227	77	.0130
12	.0833	45	.0222	78	.0128
13	.0769	46	.0217	79	.0127
14	.0714	47	.0213	80	.0125
15	.0667	48	.0208	81	.0123
16	.0625	49	.0204	82	.0122
17	.0588	50	0200	83	.0120
18	.0556	51	.0196	84	.0119
19	.0526	52	.0192	85	.0118
20	.0500	53	.0189	86	.0116
21	.0476	54	0185	87	.0115
22	.0455	55	.0182	88	.0114
23	.0435	56	.0179	89	.0112
24	.0417	57	.0175	90	.0111
25	.0400	58	.0172	91	.0110
26	.0385	59	.0169	92	.0109
27	.0370	60	.0167	93	.0108
28	.0357	61	.0164	94	.0106
29	.0345	62	.0161	95	.0105
30	.0333	63	.0159	96	.0104
31	.0323	64	.0156	97	.0103
32	.0312	65	.0154	98	.0102
33	.0303	66	.0152	99	.0101
				100	.0100

□ **Example**:

What is the period of a sine waveform having a frequency of 550 kilohertz?

In Table 2-4, the frequency is given in hertz. In that table 55 can be made to represent 550 kHz by multiplying it by 10,000 or by moving its decimal point four places to the right (550,000). However, as shown in the formula given earlier, T and f are inverse. Thus, if we move the decimal point to the right for the frequency column, we must move it to the left an equal number of places for the time column. For 55 hertz, the time is 0.0182 second. For 550,00 hertz, the time is 0.00000182 second or 1.82 μsec.

□ **Example:**

The time of a half wave is 61 microseconds. What is its frequency?

Assuming the problem involves a sine wave, first multiply 61 by 2 to get the time of a full wave. 2 × 61 equals 122. 122 microseconds corresponds to 0.000122 second. The nearest value shown in the table is 0.0122 second, and the frequency for this time value is 82 Hz. We can get 0.000122 by moving the decimal point of 0.0122 two places to the left. The decimal point for the frequency, then, should be moved an equivalent number of places to the right. This would give an answer of 8200 Hz.

Capacitance

CAPACITIVE REACTANCE

The reactance of a capacitor, or its opposition to the flow of a varying or an alternating current, varies inversely with frequency and with capacitance. Capacitive reactance, expressed in ohms, is based on the ability of a capacitor to store a charge or counter-electromotive force. This emf, acting in opposition to the applied voltage, reduces the amount of current flowing in a circuit, hence produces an effect analogous to that of a resistor. With a resistor, though, the current through it and the voltage across it are in phase. However, the counter emf of a capacitor causes the voltage to lag behind the current. Ideally, the phase angle is 90 degrees, but in practice the phase angle is less than this.

Table 3-1 gives the reactance of capacitors ranging from 0.0001 μF to 0.0005 μF for frequencies ranging from 10 to 5,000 kHz. Table 3-1 is also for capacitors having values from 0.25 to 3 μF and for frequencies from 25 to 20,000 hertz. Both tables can be extended since doubling the frequency will halve the reactance. The same effect can be obtained by doubling the capacitance. Similarly, halving the frequency or capacitance will double the reactance. Naturally, other multiplication or division factors can be used. See also Table 3-2.

☐ **Example:**
What is the reactance of a .01-μF capacitor at a frequency of

Table 3-1. Capacitive Reactance (Ohms).

Frequency (Hz)	Capacitance (μF)				
	0.25	0.5	1.0	2.0	3.0
25	25,478	12,739	6,369	3,185	2,123
30	21,231	10,616	5,308	2,654	1,769
50	12,739	6,369	3,185	1,593	1,062
60	10,616	5,308	2,654	1,327	885
75	8,492	4,246	2,123	1,062	708
100	6,369	3,185	1,592	796	531
120	5,308	2,654	1,327	664	442
150	4,246	2,123	1,062	531	354
180	3,538	1,769	885	443	295
200	3,185	1,592	796	398	265
250	2,548	1,274	637	319	212
300	2,123	1,062	531	265	177
350	1,820	910	455	228	152
400	1,592	796	398	199	133
450	1,415	708	354	177	118
500	1,274	637	319	159	106
600	1,107	531	265	133	88
700	948	455	228	114	76
800	796	398	199	99	66
900	708	354	177	89	59
1,000	637	318	159	79	53
2,000	319	159	79	39	27
3,000	213	107	53	27	18
4,000	159	79	39	20	14
5,000	127	64	32	16	11
6,000	106	53	27	14	9
7,000	91	46	23	12	8
8,000	80	40	20	10	7
9,000	71	36	18	9	6
10,000	64	32	16	8	5
12,000	53	27	14	7	4.6
14,000	46	23	12	6	4
16,000	40	20	10	5	3.3
18,000	36	18	9	4.5	3
20,000	32	16	8	4	2.6

Table 3-1. Capacitive Reactance (Ohms) (cont'd).

Frequency (kHz)	Capacitance (μF)					
	0.001	0.00015	0.0002	0.00025	0.0003	0.0005
10	159,236	106,157	79,618	63,694	53,078	31,847
20	79,618	53,079	39,809	31,848	26,539	15,924
30	53,079	35,836	26,539	21,232	17,693	10,616
40	39,809	26,540	19,905	15,924	13,270	7,962
50	31,847	21,230	15,924	12,740	10,616	6,370
60	26,539	17,693	13,270	10,616	8,847	5,308
70	22,748	15,165	11,374	9,098	7,852	4,549
80	19,905	13,270	9,953	7,962	6,635	3,981
90	17,693	11,795	8,847	7,078	5,897	3,539
100	15,924	10,615	7,962	6,370	5,308	3,185
150	10,616	7,077	5,308	4,246	3,539	2,123
200	7,962	5,308	3,981	3,186	2,654	1,593
250	6,369	4,246	3,185	2,548	2,123	1,274
300	5,308	3,538	2,654	2,124	1,770	1,062
350	4,550	3,033	2,275	1,820	1,516	910
400	3,981	2,654	1,991	1,594	1,326	797
450	3,539	2,539	1,769	1,414	1,179	707
500	3,185	2,123	1,592	1,274	1,062	637
550	2,895	1,930	1,448	1,158	965	579
600	2,654	1,769	1,327	1,062	885	531
650	2,450	1,633	1,225	980	816	490
700	2,275	1,516	1,138	910	758	455
750	2,123	1,416	1,062	850	708	425
800	1,991	1,327	896	798	663	399
850	1,873	1,249	937	750	624	375
900	1,769	1,179	885	708	589	354
950	1,676	1,117	838	670	559	335
1,000	1,592	1,062	796	637	530	319
2,000	796	531	398	319	265	159
2,500	637	425	319	255	212	127
3,000	531	354	266	212	177	106
3,500	455	303	228	182	152	91
4,000	398	265	199	159	133	80
4,500	354	254	177	141	118	71
5,000	319	212	159	127	106	64

Table 3-2. Capacitive Reactance at Spot Frequencies.

Capacitance (µF)	AUDIO FREQUENCIES					
	Reactance in Ohms					
	30 Hz	60 Hz	100 Hz	400 Hz	1000 Hz	5000 Hz
.00005	-	-	-	-	-	637.000
.0001	-	-	-	-	1.590.000	318.000
.00025	-	-	-	1.590.000	637.000	127.000
.0005	-	-	3.180.000	796.000	318.000	63.700
.001	-	2.650.000	1.590.000	398.000	159.000	31.800
.005	1.060.000	530.834	318.000	79.600	31.800	6.370
.01	531.000	265.000	159.000	39.800	15.900	3.180
.02	263.000	132.500	79.600	19.900	7.960	1.590
.05	106.000	53.083	31.800	7.960	3.180	637
.1	53.100	26.500	15.900	3.980	1.590	318
.25	21.200	10.584	6.370	1.590	637	127
.5	10.600	5.308	3.180	796	318	63.7
1	5.310	2.650	1.590	389	159	31.8
2	2.650	1.325	796	199	79.6	15.9
4	1.310	663	398	99.5	39.8	7.96
8	663	332	199	49.7	19.9	3.98
16	332	166	99.5	24.9	9.95	1.99
20	262	133	80	20	8	1.6
25	212	106	63.7	15.9	6.37	1.27
35	152	86	45.5	11.4	4.55	.910
40	131	66	39.8	9.9	3.9	.8

RADIO FREQUENCIES

Capacitance (μF)	Reactance in Ohms					
	175 Hz	465 Hz	550 Hz	1000 Hz	1500 Hz	2000 Hz
.00005	18.200	6.850	5.800	3.180	2.120	1590
.0001	9.100	3.420	2.900	1.590	1.060	795
.00025	3.640	1.370	1.160	637	424	319
.0005	1.820	685	579	318	212	159
.001	910	342	290	159	106	80
.005	182	68.5	57.9	31.8	21.2	15.9
.01	91.0	34.2	28.9	15.9	10.6	8.0
.02	45.5	17.1	14.5	7.96	5.31	3.95
.05	18.2	6.85	4.79	3.18	2.12	1.59
.1	9.10	3.42	2.89	1.59	1.06	.80
.25	3.64	1.37	1.16	.637	.424	.319
.5	1.82	.685	.579	.318	.212	.159
1	.910	.342	.289	.159	.106	.080
2	.455	.171	.145	.0796	.0531	.0398
4	.227	.0856	.0723	.0398	.0265	.0199

1000 Hz? What will happen to this reactance if the frequency is increased to 10,000 Hz?

Table 3-1 shows a frequency of 1000 Hz, but does not have a capacitance value marked 0.01. Locate 1000 in the left-hand column and move horizontally to the right until you reach the number 159 under the heading of 1.0 μF. Change 1.0 μF to 0.01 by dividing it by 100. When we do this, we must multiply 159 by 100. The answer is then 15,900 ohms. If we increase the frequency to 10,000 Hz we will be multiplying the original frequency by a factor of 10. This means the reactance should be divided by a similar factor. The answer will be 15900 divided by 10 or 1590 ohms.

☐ **Example**:

What is the reactance of a 0.00015 μF capacitor at a frequency of 3,000 kHz? The Table shows that at this frequency the reactance is 354 ohms.

CAPACITORS IN SERIES

The total capacitance of series capacitors (Fig. 3-1) is always less than that of the smallest capacitor in the series network. As a general rule of thumb, if two capacitors are in series, and one has ten or more times the value of the other, the resultant total capacitance can be considered as slightly less than or equal to the value of the smaller capacitor.

Where two series capacitors have equal values, the resultant capacitance is one-half that of either unit. When three capacitors in series have equal values, the resultant capacitance is one-third that of any of the units.

Table 3-3 gives the resultant capacitance of two capacitors in

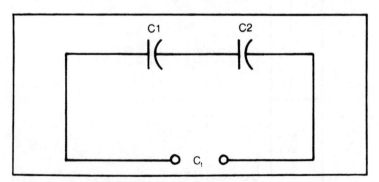

Fig. 3-1. The total capacitance (C$_t$) in a series arrangement is always less than that of any capacitor in the circuit.

Table 3-3. Capacitors in Series.

	C1				
C2	1	1.5	2	2.2	3
1	0.50	0.60	0.666	0.69	0.75
1.5	0.60	0.75	0 857	0.89	1.00
2	0.666	0.888	1.00	1.05	1.20
2.2	0.69	0.89	1.05	1.10	1.27
3	0.75	1.00	1.20	1.27	1.50
3.3	0.77	1.03	1.24	1.31	1.57
4	0.80	1.09	1.33	1.42	1.71
4.7	0.82	1.13	1.40	1.49	1.83
5	0.833	1.15	1.43	1.53	1.87
5.6	0.848	1 18	1.47	1.58	1.95
6.8	0.87	1.22	1.54	1.66	2.08
7.5	0.88	1.25	1.58	1.69	2.14
8.2	0.89	1.26	1.60	1.73	2.19
10	0 91	1.30	1.66	1.80	2.31
12	0.923	1.33	1.71	1.85	2.40
15	0.937	1.36	1.76	1.92	2.50
18	0.947	1.38	1.80	1.96	2.57
20	0.952	1.40	1.82	1.98	2.61
22	0.956	1.40	1.83	2.00	2.64
24	0.96	1.41	1.84	2.01	2.66
27	0.964	1.42	1.86	2.03	2.70
30	0.968	1.428	1.87	2.05	2.73
33	0.970	1.43	1.88	2.06	2.75
36	0.973	1.44	1.89	2.07	2.77
39	0.975	1.444	1.90	2.08	2.79
43	0.977	1.449	1.91	2.09	2.80
47	0.979	1.45	1.92	2.10	2.82
51	0.98	1.457	1.924	2.11	2.83
56	0.982	1.461	1.93	2.12	2.84
62	0.984	1.464	1.937	2.124	2.86
68	0.985	1.467	1.94	2.13	2.87
75	0.987	1.47	1.95	2.14	2.88
82	0.988	1.473	1.952	2.142	2 89
91	0.989	1.475	1 957	2.148	2.90
100	0.991	1.478	1.96	2.153	2.91

Table 3-3. Capacitors in Series (cont'd).

	C1				
C2	3.3	4	4.7	5	5.6
1	0.77	0.80	0.82	0.83	0.84
1.5	1.03	1.09	1.13	1.15	1.18
2	1.24	1.33	1.40	1.43	1.47
2.2	1.31	1.42	1.49	1.53	1.58
3	1.57	1.71	1.83	1.87	1.95
3.3	1.65	1.81	1.94	1.98	2.07
4	1.81	2.00	2.16	2.22	2.33
4.7	1.94	2.16	2.35	2.42	2.55
5	1.99	2.22	2.42	2.50	2.64
5.6	2.07	2.33	2.55	2.64	2.80
6.8	2.22	2.52	2.78	2.88	3.07
7.5	2.29	2.66	2.88	3.00	3.21
8.2	2.35	2.69	2.99	3.10	3.32
10	2.48	2.85	3.19	3.33	3.59
12	2.59	3.00	3.37	3.53	3.82
15	2.72	3.15	3.57	3.75	4.08
18	2.79	3.27	3.72	3.91	4.27
20	2.83	3.33	3.80	4.00	4.37
22	2.87	3.38	3.87	4.07	4.46
24	2.90	3.42	3.92	4.14	4.54
27	2.94	3.48	4.00	4.22	4.63
30	2.97	3.53	4.06	4.28	4.72
33	3.00	3.56	4.11	4.34	4.78
36	3.02	3.60	4.15	4.39	4.84
39	3.04	3.63	4.19	4.43	4.89
43	3.06	3.66	4.23	4.48	4.95
47	3.08	3.68	4.27	4.52	5.00
51	3.09	3.71	4.30	4.55	5.04
56	3.12	3.73	4.33	4.59	5.09
62	3.13	3.75	4.36	4.62	5.13
68	3.14	3.77	4.39	4.65	5.17
75	3.16	3.79	4.42	4.69	5.21
82	3.17	3.81	4.45	4.71	5.24
91	3.18	3.83	4.47	4.74	5.27
100	3.195	3.84	4.49	4.76	5.30

Table 3-3. Capacitors in Series (cont'd).

C1

C2	6.8	7.5	8.2	10	12
1	0.87	0.88	0.89	0.91	0.92
1.5	1.22	1.25	1.26	1.30	1.33
2	1.54	1.58	1.60	1.66	1.71
2.2	1.66	1.69	1.73	1.80	1.86
3	2.08	2.14	2.19	2.31	2.40
3.3	2.22	2.29	2.35	2.48	2.59
4	2.52	2.66	2.69	2.85	3.00
4.7	2.78	2.88	2.99	3.19	3.37
5	2.88	3.00	3.10	3.33	3.53
5.6	3.07	3.21	3.32	3.59	3.82
6.8	3.40	3.56	3.66	4.05	4.34
7.5	3.56	3.75	3.92	4.29	4.61
8.2	3.66	3.92	4.10	4.51	4.87
10	4.05	4.29	4.51	5.00	5.45
12	4.34	4.61	4.87	5.45	6.00
15	4.68	5.00	5.30	6.00	6.66
18	4.93	5.29	5.63	6.43	7.20
20	5.07	5.45	5.81	6.66	7.50
22	5.19	5.59	5.97	6.87	7.77
24	5.29	5.71	6.11	7.06	8.00
27	5.43	5.87	6.29	7.29	8.31
30	5.54	6.00	6.44	7.50	8.57
33	5.64	6.11	6.57	7.67	8.80
36	5.72	6.20	6.67	7.82	9.00
39	5.79	6.29	6.77	7.95	9.18
43	5.87	6.38	6.88	8.11	9.40
47	5.94	6.47	6.98	8.24	9.56
51	6.00	6.53	7.06	8.36	9.71
56	6.06	6.61	7.15	8.48	9.88
62	6.14	6.69	7.24	8.61	10.05
68	6.18	6.75	7.32	8.72	10.20
75	6 23	6.82	7.39	8.82	10.34
82	6.28	6.87	7.45	8.91	10.46
91	6.32	6.93	7.52	9.01	10.60
100	6.37	6.96	7.58	9.09	10.71

Table 3-3. Capacitors in Series (cont'd).

C2	C1 15	18	20	22	24
1	0 937	0.947	0.952	0.956	0.96
1.5	1.36	1.38	1.395	1.40	1.41
2	1.76	1.80	1.82	1.83	1.84
2.2	1.92	1.96	1.98	2.00	2.01
3	2.50	2.57	2.61	2.64	2.66
3.3	2.72	2.79	2.83	2.87	2.90
4	3.15	3.27	3.33	3.38	3.42
4.7	3.57	3.72	3.80	3.87	3.92
5	3.75	3.91	4.00	4.07	4.14
5.6	4.08	4.27	4.37	4.46	4.54
6.8	4.68	4.93	5.07	5.19	5.29
7.5	5.00	5.29	5.45	5.59	5.71
8.2	5.30	5.63	5.81	5.97	6.11
10	6.00	6.43	6.66	6.87	7.06
12	6.66	7.20	7.50	7.77	8.00
15	7.50	8.18	8.57	8.92	9.23
18	8.18	9.00	9.47	9.90	10.29
20	8.57	9.47	10.00	10.43	10.91
22	8.92	9.90	10.48	11.00	11.48
24	9.23	10.29	10.91	11.48	12.00
27	9.64	10.80	11.49	12.12	12.71
30	10.00	11.25	12.00	12.69	13.33
33	10.31	11.65	12.45	13.20	13.89
36	10.59	12.00	12.86	13.66	14.40
39	10 83	12.32	13.22	14.07	14.86
43	11.21	12.69	13.65	14.55	15.40
47	11.37	13.02	14.03	14.93	15.89
51	11.59	13.30	14.37	15.37	16.32
56	11.83	13.62	14.74	15.79	16.80
62	12.08	13.95	15.12	16.24	17.30
68	12.29	14.23	15.45	16.62	17.74
75	12.50	14.52	15.79	17.01	18.18
82	12.68	14.76	16.08	17.35	18.57
91	12.88	15.03	16.40	17.71	18.99
100	13.04	15.25	16.67	18.03	19.35
150	13.64	16.07	17.65	19.19	20.69
180	13.85	16.36	18.00	19.60	21.18
200	13.95	16.51	18.18	19.82	21.43
220	14.04	16.63	18.33	20.00	21.64
250	14.15	16.79	18.52	20.22	21.90

Table 3-3. Capacitors in Series (cont'd).

C2	1	1.5	2	2.2	3
1	0.50	0.60	0.666	0.69	0.75
1.5	0.60	0.75	0.857	0.89	1.00
2	0.666	0.888	1.00	1.05	1.20
2.2	0.69	0.89	1.05	1.10	1.27
3	0.75	1.00	1.20	1.27	1.50
3.3	0.77	1.03	1.24	1.31	1.57
4	0.80	1.09	1.33	1.42	1.71
4.7	0.82	1.13	1.40	1.49	1.83
5	0.833	1.15	1.43	1.53	1.87
5.6	0.848	1.18	1.47	1.58	1.95
6.8	0.87	1.22	1.54	1.66	2.08
7.5	0.88	1.25	1.58	1.69	2.14
8.2	0.89	1.26	1.60	1.73	2.19
10	0.91	1.30	1.66	1.80	2.31
12	0.923	1.33	1.71	1.85	2.40
15	0.937	1.36	1.76	1.92	2.50
18	0.947	1.38	1.80	1.96	2.57
20	0.952	1.40	1.82	1.98	2.61
22	0.956	1.40	1.83	2.00	2.64
24	0.96	1.41	1.84	2.01	2.66
27	0.964	1.42	1.86	2.03	2.70
30	0.968	1.428	1.87	2.05	2.73
33	0.970	1.43	1.88	2.06	2.75
36	0.973	1.44	1.89	2.07	2.77
39	0.975	1.444	1.90	2.08	2.79
43	0.977	1.449	1.91	2.09	2.80
47	0.979	1.45	1.92	2.10	2.82
51	0.98	1.457	1.924	2.11	2.83
56	0.982	1.461	1.93	2.12	2.84
62	0.984	1.464	1.937	2.124	2.86
68	0.985	1.467	1.94	2.13	2.87
75	0.987	1.47	1.95	2.14	2.88
82	0.988	1.473	1.952	2.142	2.89
91	0.989	1.475	1.957	2.148	2.90
100	0.991	1.478	1.96	2.153	2.91

Table 3-3. Capacitors in Series (cont'd).

C2	47	51	56	62	68	75
1	0.979	0.981	0.982	0.984	0.986	0.987
1.5	1.454	1.457	1.461	1.465	1.468	1.471
2	1.918	1.925	1.931	1.938	1.943	1.948
2.2	2.100	2.11	2.12	2.124	2.13	2.142
3	2.82	2.83	2.85	2.86	2.87	2.88
3.3	3.08	3.10	3.12	3.13	3.15	3.16
4	3.68	3.71	3.73	3.75	3.77	3.79
4.7	4.27	4.30	4.34	4.37	4.40	4.42
5	4.52	4.55	4.59	4.62	4.65	4.69
5.6	5.00	5.05	5.09	5.14	5.17	5.21
6.8	5.94	6.00	6.06	6.14	6.18	6.23
7.5	6.47	6.53	6.61	6.69	6.75	6.82
8.2	6.98	7.06	7.15	7.24	7.32	7.39
10	8.24	8.36	8.48	8.61	8.72	8.82
12	9.56	9.71	9.88	10.05	10.20	10.34
15	11.37	11.59	11.83	12.08	12.29	12.50
18	13.02	13.30	13.62	13.95	14.23	14.52
20	14.03	14.37	14.74	15.12	15.45	15.79
22	14.93	15.37	15.79	16.24	16.62	17.01
24	15.89	16.32	16.80	17.30	17.74	18.18
27	17.15	17.65	18.22	18.81	19.33	19.85
30	18.31	18.89	19.54	20.22	20.82	21.42
33	19.39	20.04	20.76	21.54	22.22	22.92
36	20.39	21.10	21.91	22.78	23.54	24.32
39	21.31	22.10	22.99	23.94	24.79	25.66
43	22.46	23.33	24.32	25.39	26.34	27.33
47	23.50	24.46	25.55	26.73	27.79	28.89
51	24.46	25.50	26.69	27.98	29.14	30.36
56	25.55	26.69	28.00	29.42	30.71	32.06
62	26.73	27 98	29.42	31.00	32.43	33.94
68	27.79	29.14	30.71	32.43	34.00	35.66
75	28.89	30.36	32.06	33.94	35.66	37.50
82	29.88	31.44	33.28	35.31	37.17	39.17
91	30.99	32.68	34.67	36.88	38.92	41.11
100	31.97	33.78	35.90	38.27	40.48	42.86

Table 3-3. Capacitors in Series (cont'd).

C2	82	91	100
1	0.988	0.989	0.990
1.5	1.473	1.476	1.478
2	1.952	1.957	1.961
2.2	2.142	2.148	2.153
3	2.89	2.90	2.91
3.3	3.17	3.19	3.20
4	3.81	3.83	3.84
4.7	4.45	4.47	4.49
5	4.71	4.74	4.76
5.6	5.24	5.28	5.30
6.8	6.28	6.32	6.37
7.5	6.87		
8.2	7.45	7.52	7.58
10	8.91	9.01	9.09
12	10.46	10.60	10.71
15	12.68	12.88	13.04
18	14.76	15.03	15.25
20	16.08	16.40	16.67
22	17.35	17.71	18.03
24	18.57	18.99	19.35
27	20.31	20.82	21.26
30	21.96	22.56	23.08
33	23.53	24.22	24.81
36	25.02	25.80	26.47
39	26.43	27.30	28.06
43	28.21	29.20	30.07
47	29.88	30.99	31.97
51	31.44	32.68	33.78
56	33.28	34.67	35.90
62	35.31	36.88	38.27
68	37.17	38.92	40.48
75	39.17	41.11	42.86
82	41.00	43.13	45.06
91	43.13	45.50	47.64
00	45.06	47.64	50.00

series. The table can be extended by moving the decimal point in the C1 and C2 columns an equal number of places to the right. The table can also be used for finding the total capacitance of three series capacitors by doing a two-step operation, i.e. determining the value of two capacitors and then combining the result with the remaining capacitor. C1 and C2 must be in similar units of μF or pF. The answers will then be in μF or pf.

□ **Example:**
What is the capacitance of two series capacitors, having values of 47 pF and 15 pF?

Locate 47 in the C2 column and move horizontally until you reach 11.37 in the C1 column headed by the number 15. The answer is 11.37 pF. You could also have solved this problem by locating 1.5 in the C2 column, and then moving across to reach 4.7 (column C1). The answer would, of course, be 1.13 pF. Moving the decimal point one place to the right supplies an answer of 11.3.

□ **Example:**
You have a number of capacitors available, but you do not have one with a capacitance of 6 pF—the value you require. What capacitor combination can you use in series to give you 6 pF?

A value of exactly 6 pF is shown in the table. It can be made by connecting a 10 pf and a 15 pF in series. Another combination which would result in a capacitance fairly close to 6 pF would be 33 pF and 7.5 pF, giving a total capacitance of 6.11 pF. Or you could use a pair of capacitors each having a value of 12 pF.

CAPACITORS IN PARALLEL OR SERIES

For capacitors connected in parallel, use the following formula:

$$C_t = C1 + C2 + C3 \ldots$$

C_t is the total capacitance, C1, C2, C3, etc., are the parallel (shunt) connected capacitors. To use the formula, the capacitors must be in the same units, i.e., microfarads or picofarads. If not, their values must be changed accordingly, that is, microfarads to picofarads, or vice versa. The total capacitance, C_t, of capacitors in parallel is always larger than that of any individual capacitor in the network (Fig. 3-2).

For capacitors connected in series, use the following equation:

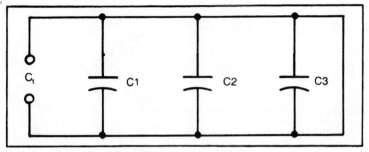

Fig. 3-2. Capacitors wired in parallel.

$$C_t = \frac{C1 \times C2}{C1 + C2}$$

As in the case of the formula for parallel capacitors, all the capacitors indicated in the series formula must be in the same capacitance units.

The formula given here is intended only for two capacitors. For any number of series connected capacitors, use this form:

$$\frac{1}{C_t} = \frac{1}{C1} + \frac{1}{C2} + \frac{1}{C3} \ldots$$

Alternatively, the following formula can be used:

$$C_t = \frac{1}{\dfrac{1}{C1} + \dfrac{1}{C2} + \dfrac{1}{C3}}$$

For capacitors in series, the total capacitance, C_t, is always less than the unit having the lowest value of capacitance in the series connected group.

Inductance

UNITS OF INDUCTANCE

The basic unit of inductance is the henry. Submultiples are the millihenry and the microhenry. Use Table 4-1 to convert from one unit to another.

VALUES OF INDUCTANCE CONVERSIONS

When working with formulas involving inductance, it is sometimes necessary to move from the basic unit of inductance to submultiples. Table 4-1 supplies the multiplication factors for making these conversions.

INDUCTIVE REACTANCE

The reactance of a coil (inductor) varies directly with frequency and with inductance. Like capacitive reactance, coil reactance or inductive reactance is measured in ohms. Inductive reactance is an effect produced by the counter-electromotive force induced across the coil. This voltage, acting in opposition to the applied voltage, reduces the amount of circuit current.

As in the case of capacitors, the coil produces a phase shift. The current lags the voltage and in a hypothetical coil (one containing no resistance) the phase angle would be 90 degrees. In practice, the phase angle is less than 90 degrees.

Table 4-1. Inductance Conversions.

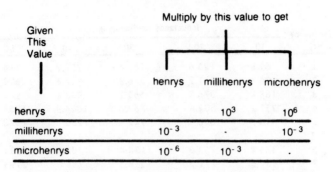

Given This Value	Multiply by this value to get		
	henrys	millihenrys	microhenrys
henrys		10^3	10^6
millihenrys	10^{-3}	.	10^{-3}
microhenrys	10^{-6}	10^{-3}	.

Table 4-2 supplies the inductive reactance of coils ranging from 10 to 100 millihenrys at frequencies ranging from 1 to 1000 kHz. The table also supplies the reactance of coils from 0.001 henry to 10 henrys from frequencies of 25 Hz to 1000 Hz.

The behavior of an inductor is opposite that of a capacitor. Increasing the frequency or the inductance increases the reactance proportionately, and vice versa. Thus, doubling either the frequency or the inductance will double the reactance. See also Table 4-3.

☐ **Example**:

What is the reactance of a coil having an inductance of 0.005 henry at a frequency of 1 kHz?

Locate 1000 hertz in the frequency column. Move to the right to the column headed by 0.005. The inductive reactance is 31.40 ohms.

☐ **Example**:

What is the reactance of a 5-henry choke coil at a frequency of 60 Hz.

Locate 60 Hz in the frequency column and move to the right, finding an inductive reactance of 1884 ohms in the column headed by the number 5. Note that the reactance of this coil at twice the frequency (120 Hz) is 3768 ohms, double its original value.

☐ **Example**:

What is the reactance of a 1-henry coil at a frequency of 150 Hz?

A frequency of 150 Hz is not listed in the table. Locate 50 Hz, find the corresponding reactance of 314 ohms and multiply this

Table 4-2. Inductive Reactance (ohms).

Frequency (kHz)	Inductance (millihenrys)				
	10	20	30	40	50
1	62.8	125.6	188.4	251.2	314
2	125.6	251.2	376.8	502.4	628
3	188.4	376.8	565.2	753.6	942
4	251.2	502.4	753.6	1004.8	1256
5	314	628	942	1256	1570
6	376.8	753.6	1130.4	1507.2	1884
7	439.6	879.2	1318.8	1758.4	2198
8	502.4	1004.8	1507.2	2009.6	2512
9	565.2	1130.4	1695.6	2260.8	2826
10	628	1256	1884	2512	3140
20	1256	2512	3768	5024	6280
25	1570	3140	4710	6280	7850
30	1884	3768	5652	7536	9420
40	2512	5024	7536	10048	12560
50	3140	6280	9420	12560	15700
60	3768	7536	11304	15072	18840
70	4396	8792	13188	17584	21980
80	5024	10048	15072	20096	25120
90	5652	11304	16956	22608	28260
100	6280	12560	18840	25120	31400
150	9420	18840	28260	37680	47100
200	12560	25120	37680	50240	62800
250	15700	31400	47100	62800	78500
300	18840	37680	56520	75360	94200
350	21980	43960	65940	87920	109900
400	25120	50240	75360	100480	125600
450	28260	56520	84780	113040	141300
500	31400	62800	94200	125600	157000
550	34540	69080	103620	138160	172700
600	37680	75360	113040	150720	188400
650	40820	81640	122460	163280	204100
700	43960	87920	131880	175840	219800
800	50240	100480	150720	200960	251200
900	56520	113040	168560	226080	282600
1000	62800	125600	188400	251200	314000

Table 4-2. Inductive Reactance (ohms) (cont'd).

Frequency (kHz)	Inductance (millihenrys)				
	60	70	80	90	100
1	376.8	439.6	502.4	565.2	628
2	753.6	879.2	1004.8	1130.4	1256
3	1130.4	1318.8	1507.2	1695.6	1884
4	1507.2	1758.4	2009.6	2260.8	2512
5	1884	2198	2512	2826	3140
6	2260.8	2637.6	3014.4	3391.2	3768
7	2637.6	3077.2	3516.8	3956.4	4396
8	3014.4	3516.8	4019.2	4521.6	5024
9	3391.2	3856.4	4521.6	5086.8	5652
10	3768	4396	5024	5652	6280
20	7536	8792	10048	11304	12560
25	9420	10990	12560	14130	15700
30	11304	13188	15072	16956	18840
40	15072	17584	20096	22608	25120
50	18840	21980	25120	28260	31400
60	22608	26376	30144	33912	37680
70	26376	30772	35168	39564	43960
80	30144	35168	40192	45216	50240
90	33912	38564	45216	50868	56520
100	37680	43960	50240	56520	62800
150	56520	65940	75360	84780	94200
200	75360	87920	100480	113040	125600
250	94200	109900	125600	141300	157000
300	113040	131880	150720	169560	188400
350	131880	153860	175840	197820	219800
400	150720	175840	200960	226080	251200
450	169560	192820	226080	254340	282600
500	188400	219800	251200	282600	314000
550	207240	241780	276320	310860	345400
600	226080	263760	301440	339120	376800
650	244920	285740	326560	367380	408200
700	263760	307720	351680	395640	439600
800	301440	351680	401920	452160	502400
900	339120	385640	452160	508680	565200
1000	376800	439600	502400	565200	628000

Table 4-2. Inductive Reactance (ohms) (cont'd).

Frequency (Hz)	Inductance (henrys)				
	1	2	3	4	5
25	157.0	314.0	471.0	628.0	785
30	188.4	376.8	565.2	753.6	942
35	219.8	439.6	659.4	879.2	1099
40	251.2	502.4	753.6	1004.8	1256
45	282.6	565.2	847.8	1130.4	1413
50	314.0	628.0	942.0	1256.0	1570
55	345.4	690.8	1036.2	1318.6	1727
60	376.8	753.6	1130.4	1507.2	1884
65	408.2	816.4	1224.6	1632.8	2041
70	439.6	879.2	1318.8	1758.4	2198
75	471.0	942.0	1413.0	1884.0	2355
80	502.4	1004.8	1507.2	2009.6	2512
85	533.8	1067.6	1601.4	2135.2	2669
90	565.2	1130.4	1695.6	2260.8	2826
95	596.6	1193.2	1789.8	2386.4	2983
100	628.0	1256.0	1884.0	2512.0	3140
120	753.6	1507.2	2260.8	3014.4	3768
200	1256.0	2512.0	3768.0	5024.0	6280
240	1507.2	3014.4	4521.6	6028.8	7536
250	1570.0	3140.0	4710.0	6280.0	7850
300	1884	3768	5652	7536	9420
350	2198	4396	6594	8792	10990
360	2260	4522	6783	9044	11305
400	2512	5024	7536	10048	12560
450	2826	5652	8478	11304	14130
500	3140	6280	9420	12560	15700
550	3454	6908	10362	13816	17270
600	3768	7536	11304	15072	18840
650	4082	8164	12246	16328	20410
700	4396	8792	13188	17584	21980
750	4710	9420	14130	18840	23550
800	5024	10048	15072	20096	25120
850	5338	10676	16014	21352	26690
900	5652	11304	16956	22608	28260
1000	6280	12560	18840	25120	31400

Table 4-2. Inductive Reactance (ohms) (cont'd).

Frequency (Hz)	Inductance (henrys)				
	.001	.002	.003	.005	.01
25	.1570	.3140	.4710	.785	1.570
30	.1884	.3768	.5652	.942	1.884
40	.2512	.5024	.7536	1.256	2.512
45	.2826	.5652	.8478	1.413	2.826
50	.3140	.6280	.9420	1.570	3.140
55	.3454	.6908	1.0362	1.727	3.454
60	.3768	.7536	1.1304	1.884	3.768
65	.4082	.8164	1.2246	2.041	4.082
70	.4396	.8792	1.3188	2.198	4.396
75	.4710	.9420	1.4130	2.355	4.710
80	.5024	1.0048	1.5072	2.512	5.024
85	.5338	1.0676	1.6014	2.669	5.338
90	.5652	1.1304	1.6956	2.826	5.652
95	.5966	1.1932	1.7898	2.983	5.966
100	.6280	1.2560	1.8840	3.140	6.280
110	.6908	1.3816	2.0724	3.454	6.908
120	.7536	1.5072	2.2608	3.768	7.536
150	.942	1.8840	2.826	4.710	9.420
175	1.099	2.1980	3.297	5.495	10.990
200	1.256	2.5120	3.768	6.280	12.560
250	1.570	3.140	4.710	7.85	15.70
300	1.884	3.768	5.652	9.42	18.84
350	2.198	4.396	6.594	10.99	21.98
400	2.512	5.024	7.536	12.56	25.12
500	3.140	6.280	9.420	15.70	31.40
550	3.454	6.908	10.362	17.27	34.54
600	3.768	7.536	11.304	18.84	37.68
650	4.082	8.164	12.246	20.41	40.82
700	4.396	8.792	13.188	21.98	43.96
750	4.710	9.420	14.130	23.55	47.10
800	5.024	10.048	15.072	25.12	50.24
850	5.338	10.676	16.014	26.69	53.38
900	5.652	11.304	16.956	28.26	56.52
950	5.966	11.932	17.898	29.83	59.66
1000	6.280	12.560	18.840	31.40	62.80

Table 4-2. Inductive Reactance (ohms) (cont'd).

Frequency (Hz)	Inductance (henrys)				
	6	7	8	9	10
25	942.0	1099.0	1256.0	1413.0	1570
30	1130.4	1318.8	1507.2	1695.6	1884
35	1318.8	1538.6	1758.4	1978.2	2198
40	1507.2	1758.4	2009.6	2260.8	2512
45	1695.6	1978.2	2260.8	2543.4	2826
50	1884.0	2198.0	2512.0	2826.0	3140
55	2072.4	2417.8	2763.2	3108.6	3454
60	2260.8	2637.6	3014.4	3391.2	3768
65	2449.2	2857.4	3265.6	3673.8	4082
70	2637.6	3077.2	3516.8	3956.4	4396
75	2826.0	3297.0	3768.0	4239.0	4710
80	3014.4	3516.8	4019.2	4521.6	5024
85	3202.8	3736.6	4270.4	4804.2	5338
90	3391.2	3956.4	4521.6	5086.8	5652
95	3579.6	4176.2	4772.8	5396.4	5966
100	3768.0	4396.0	5024.0	5652.0	6280
120	4521.6	5275.2	6028.8	6782.4	7536
200	7536.0	8792.0	10048.0	11304	12560
240	9043.2	10550.4	12057.6	13565	15072
250	9420.0	10990.0	12460.0	14130	15700
300	11304	13188	15072	16956	18840
350	13188	15386	17584	19782	21980
360	13414	15826	18086	20347	22608
400	15072	17584	20096	22608	25120
450	16956	19782	22608	25434	28260
500	18840	21980	25120	28260	31400
550	20724	24178	27632	31086	34540
600	22608	26376	30124	33912	37680
650	24492	28574	32656	36738	40820
700	26376	30772	35168	39564	43960
750	28260	32970	37680	42390	47100
800	30144	35168	40192	45216	50240
850	32028	37366	42704	48042	53380
900	33912	39564	45216	50868	56520
1000	37680	43960	50240	56520	62800

Table 4-3. Inductive Reactance at Spot Frequencies.

	50 Hz	100 Hz	1 kHz	10 kHz	100 kHz	1 MHz	10 MHz	100 MHz
1 µH	—	—	—	—	0.63	6.3	63	630
5 µH	—	—	—	0.31	3.1	31	310	3.1 K
10 µH	—	—	—	0.63	6.3	63	630	6.3 K
50 µH	—	—	0.31	3.1	31	310	3.1 K	31 K
100 µH	—	—	0.63	6.3	63	630	6.3 K	63 K
250 µH	—	0.16	1.6	16	160	1.6 K	16 K	160 K
1 mH	—	0.63	6.3	63	630	6.3 K	63 K	630 K
2.5 mH	0.8	1.6	16	160	1.6 K	16 K	160 K	1.6 M
10 mH	3.1	6.3	63	630	6.3 K	63 K	630 K	6.3 M
25 mH	8	16	160	1.6 K	16 K	160 K	1.6 M	—
100 mH	31	63	630	6.3 K	63 K	630 K	6.3 M	—
1 H	310	630	6.3 K	63 K	630 K	6.3 M	—	—
5 H	1.5 K	3.1 K	31 K	310 K	3.1 M	—	—	—
10 H	3.1 K	6.3 K	63 K	630 K	6.3 M	—	—	—
100 H	31 K	63 K	630 K	6.3 M	—	—	—	—

Values in ohms.

value by 3 to get your answer. You could also get the same result by locating 75 Hz in the Table, finding the reactance of 471 ohms and multiplying this result by 2. In either case the answer is 942 ohms.

LC PRODUCT FOR RESONANCE

Table 4-4 shows the relationship between the wavelength in meters, the frequency in kilohertz and the inductance-capacitance product (L × C) required to produce resonance. The inductance is in microhenrys and the capacitance is in microfarads.

In an LC circuit a condition of resonance is reached when the reactive elements are equal—that is, when the inductive and capacitive reactances are identical. When the inductance and the capacitance are both in microunits (microhenrys and microfarads), the resonant frequency will be in kilohertz.

To find the resonant frequency without resorting to formulas, multiply the values of L and C, first converting these to microhenrys and microfarads. Knowing the LC product, find the resonant frequency by using Table 4-4. At the same time the table also supplies the wavelength in meters.

☐ **Example:**

What is the wavelength in meters of a circuit when the inductance is 221 µh and the capacitance is 100 pf?

Table 4-4 requires that the capacitance be in microfarads. 100 µF is equivalent to 0.0001 µF. Multiply 221 by .0001. The answer

Table 4-4. LC Product for Resonance.

Wavelength (meters)	Frequency (kHz)	L × C	Wavelength (meters)	Frequency (kHz)	L × C
1	300,000	0.0000003	270	1, 1	0.0205
2	150,000	0.0000111	280	1,071	0.0221
3	100,000	0.0000018	290	1,034	0.0237
4	75,000	0.0000045	300	1,000	0.0253
5	60,000	0.0000057	310	968	0.0270
6	50,000	0.0000101	320	938	0.0288
7	42,900	0.0000138	330	909	0.0306
8	37,500	0.0000180	340	883	0.0325
9	33,333	0.0000228	350	857	0.0345
10	30,000	0.0000282	360	834	0.0365
20	15,000	0.0001129	370	811	0.0385
30	10,000	0.0002530	380	790	0.0406
40	7,500	0.0004500	390	769	0.0428
50	6,000	0.0007040	400	750	0.0450
60	5,000	0.0010140	410	732	0.0473
70	4,290	0.0013780	420	715	0.0496
80	3.750	0.0018010	430	698	0.0520
90	3,333	0.0022800	440	682	0.0545
100	3,000	0.00282	450	667	0.0570
110	2,727	0.00341	460	652	0.0596
120	2,500	0.00405	470	639	0.0622
130	2,308	0.00476	480	625	0.0649
140	2,143	0.00552	490	612	0.0676
150	2,000	0.00633	500	600	0.0704
160	1,875	0.00721	505	594	0.0718
170	1,764	0.00813	510	588	0.0732
180	1,667	0.00912	515	583	0.0747
190	1,579	0.01015	520	577	0.0761
200	1,500	0.01126	525	572	0.0776
210	1,429	0.01241	530	566	0.0791
220	1,364	0.01362	535	561	0.0806
230	1,304	0.01489	540	556	0.0821
240	1,250	0.01621	545	551	0.0836
250	1,200	0.01759	550	546	0.0852
260	1,154	0.01903	555	541	0.0867

Table 4-4. LC Product for Resonance (cont'd).

Wave-length (meters)	Frequency (kHz)	L × C	Wave-length (meters)	Frequency (kHz)	L × C
560	536	0.0883	735	408	0.1521
565	531	0.0899	740	405	0.1541
570	527	0.0915	745	403	0.1562
575	522	0.0931	750	400	0.1583
580	517	0.0947	755	397	0.1604
585	513	0.0963	760	395	0.1626
590	509	0.0980	765	392	0.1647
595	504	0.0996	770	390	0.1669
600	500	0.1013	775	387	0.1690
605	496	0.1030	780	385	0.1712
610	492	0.1047	785	382	0.1734
615	488	0.1065	790	380	0.1756
620	484	0.1082	795	377	0.1779
625	480	0.1100	800	375	0.1801
630	476	0.1117	805	373	0.1824
635	472	0.1135	810	370	0.1847
640	469	0.1153	815	368	0.1870
645	465	0.1171	820	366	0.1893
650	462	0.1189	825	364	0.1916
655	458	0.1208	830	361	0.1939
660	455	0.1226	835	359	0.1962
665	451	0.1245	840	357	0.1986
670	448	0.1264	845	355	0.201
675	444	0.1283	850	353	0.203
680	441	0.1302	855	351	0.206
685	438	0.1321	860	349	0.208
690	435	0.1340	865	347	0.211
695	432	0.1360	870	345	0.213
700	429	0.1379	875	343	0.216
705	426	0.1399	880	341	0.218
710	423	0.1419	885	339	0.220
715	420	0.1439	890	337	0.223
720	417	0.1459	895	335	0.225
725	414	0.1479	900	333	0.228
730	411	0.1500	905	331	0.231

73

Table 4-4. LC Product for Resonance (cont'd).

Wave-length (meters)	Frequency (kHz)	L × C	Wave-length (meters)	Frequency (kHz)	L × C
910	330	0.233	955	314	0.257
915	328	0.236	960	313	0.260
920	326	0.238	965	311	0.262
925	324	0.241	970	309	0.265
930	323	0.243	975	308	0.268
935	321	0.246	980	306	0.270
940	319	0.249	985	305	0.273
945	317	0.251	990	303	0.276
950	316	0.254	995	302	0.279
			1000	300	0.282

will be 0.0221. Locate this value in the L × C column in the table. You will find that the frequency in kilohertz is 1071 and the corresponding wavelength is 280 meters.

□ **Example**:

What is the resonant frequency of a circuit whose capacitance has a value of 250 pF and an inductance of 136 μH?

The inductance value can be used as it is, but the capacitance must be changed to microfarads. 250 pF is equivalent to 0.00025 μF. The product of 0.00025 and 136 is 0.0340. The table does not list such a value, but it does have a value that is close. The L × C column shows this to be 0.0345 with a corresponding frequency in kilohertz of 857. The wavelength is 350 meters.

□ **Example**:

You want a circuit that will be resonant at 5 megahertz. What must be the value of the LC product?

5 megahertz is equal to 5000 kilohertz. The table shows that the LC product is 0.0010140. Any combination of inductance and capacitance producing this product will be resonant at 5 MHz.

5

Impedance

IMPEDANCE VECTORS

The phase relationships of resistive and reactive components (such as inductors and capacitors) can be indicated by lines (vectors) terminating in arrows. For series resistors, as shown in Fig. 5-1A, the phase difference is zero. Voltages appearing across these resistors will be in phase.

For a coil and resistor, Fig. 5-1B, the phase angle is a maximum of 90°. This is indicated by drawing the vectors for the two components at a right angle to each other. The length of the vector line should be comparable to the value, in ohms, of each component. Thus, if the resistor is 10 ohms and the inductive reactance (X_L) of the coil is 20 ohms, the vector line for the coil should have twice the length of the line for the resistor.

Vectors can also be drawn for a series circuit consisting of R and C, as indicated in Fig. 5-1C. For a series circuit consisting of R, C, and L, as shown in Fig. 5-1D, the final vector diagram indicates the difference between X_L and X_C. If X_L is larger than X_C, then the amount of capacitive reactance is subtracted from inductive reactance, reducing it accordingly as shown in Fig. 5-1D. The capacitive reactance can be larger than the inductive reactance, though, in which case X_L is subtracted from X_C.

If you have both types of reactance, then, subtract one from the other to get the value of X. It makes no difference which reactive

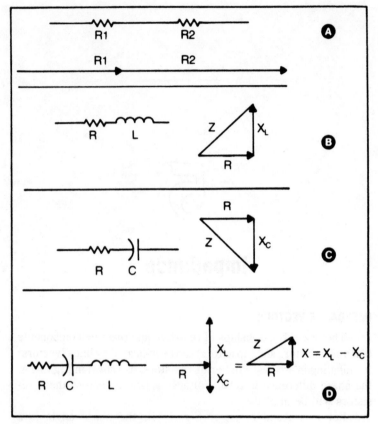

Fig. 5-1. Vector relationships in a series circuit.

component (inductance or capacitance) has the larger reactance. Subtract the value of the smaller reactance from that of the larger to obtain the value of X.

When X_L is equal to X_C, the two reactances effectively cancel and the impedance is equal to R.

$$X = X_L - X_C$$
$$\text{or}$$
$$X = X_C - X_L$$
$$Z = \sqrt{R^2 + X^2}$$

The impedance of a series RC, RL, or RLC circuit is the vector sum of the individual reactances and resistance. Table 5-1 supplies the impedance in ohms, when the values of R and X are known. X

Table 5-1. Impedance (ohms) for Series R and X.

X	R 1	2	3	4	5	6
1	1.41	2.24	3.16	4.12	5.10	6.08
2	2.24	2.83	3.61	4.47	5.39	6.32
3	3.16	3.61	4.24	5.00	5.83	6.71
4	4.12	4.47	5.00	5.57	6.40	7.21
5	5.10	5.39	5.83	6.40	7.07	7.81
6	6.08	6.32	6.71	7.21	7.81	8.48
7	7.07	7.28	7.62	8.06	8.60	9.22
8	8.06	8.25	8.54	8.94	9.43	10.00
9	9.06	9.22	9.49	9.85	10.29	10.81
10	10.05	10.19	10.44	10.77	11.18	11.66
11	11.04	11.18	11.40	11.70	12.08	12.52
12	12.04	12.16	12.36	12.64	13.00	13.41
13	13.03	13.15	13.34	13.60	13.92	14.31
14	14.03	14.14	14.31	14.56	14.86	15.23
15	15.03	15.13	15.29	15.52	15.81	16.15
16	16.03	16.12	16.28	16.49	16.76	17.09
17	17.03	17.12	17.26	17.46	17.72	18.03
18	18.03	18.11	18.25	18.44	18.68	18.97
19	19.03	19.10	19.24	19.42	19.65	19.92
20	20.02	20.05	20.22	20.40	20.62	20.88
21	21.02	21.10	21.21	21.38	21.59	21.84
22	22.02	22.09	22.20	22.36	22.56	22.80
23	23.02	23.09	23.19	23.35	23.54	23.75
24	24.02	24.08	24.19	24.33	24.52	24.74
25	25.02	25.08	25.18	25.32	25.50	25.71
26	26.02	26.08	26.17	26.31	26.48	26.68
27	27.02	27.07	27.16	27.29	27.46	27.66
28	28.02	28.07	28.16	23.28	28.44	28.64
29	29.02	29.07	29.15	29.27	29.43	29.61
30	30.02	30.07	30.15	30.27	30.41	30.59
31	31.02	31.06	31.14	31.26	31.40	31.58
32	32.01	32.06	32.14	32.24	32.39	32.55
33	33.01	33.06	33.14	33.24	33.38	33.54
34	34.01	34.06	34.14	34.24	34.37	34.53
35	35.01	35.06	35.13	35.23	35.36	35.51

Table 5-1. Impedance (ohms) for Series R and X (cont'd).

X	7	8	9	10	11	12
1	7.07	8.06	9.06	10.05	11.04	12.04
2	7.28	8.25	9.22	10.19	11.18	12.16
3	7.62	8.54	9.49	10.44	11.40	12.36
4	8.06	8.94	9.85	10.77	11.70	12.64
5	8.60	9.43	10.29	11.18	12.08	13.00
6	9.22	10.00	10.81	11.66	12.52	13.41
7	9.89	10.63	11.40	12.20	13.03	13.89
8	10.63	11.31	12.04	12.80	13.60	14.42
9	11.40	12.04	12.72	13.45	14.21	15.00
10	12.20	12.80	13.45	14.14	14.86	15.62
11	13.03	13.60	14.21	14.86	15.55	16.27
12	13.89	14.42	15.00	15.62	16 27	16.97
13	14.76	15.26	15.81	16 40	17 02	17.69
14	15.65	16.12	16.64	17.20	17.80	18.43
15	16.55	17.00	17.49	18.02	18.60	19.20
16	17.46	17.89	18.36	18.87	19.42	20.00
17	18.38	18.79	19.24	19.72	20.25	20 81
18	19.31	19.70	20.12	20.59	21.10	21.63
19	20.25	20.62	21.02	21.47	21.95	22.47
20	21.19	21.54	21.93	22.36	22.83	23.32
21	22.14	22.47	22.85	23 26	23.71	24.19
22	23.09	23.41	23.77	24.17	24.60	25.06
23	24.04	24.35	24.70	25 08	25.50	25 94
24	25 00	25.30	25.63	26 00	26.40	26 83
25	25.96	26.25	26.57	26.93	27.31	27 73
26	26.93	27.20	27.51	27.86	28.23	28.64
27	27.89	28.16	28.46	28.79	29.15	29.55
28	28 86	29.12	29.41	29 73	30 08	30 46
29	29.83	30.08	30.36	30 68	31.02	31 38
30	30.81	31.05	31.32	31.62	31 96	32.31
31	31.78	31.93	32.25	32.56	32 90	33 24
32	32.75	32.98	33.24	33.52	33.85	34.18
33	33.74	33.96	34.21	34.49	34.79	35 12
34	34.72	34.93	35.17	35.44	35 74	36 06
35	35.70	35.90	36.14	36.40	36 69	37.00

Table 5-1. Impedance (ohms) for Series R and X (cont'd).

X	R					
	13	14	15	16	17	18
1	13.03	14.03	15.03	16.03	17.03	18.03
2	13.15	14.14	15.13	16.12	17.12	18.11
3	13.34	14.31	15.29	16.28	17.26	18.25
4	13.60	14.56	15.52	16.49	17.46	18.44
5	13.92	14.86	15.81	16.76	7.72	18.68
6	14.31	15.23	16.15	17.09	18.03	18.97
7	14.76	15.65	16.55	17.47	18.38	19.31
8	15.26	16.12	17.00	17.89	18.79	19.70
9	15.81	16.64	17.49	18.36	19.24	20.12
10	16.40	17.20	18.02	18.87	19.72	20.59
11	17.02	17.80	18.60	19.42	20.25	21.10
12	17.69	18.43	19.20	20.00	20.81	21.63
13	18.38	19.10	19.84	20.62	21.40	22.20
14	19.10	19.79	20.51	21.26	22.02	22.80
15	19.84	20.51	21.21	21.93	22.67	23.43
16	20.62	21.26	21.93	22.63	23.35	24.08
17	21.40	22.02	22.67	23.35	24.04	24.76
18	22.20	22.80	23.43	24.08	24.76	25.46
19	23.02	23.60	24.20	24.84	25.50	26.17
20	23.85	24.41	25.00	25.61	26.25	26.91
21	24.70	25.24	25.81	26.40	27.02	27.66
22	25.55	26.08	26.63	27.20	27.80	28.43
23	26.42	26.93	27.46	28.02	28.60	29.21
24	27.29	27.78	28.30	28.84	29.41	30.00
25	28.18	28.65	29.15	29.68	30.23	30.81
26	29.07	29.53	30.02	30.53	31.06	31.62
27	29.97	30.41	30.89	31.38	31.91	32.45
28	30.87	31.30	31.77	32.35	32.76	33.29
29	31.78	32.21	32.65	33.12	33.62	34.14
30	32.70	33.11	33.54	34.00	34.48	34.99
31	33.62	34.01	34.44	34.89	35.36	35.85
32	34.54	35.50	35.90	36.33	36.78	37.26
33	35.47	35.85	36.25	36.67	37.12	37.59
34	36.40	36.77	37.16	37.58	38.01	38.47
35	37.34	37.60	38.08	38.48	38.91	39.36

Table 5-1. Impedance (ohms) for Series R and X (cont'd).

X	R 19	20	21	22	23	24
1	19.03	20.02	21.02	22.02	23.02	24.02
2	19.10	20.10	21.10	22.09	23.09	24.08
3	19.24	20.22	21.21	22.20	23.19	24.19
4	19.42	20.40	21.38	22.36	23.35	24.33
5	19.65	20.62	21.59	22.56	23.54	24.52
6	19.92	20.88	21.84	22.80	23.77	24.74
7	20.25	21.18	22.14	23.09	24.04	25.00
8	20.62	21.54	22.47	23.41	24.35	25.30
9	21.02	21.93	22.85	23.77	24.70	25.63
10	21.47	22.36	23.26	24.17	25.08	26.00
11	21.95	22.82	23.71	24.60	25.50	26.40
12	22.47	23.32	24.19	25.06	25.94	26.83
13	23.02	23.85	24.70	25.55	26.42	27.29
14	23.60	24.41	25.24	26.08	26.93	27.78
15	24.20	25.00	25.81	26.63	27.46	28.30
16	24.84	25.61	26.40	27.20	28.02	28.84
17	25.50	26.25	27.02	27.80	28.60	29.41
18	26.17	26.91	27.66	28.43	29.21	30.00
19	26.86	27.59	28.32	29.07	29.83	30.61
20	27.59	28.28	29.00	29.73	30.48	31.24
21	28.32	29.00	29.70	30.41	31.14	31.89
22	29.07	29.73	30.41	31.11	31.83	32.56
23	29.83	30.48	31.14	31.83	32.52	33.24
24	30.61	31.24	31.89	32.56	33.24	33.94
25	31.40	32.02	32.65	33.30	33.97	34.66
26	32.20	32.80	33.42	34.06	34.71	35.38
27	33.02	33.60	34.21	34.83	35.47	36.13
28	33.85	34.41	35.01	35.61	36.24	36.88
29	34.67	35.23	35.81	36.40	37.01	37.64
30	35.51	36.06	36.62	37.20	37.80	38.42
31	36.36	36.89	37.44	38.01	38.60	39.21
32	37.22	37.73	38.28	38.83	39.41	40.00
33	38.08	38.59	39.12	39.66	40.22	40.81
34	38.95	39.45	39.96	40.40	41.05	41.62
35	39.82	40.31	40.82	41.34	41.88	42.44

Table 5-1. Impedance (ohms) for Series R and X (cont'd).

X	R					
	25	26	27	28	29	30
1	25.02	26.02	27.02	28.02	29.02	30.02
2	25.08	26.08	27.07	28.07	29.07	30.07
3	25.18	26.17	27.17	28.16	29.15	30.15
4	25.32	26.31	27.29	28.28	29.27	30.27
5	25.50	26.46	27.46	28.44	29.43	30.41
6	25.71	26.68	27.66	28.64	29.61	30.59
7	25.98	26.93	27.89	28.86	29.83	30.81
8	26.25	27.20	28.16	29.12	30.08	31.05
9	26.57	27.51	28.46	29.41	30.36	31.32
10	26.93	27.86	28.79	29.73	30.68	31.62
11	27.31	28.23	29.15	30.08	31.02	31.95
12	27.73	28.64	29.55	30.46	31.38	32.31
13	28.18	29.07	29.97	30.87	31.78	32.60
14	28.65	29.53	30.41	31.30	32.20	33.11
15	29.15	30.02	30.89	31.77	32.64	33.54
16	29.68	30.53	31.38	32.25	33.12	33.80
17	30.23	31.06	31.91	32.76	33.62	34.48
18	30.81	31.62	32.45	33.29	34.13	34.99
19	31.40	32.20	33.02	33.85	34.67	35.51
20	32.02	32.80	33.60	34.41	35.23	36.06
21	32.65	33.42	34.21	35.01	35.81	36.62
22	33.30	34.06	34.83	35.61	36.40	37.20
23	33.97	34.71	35.47	36.24	37.01	37.80
24	34.66	35.38	36.13	36.88	37.64	38.42
25	35.36	36.07	36.70	37.54	38.29	39.05
26	36.07	36.77	37.48	38.21	38.95	39.60
27	36.70	37.48	38.18	38.80	39.62	40.36
28	37.54	38.21	38.71	39.50	40.31	41.04
29	38.29	38.95	39.62	40.31	41.01	41.73
30	39.05	39.60	40.36	41.04	41.73	42.43
31	39.81	40.46	41.11	41.77	42.44	43.14
32	40.61	41.23	41.87	42.52	43.19	43.86
33	41.40	42.01	42.64	43.28	43.93	44.50
34	42.20	42.80	43.42	44.05	44.69	45.34
35	43.01	43.60	44.20	44.82	45.45	46.00

Table 5-1. Impedance (ohms) for Series R and X (cont'd).

X	R				
	31	32	33	34	35
1	31.01	32.02	33.02	34.02	35.01
2	31.06	32.06	33.06	34.06	35.06
3	31.14	32.14	33.14	34.13	35.13
4	31.26	32.25	33.24	34.23	35.23
5	31.40	32.39	33.38	34.37	35.36
6	31.57	32.56	33.54	34.53	35.51
7	31.78	32.76	33.74	34.71	35.69
8	32.02	32.99	33.96	34.93	35.90
9	32.28	33.24	34.21	35.17	36.14
10	32.57	33.53	34.48	35.36	36.40
11	32.89	33.85	34.79	35.74	36.69
12	33.24	34.18	35.11	36.06	36.80
13	33.62	34.53	35.48	36.40	37.34
14	34.02	34.93	35.85	36.77	37.60
15	34.44	35.34	36.25	37.16	38.08
16	34.89	35.78	36.67	37.58	38.48
17	35.37	36.24	37.12	38.01	38.90
18	35.85	36.72	37.59	38.47	39.36
19	36.36	37.22	38.08	38.95	39.82
20	36.89	37.74	38.59	39.46	40.31
21	37.44	38.28	39.12	39.96	40.82
22	38.01	38.83	39.66	40.40	41.34
23	38.60	39.41	40.22	41.05	41.88
24	39.21	40.00	40.80	41.63	42.44
25	39.81	40.61	41.40	42.20	43.01
26	40.46	41.23	42.01	42.80	43.60
27	41.11	41.87	42.64	43.42	44.20
28	41.77	42.52	43.28	44.05	44.82
29	42.44	43.19	43.93	44.69	45.45
30	43.14	43.86	44.50	45.34	46.00
31	43.84	44.55	45.28	46.01	46.76
32	44.55	45.26	45.97	46.69	47.42
33	45.28	45.98	46.68	47.38	48.10
34	46.01	46.69	47.38	48.08	48.70
35	46.75	47.42	48.10	48.70	49.40

can represent either inductive or capacitive reactance, or X may be the vector sum of these reactive components when both are present in the circuit.

Table 5-1 covers a range of X and of R from 1 ohm to 35 ohms. The table can be extended by moving the decimal point an equal number of places for R, X, and the answer. Thus, if you had a resistor with a value of 30 ohms and a reactance with a value of 90 ohms, you could consider 3 in the R column as 30, and 9 in the X column as 90. The value of impedance would be shown in the table as 9.49 ohms, but moving the decimal point one place to the right gives an impedance of 94.9 ohms. Table 5-2 lists common impedance formulas.

□Example:

A circuit consists of a 5-ohm resistor, a coil having an inductive reactance of 35 ohms and a capacitor having a capacitive reactance of 21 ohms. What is the impedance of this circuit?

Subtract the value of the smaller reactance from that of the larger. 35−21 equals 14 ohms. This is the net reactance and constitutes the value of X. Using this value, locate 14 in the X column. Move horizontally to reach the 5 column (R). The impedance is shown as 14.86 ohms.

□Example:

You require an impedance of 30 ohms for an RLC circuit. What values of R and X can you use to obtain this impedance?

The table shows you can get this impedance by using an 18-ohm resistor and a reactance having a value of 24 ohms. This reactance may be a coil having an inductive reactance of this value, or a capacitor, or both components whose net reactance is 24 ohms. Of course you could also use a 24-ohm resistor and a reactance of 18 ohms. There are many other combinations in the table that are very close to this desired 30-ohm impedance.

VECTOR CONVERSION

Table 5-3 can be used to change the form to vector quantities from j-notation to polar notation, or from polar to j-notation. The first column is the ratio of reactance, X, to resistance, R. The second column is the phase angle of the polar vector and is the angle whose tangent is X/R. The third column, Z, is the absolute magnitude of the vector in terms of X.

Table 5-2. Impedance Formulas.

Circuit	Impedance	Phase Angle
X_L R	$Z = \sqrt{R^2 + X_L^2}$	$\phi = \tan^{-1} \dfrac{X_L}{R}$
X_C R	$Z = \sqrt{R^2 + X_C^2}$	$\phi = \tan^{-1} - \dfrac{X_C}{R}$
X_C X_L	$Z = X_L - X_C$ $Z = 0$ when $X_L = X_C$	$\phi = 0$ when $X_L = X_C$ $\phi = -90°$ when $X_L < X_C$ $\phi = +90°$ when $X_L > X_C$
X_C X_L R	$Z = \sqrt{R^2 + (X_L - X_C)^2}$ $Z = R$ when $X_L = X_C$	$\phi = \tan^{-1} \dfrac{X_L - X_C}{R}$ $\phi = 0$ when $X_L = X_C$
R X_L	$Z = \sqrt{\dfrac{R X_L}{R^2 + X_L^2}}$	$\phi = \tan^{-1} + \dfrac{R}{X_L}$
R X_C	$Z = \sqrt{\dfrac{R X_C}{R^2 + X_C^2}}$	$\phi = \tan^{-1} - \dfrac{R}{X_C}$
X_C X_L	$Z = \dfrac{L}{C (X_L - X_C)}$ $Z = \infty$ when $X_L = X_C$	$\phi = 0$ when $X_L = X_C$ $\phi = +90°$ when $X_L < X_C$ $\phi = -90°$ when $X_L > X_C$
R X_C X_L	$Z = \dfrac{R X_L X_C}{\sqrt{(R X_L - R X_C)^2 + X_L^2 X_C^2}}$ $Z = R$ when $X_L = X_C$	$\phi = \tan^{-1} - \dfrac{R X_L - R X_C}{X_L X_C}$ $\phi = 0$ when $X_L = X_C$

Where: Z = equivalent or resultant impedance in ohms
R = resistance in ohms
X_L = inductive reactance in ohms (magnitude only)
X_C = capacitive reactance in ohms (magnitude only)
ϕ = phase angle in degrees

Table 5-2. Impedance Formulas (cont'd).

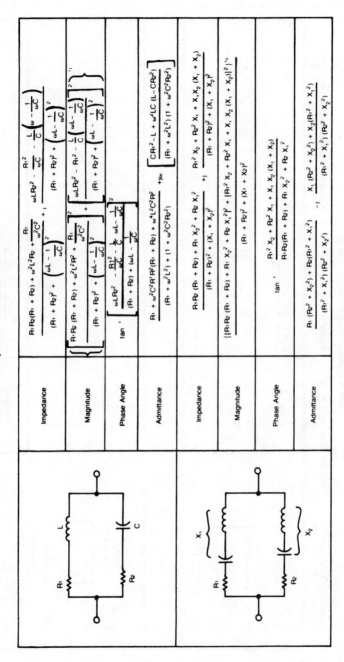

	Impedance	$\dfrac{R_1 R_2 (R_1 + R_2) + \omega^2 L^2 R_2 + \dfrac{R_1}{\omega^2 C^2}}{(R_1 + R_2)^2 + \left(\omega L - \dfrac{1}{\omega C}\right)^2} + j\omega \left[\dfrac{\omega L R_2^2 - \dfrac{R_1^2}{\omega C} - \dfrac{L}{C}\left(\omega - \dfrac{1}{\omega C}\right)}{(R_1 + R_2)^2 + \left(\omega L - \dfrac{1}{\omega C}\right)^2}\right]$
	Magnitude	$\left\{\dfrac{\left[R_1 R_2 (R_1 + R_2) + \omega^2 L^2 R_2 + \dfrac{R_1}{\omega^2 C^2}\right]^2 + \left[\omega L R_2^2 - \dfrac{R_1^2}{\omega C} - \dfrac{L}{C}\left(\omega - \dfrac{1}{\omega C}\right)\right]^2}{\left[(R_1 + R_2)^2 + \left(\omega L - \dfrac{1}{\omega C}\right)^2\right]^2}\right\}^{1/2}$
	Phase Angle	$\tan^{-1} \dfrac{\omega L R_2^2 - \dfrac{R_1^2}{\omega C} - \dfrac{L}{C}\left(\omega L - \dfrac{1}{\omega C}\right)}{R_1 R_2^2 (R_1 + R_2) + \omega^2 L^2 R_2 + \dfrac{R_1}{\omega^2 C^2}}$
	Admittance	$\dfrac{R_1 + \omega^2 C^2 R_1^2 R^2 (R_1 + R_2) + \omega^4 L^2 C^2 R^2}{(R_1 + \omega^2 L^2) (1 + \omega^2 C^2 R_2^2)} + j\omega \left[\dfrac{CR^2 - L + \omega^4 LC (L - CR_2^2)}{(R_1 + \omega^2 L^2)(1 + \omega^2 C^2 R_2^2)}\right]$
	Impedance	$\dfrac{R_1 R_2 (R_1 + R_2) + R_1 X_2^2 + R_2 X_1^2}{(R_1 + R_2)^2 + (X_1 + X_2)^2} + j\left[\dfrac{R_1^2 X_2 + R_2^2 X_1 + X_1 X_2 (X_1 + X_2)}{(R_1 + R_2)^2 + (X_1 + X_2)^2}\right]$
	Magnitude	$\dfrac{\left\{[R_1 R_2 (R_1 + R_2) + R_1 X_2^2 + R_2 X_1^2]^2 + [R_1^2 X_2 + R_2^2 X_1 + X_1 X_2 (X_1 + X_2)]^2\right\}^{1/2}}{(R_1 + R_2)^2 + (X_1 + X_2)^2}$
	Phase Angle	$\tan^{-1} \dfrac{R_1^2 X_2 + R_2^2 X_1 + X_1 X_2 (X_1 + X_2)}{R_1 R_2 (R_1 + R_2) + R_1 X_2^2 + R_2 X_1^2}$
	Admittance	$\dfrac{R_1 (R_2^2 + X_2^2) + R_2 (R_1^2 + X_1^2)}{(R_1^2 + X_1^2) (R_2^2 + X_2^2)} - j\dfrac{X_1 (R_2^2 + X_2^2) + X_2 (R_1^2 + X_1^2)}{(R_1^2 + X_1^2) (R_2^2 + X_2^2)}$

Table 5-2. Impedance Formulas (cont'd).

Impedance $Z = R + jX$ ohms
Magnitude $Z = |R^2 + X^2|^{1/2}$ ohms

Phase Angle $\phi = \tan^{-1}\frac{X}{R}$

Admittance $Y = \frac{1}{Z}$ mhos

Phase Angle of the Admittance is $-\tan^{-1}\frac{X}{R}$

	Impedance	$\dfrac{R + j\omega\,[L(1-\omega^2 LC2)-CR^2]}{(1-\omega^2 LC)^2 + \omega^2 C^2 R^2}$
	Magnitude	$\left[\dfrac{R^2 + \omega^2 L^2}{(1-\omega^2 LC)^2 + \omega^2 C^2 R^2}\right]^{\frac{1}{2}}$
	Phase Angle	$\tan^{-1}\dfrac{\omega\,[L(1-\omega^2 LC)-CR^2]}{R}$
	Admittance	$\dfrac{R- j\omega\,[L(1-\omega^2 LC)-CR^2]}{R^2 + \omega^2 L^2}$
	Impedance	$\dfrac{X_1 R_2 + j\,[R_2^2 + X_2(X_1 + X_2)]}{R_2^2 + (X_1 + X_2)^2}\,X_1$
	Magnitude	$X_1\dfrac{\{X_1{}^2 R_2{}^2 + [R_2^2 + X_2(X_1 + X_2)]^2\}^{.5}}{R_2^2 + (X_1 + X_2)^2}$
	Phase Angle	$\tan^{-1}\dfrac{R_2^2 + X_2(X_1 + X_2)}{X_1 R_2}$
	Admittance	$\dfrac{R_2 X_1 - j[R_2^2 + X_2^2 + X_1 X_2)}{X_1(R_2^2 + X_2^2)}$

Table 5-3. Vector Conversion.

$\frac{X}{R}$	θ degrees	Z	$\frac{X}{R}$	θ degrees	Z
0.00000	0.0	Z = R	0.06116	3.5	16.380 X
.00175	.1	572.96 X	.06291	3.6	15.926 X
.00349	.2	286.48 X	.06467	3.7	15.496 X
.00524	.3	190.98 X	.06642	3.8	15.089 X
.00698	.4	143.24 X	.06817	3.9	14.702 X
0.00873	0.5	114.59 X	0.06993	4.0	14.335 X
0.01047	0.6	95.495 X	0.07168	4.1	13.986 X
.01222	0.7	81.853 X	.07344	4.2	13.654 X
.01396	0.8	71.622 X	.07519	4.3	13.337 X
.01571	0.9	63.664 X	.07694	4.4	13.034 X
.01746	1.0	57.299 X	0.07870	4.5	12.745 X
0.01920	1.1	52.090 X			
0.02095	1.2	47.750 X	0.08046	4.6	12.469 X
.02269	1.3	44.077 X	.08221	4.7	12.204 X
.02444	1.4	40.930 X	.08397	4.8	11.950 X
.02618	1.5	38.201 X	.08573	4.9	11.707 X
.02793	1.6	35.814 X	.08749	5.0	11.474 X
0.02968	1.7	33.708 X	0.08925	5.1	11.249 X
0.03143	1.8	31.836 X	0.09101	5.2	11.033 X
.03317	1.9	30.161 X	.09277	5.3	10.826 X
.03492	2.0	28.654 X	.09453	5.4	10.626 X
.03667	2.1	27.290 X	.09629	5.5	10.433 X
0.03842	2.2	26.050 X	.09805	5.6	10.248 X
0.04016	2.3	24.918 X	0.09981	5.7	10.068 X
.04191	2.4	23.880 X	0.10158	5.8	9.8955 X
.04366	2.5	22.925 X	.10344	5.9	9.7283 X
.04541	2.6	22.044 X	.10510	6.0	9.5668 X
.04716	2.7	21.228 X	.10687	6.1	9.4105 X
0.04891	2.8	20.471 X	0.10863	6.2	9.2593 X
0.05066	2.9	19.766 X	0.11040	6.3	9.1129 X
.05241	3.0	19.107 X	.11217	6.4	8.9711 X
.05416	3.1	18.491 X	.11394	6.5	8.8337 X
.05591	3.2	17.914 X	.11570	6.6	8.7004 X
.05766	3.3	17.372 X	.11747	6.7	8.5711 X
0.05941	3.4	16.861 X	0.11924	6.8	8.4457 X

Table 5-3. Vector Conversion (cont'd).

$\frac{X}{R}$	θ degrees	Z	$\frac{X}{R}$	θ degrees	Z
0.12107	6.9	8.3238 X	0.18173	10.3	5.5928 X
.12278	7.0	8.2055 X	.18353	10.4	5.5396 X
.12456	7.1	8.0905 X	.18534	10.5	5.4874 X
.12663	7.2	7.9787 X	.18714	10.6	5.4362 X
.12810	7.3	7.8700 X	0.18895	10.7	5.3860 X
0.12988	7.4	7.7642 X	0.19076	10.8	5.3367 X
0.13165	7.5	7.6613 X	.19257	10.9	5.2883 X
.13343	7.6	7.5611 X	.19438	11.0	5.2408 X
.13521	7.7	7.4634 X	.19619	11.1	5.1942 X
.13698	7.8	7.3683 X	.19800	11.2	5.1484 X
0.13876	7.9	7.2756 X	0.19982	11.3	5.1034 X
0.14054	8.0	7.1853 X	0.20163	11.4	5.0593 X
.14232	8.1	7.0972 X	.20345	11.5	5.0158 X
.14410	8.2	7.0112 X	.20527	11.6	5.9732 X
.14588	8.3	6.9273 X	.20709	11.7	5.9313 X
.14767	8.4	6.8454 X	0.20891	11.8	5.8901 X
0.14945	8.5	6.7755 X			
0.15124	8.6	6.6874 X	0.21073	11.9	5.8496 X
.15302	8.7	6.6111 X	.21256	12.0	4.8097 X
.15481	8.8	6.5365 X	.21438	12.1	4.7706 X
.15660	8.9	6.4637 X	.21621	12.2	4.7320 X
0.15838	9.0	6.3924 X	.21803	12.3	4.6942 X
0.16017	9.1	6.3228 X	0.21986	12.4	4.6569 X
.16196	9.2	6.2546 X	0.22169	12.5	4.6201 X
.16376	9.3	6.1880 X	.22353	12.6	4.5841 X
.16555	9.4	6.1227 X	.22536	12.7	4.5486 X
.16734	9.5	6.0588 X	.22719	12.8	4.5137 X
0.16914	9.6	5.9963 X	0.22903	12.9	4.4793 X
0.17093	9.7	5.9351 X	0.23087	13.0	4.4454 X
.17273	9.8	5.8751 X	.23270	13.1	4.4121 X
.17453	9.9	5.8163 X	.23455	13.2	4.3792 X
.17633	10.0	5.7588 X	.23693	13.3	4.3469 X
.17813	10.1	5.7023 X	0.23823	13.4	4.3150 X
0.17993	10.2	5.6470 X			

Table 5-3. Vector Conversion (cont'd).

$\frac{X}{R}$	θ degrees	Z	$\frac{X}{R}$	θ degrees	Z
0.24008	13.5	4.2836 X	0.30001	16.7	3.4799 X
.24192	13.6	4.2527 X	.30192	16.8	3.4598 X
.24377	13.7	4.2223 X	.30382	16.9	3.4399 X
.24562	13.8	4.1923 X	.30573	17.0	3.4203 X
.24747	13.9	4.1627 X	.30764	17.1	3.4009 X
0.24933	14.0	4.1336 X	0.30995	17.2	3.3817 X
0.25118	14.1	4.1048 X	0.31146	17.3	3.3627 X
.25304	14.2	4.0765 X	.31338	17.4	3.3440 X
.25490	14.3	4.0486 X	.31530	17.5	3.3255 X
.25676	14.4	4.0211 X	.31722	17.6	3.3072 X
0.25862	14.5	3.9939 X	0.31914	17.7	3.2891 X
0.26048	14.6	3.9672 X	0.32106	17.8	3.2712 X
.26234	14.7	3.9408 X	.32299	17.9	3.2535 X
.26421	14.8	3.9147 X	.32492	18.0	3.2361 X
.26608	14.9	3.8890 X	.32685	18.1	3.2188 X
.26795	15.0	3.8637 X	0.32878	18.2	3.2017 X
0.26982	15.1	3.8387 X			
0.27169	15.2	3.8140 X	0.33072	18.3	3.1848 X
.27357	15.3	3.7897 X	.33265	18.4	3.1681 X
.27554	15.4	3.7657 X	.33459	18.5	3.1515 X
.27732	15.5	3.7420 X	.33654	18.6	3.1352 X
0.27920	15.6	3.7186 X	0.33848	18.7	3.1190 X
0.28109	15.7	3.6955 X	0.34043	18.8	3.1030 X
.28297	15.8	3.6727 X	.34238	18.9	3.0872 X
.28486	15.9	3.6502 X	.34433	19.0	3.0715 X
.28674	16.0	3.6279 X	.34628	19.1	3.0561 X
0.28863	16.1	3.6060 X	0.34824	19.2	3.0407 X
0.29053	16.2	3.5843 X	0.35019	19.3	3.0256 X
.29242	16.3	3.5629 X	.35215	19.4	3.0106 X
.29432	16.4	3.5418 X	.35412	19.5	2.9957 X
.29621	16.5	3.5209 X	.35608	19.6	2.9810 X
0.29811	16.6	3.5003 X	0.35805	19.7	2.9665 X

Table 5-3. Vector Conversion (cont'd).

$\frac{X}{R}$	θ degrees	Z	$\frac{X}{R}$	θ degrees	Z
0.36002	19.8	2.9521 X	0.42036	22.8	2.5805 X
.36199	19.9	2.9379 X	.42242	22.9	2.5699 X
.36397	20.0	2.9238 X	.42447	23.0	2.5593 X
.36595	20.1	2.9098 X	.42654	23.1	2.5488 X
.36793	20.2	2.8960 X	0.42860	23.2	2.5384 X
0.36991	20.3	2.8824 X			
0.37190	20.4	2.8688 X	0.43067	23.3	2.5281 X
.37388	20.5	2.8554 X	.43274	23.4	2.5179 X
.37587	20.6	2.8422 X	.43481	23.5	2.5078 X
.37787	20.7	2.8290 X	.43689	23.6	2.4978 X
0.37986	20.8	2.8160 X	0.43897	23.7	2.4879 X
0.38186	20.9	2.8032 X	0.44105	23.8	2.4780 X
.38386	21.0	2.7904 X	.44314	23.9	2.4683 X
.38587	21.1	2.7778 X	.44523	24.0	2.4586 X
.38787	21.2	2.7653 X	.44732	24.1	2.4490 X
0.38988	21.3	2.7529 X	0.44942	24.2	2.4395 X
0.39189	21.4	2.7406 X	0.45152	24.3	2.4300 X
.39391	21.5	2.7285 X	.45362	24.4	2.4207 X
.39593	21.6	2.7165 X	.45573	24.5	2.4114 X
.39795	21.7	2.7045 X	.45783	24.6	2.4022 X
0.39997	21.8	2.6927 X	0.45995	24.7	2.3931 X
0.40200	21.9	2.6810 X	0.46206	24.8	2.3841 X
.40403	22.0	2.6695 X	.46418	24.9	2.3751 X
.40606	22.1	2.6580 X	.46631	25.0	2.3662 X
0.40809	22.2	2.6466 X	0.46843	25.1	2.3574 X
0.41013	22.3	2.6353 X	0.47056	25.2	2.3486 X
.41217	22.4	2.6242 X	.47270	25.3	2.3399 X
.41421	22.5	2.6131 X	.47483	25.4	2.3313 X
.41626	22.6	2.6022 X	.47697	25.5	2.3228 X
0.41831	22.7	2.5913 X	0.47912	25.6	2.3143 X

Table 5-3. Vector Conversion (cont'd).

$\dfrac{X}{R}$	θ degrees	Z	$\dfrac{X}{R}$	θ degrees	Z
0.48127	25.7	2.3059 X	0.55203	28.9	2.0692 X
.48342	25.8	2.2976 X	.55431	29.0	2.0627 X
.48557	25.9	2.2894 X	.55659	29.1	2.0562 X
.48773	26.0	2.2812 X	0.55888	29.2	2.0498 X
0.48989	26.1	2.2730 X			
0.49206	26.2	2.2650 X	0.56117	29.3	2.0434 X
.49423	26.3	2.2570 X	.56347	29.4	2.0370 X
.49640	26.4	2.2490 X	.56577	29.5	2.0308 X
0.49858	26.5	2.2411 X	0.56808	29.6	2.0245 X
0.50076	26.6	2.2333 X	0.57039	29.7	2.0183 X
.50295	26.7	2.2256 X	.57270	29.8	2.0122 X
.50514	26.8	2.2179 X	.57502	29.9	2.0061 X
.50733	26.9	2.2103 X	.57735	30.0	2.0000 X
0.50952	27.0	2.2027 X	0.57968	30.1	1.9940 X
0.51172	27.1	2.1952 X	0.58201	30.2	1.9880 X
.51393	27.2	2.1877 X	.58435	30.3	1.9820 X
.51614	27.3	2.1803 X	.58670	30.4	1.9761 X
0.51835	27.4	2.1730 X	0.58904	30.5	1.9703 X
0.52057	27.5	2.1657 X	0.59140	30.6	1.9645 X
.52279	27.6	2.1584 X	.59376	30.7	1.9587 X
.52501	27.7	2.1513 X	.59612	30.8	1.9530 X
.52724	27.8	2.1441 X	0.59849	30.9	1.9473 X
0.52947	27.9	2.1371 X			
0.53171	28.0	2.1300 X	0.60086	31.0	1.9416 X
.53395	28.1	2.1231 X	.60324	31.1	1.9360 X
.53619	28.2	2.1162 X	.60562	31.2	1.9304 X
0.53844	28.3	2.1093 X	0.60801	31.3	1.9248 X
0.54070	28.4	2.1025 X	0.61040	31.4	1.9193 X
.54295	28.5	2.0957 X	.61280	31.5	1.9139 X
.54522	28.6	2.0890 X	.61520	31.6	1.9084 X
.54748	28.7	2.0824 X	0.61761	31.7	1.9030 X
0.54975	28.8	2.0757 X			

Table 5-3. Vector Conversion (cont'd).

$\frac{X}{R}$	θ degrees	Z	$\frac{X}{R}$	θ degrees	Z
0.62003	31.8	1.8977 X	0.69243	34.7	1.7566 X
.62244	31.9	1.8924 X	.69502	34.8	1.7522 X
.62487	32.0	1.8871 X	0.69761	34.9	1.7478 X
.62730	32.1	1.8818 X			
0.62973	32.2	1.8766 X			
0.63217	32.3	1.8714 X	0.70021	35.0	1.7434 X
.63462	32.4	1.8663 X	.70281	35.1	1.7391 X
.63707	32.5	1.8611 X	.70542	35.2	1.7348 X
0.63953	32.6	1.8561 X	0.70804	35.3	1.7305 X
0.64199	32.7	1.8510 X	0.71066	35.4	1.7263 X
.64466	32.8	1.8460 X	.71329	35.5	1.7220 X
.64693	32.9	1.8410 X	.71593	35.6	1.7178 X
0.64941	33.0	1.8361 X	0.71857	35.7	1.7137 X
0.65189	33.1	1.8311 X	0.72122	35.8	1.7095 X
.65438	33.2	1.8263 X	.72388	35.9	1.7054 X
.65688	33.3	1.8214 X	.72654	36.0	1.7013 X
0.65938	33.4	1.8166 X	0.72921	36.1	1.6972 X
0.66188	33.5	1.8118 X	0.73189	36.2	1.6932 X
.66440	33.6	1.8070 X	.73457	36.3	1.6891 X
.66692	33.7	1.8023 X	.73726	36.4	1.6851 X
0.66944	33.8	1.7976 X	0.73996	36.5	1.6812 X
0.67197	33.9	1.7929 X	0.74266	36.6	1.6772 X
.67451	34.0	1.7883 X	.74538	36.7	1.6733 X
.67705	34.1	1.7837 X	0.74809	36.8	1.6694 X
0.67160	34.2	1.7791 X			
0.68215	34.3	1.7745 X	0.75082	36.9	1.6655 X
.68471	34.4	1.7700 X	.75355	37.0	1.6616 X
.68728	34.5	1.7655 X	.75629	37.1	1.6578 X
0.68985	34.6	1.7610 X	0.75904	37.2	1.6540 X

Table 5-3. Vector Conversion (cont'd).

$\frac{X}{R}$	θ degrees	Z	$\frac{X}{R}$	θ degrees	Z
0.76179	37.3	1.6502 X	0.85107	40.4	1.5429 X
.76546	37.4	1.6464 X	.85408	40.5	1.5398 X
0.76733	37.5	1.6427 X	0.85710	40.6	1.5366 X
0.77010	37.6	1.6389 X	0.86013	40.7	1.5335 X
.77289	37.7	1.6352 X	.86318	40.8	1.5304 X
.77568	37.8	1.6316 X	.86623	40.9	1.5273 X
0.77848	37.9	1.6279 X	0.86929	41.0	1.5242 X
0.78128	38.0	1.6243 X	0.87235	41.1	1.5212 X
.78410	38.1	1.6206 X	.87543	41.2	1.5182 X
.78692	38.2	1.6170 X	0.87852	41.3	1.5151 X
0.78975	38.3	1.6135 X			
0.79259	38.4	1.6099 X	0.88162	41.4	1.5121 X
.79543	38.5	1.6064 X	.88472	41.5	1.5092 X
0.79829	38.6	1.6029 X	0.88784	41.6	1.5062 X
0.80115	38.7	1.5994 X	0.89097	41.7	1.5032 X
.80402	38.8	1.5959 X	.89410	41.8	1.5003 X
.80690	38.9	1.5924 X	0.89725	41.9	1.4974 X
0.80978	39.0	1.5890 X			
0.81268	39.1	1.5856 X	0.90040	42.0	1.4945 X
.81558	39.2	1.5822 X	.90357	42.1	1.4916 X
0.81849	39.3	1.5788 X	.90674	42.2	1.4887 X
			0.90993	42.3	1.4858 X
0.82141	39.4	1.5755 X	0.91312	42.4	1.4830 X
.82434	39.5	1.5721 X	.91633	42.5	1.4802 X
0.82727	39.6	1.5688 X	0.91955	42.6	1.4774 X
0.83022	39.7	1.5655 X	0.92277	42.7	1.4746 X
.83317	39.8	1.5622 X	.92601	42.8	1.4718 X
.83613	39.9	1.5590 X	0.92926	42.9	1.4690 X
0.83910	40.0	1.5557 X			
0.84208	40.1	1.5525 X	0.93251	43.0	1.4663 X
.84506	40.2	1.5493 X	.93578	43.1	1.4635 X
0.84806	40.3	1.5461 X	0.93906	43.2	1.4608 X

Table 5-3. Vector Conversion (cont'd).

$\frac{X}{R}$	θ degrees	Z	$\frac{X}{R}$	θ degrees	Z
0.94235	43.3	1.4581 X	1.0538	46.5	1.3786 X
.94565	43.4	1.4554 X	1.0575	46.6	1.3763 X
0.94896	43.5	1.4527 X	1.0612	46.7	1.3740 X
0.95229	43.6	1.4501 X	1.0649	46.8	1.3718 X
.95562	43.7	1.4474 X	1.0686	46.9	1.3695 X
0.95896	43.8	1.4448 X	1.0724	47.0	1.3673 X
0.96232	43.9	1.4422 X	1.0761	47.1	1.3651 X
.96569	44.0	1.4395 X	1.0799	47.2	1.3629 X
0.96907	44.1	1.4370 X	1.0837	47.3	1.3607 X
0.97246	44.2	1.4344 X	1.0875	47.4	1.3585 X
.97586	44.3	1.4318 X	1.0913	47.5	1.3563 X
0.97927	44.4	1.4292 X	1.0951	47.6	1.3542 X
0.98270	44.5	1.4267 X	1.0990	47.7	1.3520 X
.98613	44.6	1.4242 X	1.1028	47.8	1.3499 X
0.98958	44.7	1.4217 X	1.1067	47.9	1.3477 X
0.99304	44.8	1.4192 X	1.1106	48.0	1.3456 X
0.99651	44.9	1.4167 X	1.1145	48.1	1.3435 X
1.0000	45.0	1.4142 X	1.1184	48.2	1.3414 X
1.0035	45.1	1.4117 X	1.1224	48.3	1.3393 X
1.0070	45.2	1.4093 X	1.1263	48.4	1.3372 X
1.0105	45.3	1.4069 X	1.1303	48.5	1.3352 X
1.0141	45.4	1.4040 X	1.1343	48.6	1.3331 X
1.0176	45.5	1.4020 X	1.1383	48.7	1.3311 X
1.0212	45.6	1.3996 X	1.1423	48.8	1.3290 X
1.0247	45.7	1.3972 X	1.1463	48.9	1.3270 X
1.0283	45.8	1.3949 X	1.1504	49.0	1.3250 X
1.0319	45.9	1.3925 X	1.1544	49.1	1.3230 X
1.0355	46.0	1.3902 X	1.1585	49.2	1.3210 X
1.0391	46.1	1.3878 X	1.1626	49.3	1.3190 X
1.0428	46.2	1.3855 X	1.1667	49.4	1.3170 X
1.0464	46.3	1.3832 X	1.1708	49.5	1.3151 X
1.0501	46.4	1.3809 X	1.1750	49.6	1.3131 X

Table 5-3. Vector Conversion (cont'd).

$\frac{X}{R}$	θ degrees	Z	$\frac{X}{R}$	θ degrees	Z
1.1791	49.7	1.3112 X	1.3367	53.2	1.2488 X
1.1833	49.8	1.3092 X	1.3416	53.3	1.2472 X
1.1875	49.9	1.3073 X	1.3465	53.4	1.2456 X
1.1917	50.0	1.3054 X	1.3514	53.5	1.2440 X
1.1960	50.1	1.3035 X	1.3564	53.6	1.2424 X
1.2002	50.2	1.3016 X	1.3613	53.7	1.2408 X
1.2045	50.3	1.2997 X	1.3663	53.8	1.2392 X
1.2088	50.4	1.2978 X	1.3713	53.9	1.2376 X
1.2131	50.5	1.2960 X	1.3764	54.0	1.2361 X
1.2174	50.6	1.2941 X	1.3814	54.1	1.2345 X
1.2218	50.7	1.2922 X	1.3865	54.2	1.2329 X
1.2261	50.8	1.2904 X	1.3916	54.3	1.2314 X
1.2305	50.9	1.2886 X	1.3968	54.4	1.2298 X
1.2349	51.0	1.2867 X	1.4019	54.5	1.2283 X
1.2393	51.1	1.2849 X	1.4071	54.6	1.2268 X
1.2437	51.2	1.2831 X	1.4123	54.7	1.2253 X
1.2482	51.3	1.2813 X	1.4176	54.8	1.2238 X
1.2527	51.4	1.2795 X	1.4228	54.9	1.2223 X
1.2572	51.5	1.2778 X	1.4281	55.0	1.2208 X
1.2617	51.6	1.2760 X	1.4335	55.1	1.2193 X
1.2662	51.7	1.2742 X	1.4388	55.2	1.2178 X
1.2708	51.8	1.2725 X	1.4442	55.3	1.2163 X
1.2753	51.9	1.2707 X	1.4496	55.4	1.2149 X
1.2799	52.0	1.2690 X	1.4550	55.5	1.2134 X
1.2845	52.1	1.2673 X	1.4605	55.6	1.2119 X
1.2892	52.2	1.2656 X	1.4659	55.7	1.2105 X
1.2938	52.3	1.2639 X	1.4714	55.8	1.2091 X
1.2985	52.4	1.2622 X	1.4770	55.9	1.2076 X
1.3032	52.5	1.2605 X	1.4826	56.0	1.2062 X
1.3079	52.6	1.2588 X	1.4881	56.1	1.2048 X
1.3127	52.7	1.2571 X	1.4938	56.2	1.2034 X
1.3174	52.8	1.2554 X	1.4994	56.3	1.2020 X
1.3222	52.9	1.2538 X	1.5051	56.4	1.2006 X
1.3270	53.0	1.2521 X	1.5108	56.5	1.1992 X
1.3319	53.1	1.2505 X	1.5166	56.6	1.1978 X

Table 5-3. Vector Conversion (cont'd).

$\frac{X}{R}$	θ degrees	Z	$\frac{X}{R}$	θ degrees	Z
1.5223	56.7	1.1964 X	1.7461	60.2	1.1524 X
1.5282	56.8	1.1951 X	1.7532	60.3	1.1512 X
1.5340	56.9	1.1937 X	1.7603	60.4	1.1501 X
1.5399	57.0	1.1924 X	1.7675	60.5	1.1489 X
1.5458	57.1	1.1910 X	1.7747	60.6	1.1478 X
1.5517	57.2	1.1897 X	1.7820	60.7	1.1467 X
1.5577	57.3	1.1883 X	1.7893	60.8	1.1456 X
1.5636	57.4	1.1870 X	1.7966	60.9	1.1445 X
1.5697	57.5	1.1857 X	1.8040	61.0	1.1433 X
1.5757	57.6	1.1844 X	1.8115	61.1	1.1422 X
1.5818	57.7	1.1831 X	1.8190	61.2	1.1411 X
1.5880	57.8	1.1818 X	1.8265	61.3	1.1401 X
1.5941	57.9	1.1805 X	1.8341	61.4	1.1390 X
1.6003	58.0	1.1792 X	1.8418	61.5	1.1379 X
1.6006	58.1	1.1779 X	1.8495	61.6	1.1368 X
1.6128	58.2	1.1766 X	1.8572	61.7	1.1357 X
1.6191	58.3	1.1753 X	1.8650	61.8	1.1347 X
1.6255	58.4	1.1741 X	1.8728	61.9	1.1336 X
1.6318	58.5	1.1728 X	1.8807	62.0	1.1326 X
1.6383	58.6	1.1716 X	1.8887	62.1	1.1315 X
1.6447	58.7	1.1703 X	1.8967	62.2	1.1305 X
1.6512	58.8	1.1691 X	1.9047	62.3	1.1294 X
1.6577	58.9	1.1678 X	1.9128	62.4	1.1284 X
1.6643	59.0	1.1666 X	1.9210	62.5	1.1274 X
1.6709	59.1	1.1654 X	1.9292	62.6	1.1264 X
1.6775	59.2	1.1642 X	1.9375	62.7	1.1253 X
1.6842	59.3	1.1630 X	1.9458	62.8	1.1243 X
1.6909	59.4	1.1618 X	1.9542	62.9	1.1233 X
1.6977	59.5	1.1606 X	1.9626	63.0	1.1223 X
1.7044	59.6	1.1594 X	1.9711	63.1	1.1213 X
1.7113	59.7	1.1582 X	1.9797	63.2	1.1203 X
1.7182	59.8	1.1570 X	1.9883	63.3	1.1193 X
1.7251	59.9	1.1559 X	1.9969	63.4	1.1184 X
1.7320	60.0	1.1547 X	2.0057	63.5	1.1174 X
1.7390	60.1	1.1535 X	2.0145	63.6	1.1164 X

Table 5-3. Vector Conversion (cont'd).

$\frac{X}{R}$	θ degrees	Z	$\frac{X}{R}$	θ degrees	Z
2.0233	63.7	1.1155 X	2.3789	67.2	1.0847 X
2.0323	63.8	1.1145 X	2.3906	67.3	1.0840 X
2.0412	63.9	1.1135 X	2.4023	67.4	1.0832 X
2.0503	64.0	1.1126 X	2.4142	67.5	1.0824 X
2.0594	64.1	1.1116 X	2.4262	67.6	1.0816 X
2.0686	64.2	1.1107 X	2.4382	67.7	1.0808 X
2.0778	64.3	1.1098 X	2.4504	67.8	1.0801 X
2.0872	64.4	1.1088 X	2.4627	67.9	1.0793 X
2.0965	64.5	1.1079 X	2.4751	68.0	1.0785 X
2.1060	64.6	1.1070 X	2.4876	68.1	1.0778 X
2.1115	64.7	1.1061 X	2.5002	68.2	1.0770 X
2.1251	64.8	1.1052 X	2.5129	68.3	1.0763 X
2.1348	64.9	1.1043 X	2.5257	68.4	1.0755 X
2.1445	65.0	1.1034 X	2.5386	68.5	1.0748 X
2.1543	65.1	1.1025 X	2.5517	68.6	1.0740 X
2.1642	65.2	1.1016 X	2.5649	68.7	1.0733 X
2.1741	65.3	1.1007 X	2.5781	68.8	1.0726 X
2.1842	65.4	1.0998 X	2.5916	68.9	1.0719 X
2.1943	65.5	1.0989 X	2.6051	69.0	1.0711 X
2.2045	65.6	1.0981 X	2.6187	69.1	1.0704 X
2.2147	65.7	1.0972 X	2.6325	69.2	1.0697 X
2.2251	65.8	1.0963 X	2.6464	69.3	1.0690 X
2.2355	65.9	1.0955 X	2.6604	69.4	1.0683 X
2.2460	66.0	1.0946 X	2.6746	69.5	1.0676 X
2.2566	66.1	1.0938 X	2.6889	69.6	1.0669 X
2.2673	66.2	1.0929 X	2.7033	69.7	1.0662 X
2.2781	66.3	1.0921 X	2.7179	69.8	1.0655 X
2.2889	66.4	1.0913 X	2.7326	69.9	1.0648 X
2.2998	66.5	1.0904 X	2.7475	70.0	1.0642 X
2.3109	66.6	1.0896 X	2.7625	70.1	1.0635 X
2.3220	66.7	1.0888 X	2.7776	70.2	1.0628 X
2.3332	66.8	1.0880 X	2.7929	70.3	1.0622 X
2.3445	66.9	1.0872 X	2.8083	70.4	1.0615 X
2.3558	67.0	1.0864 X	2.8239	70.5	1.0608 X
2.3673	67.1	1.0855 X	2.8396	70.6	1.0602 X

Table 5-3. Vector Conversion (cont'd).

$\frac{X}{R}$	θ degrees	Z	$\frac{X}{R}$	θ degrees	Z
2.8555	70.7	1.0595 X	3.5339	74.2	1.0393 X
2.8716	70.8	1.0589 X	3.5576	74.3	1.0387 X
2.8878	70.9	1.0582 X	3.5816	74.4	1.0382 X
2.9042	71.0	1.0576 X	3.6059	74.5	1.0377 X
2.9208	71.1	1.0570 X	3.6305	74.6	1.0372 X
2.9375	71.2	1.0563 X	3.6554	74.7	1.0367 X
2.9544	71.3	1.0557 X	3.6806	74.8	1.0362 X
2.9714	71.4	1.0551 X	3.7062	74.9	1.0358 X
2.9887	71.5	1.0545 X	3.7320	75.0	1.0353 X
3.0061	71.6	1.0539 X	3.7583	75.1	1.0348 X
3.0237	71.7	1.0533 X	3.7848	75.2	1.0343 X
3.0415	71.8	1.0527 X	3.8118	75.3	1.0338 X
3.0595	71.9	1.0521 X	3.8390	75.4	1.0334 X
3.0777	72.0	1.0515 X	3.8667	75.5	1.0329 X
3.0960	72.1	1.0509 X	3.8947	75.6	1.0324 X
3.1146	72.2	1.0503 X	3.9231	75.7	1.0320 X
3.1334	72.3	1.0497 X	3.9520	75.8	1.0315 X
3.1524	72.4	1.0491 X	3.9812	75.9	1.0311 X
3.1716	72.5	1.0485 X	4.0108	76.0	1.0306 X
3.1910	72.6	1.0479 X	4.0408	76.1	1.0302 X
3.2106	72.7	1.0474 X	4.0713	76.2	1.0297 X
3.2305	72.8	1.0468 X	4.1022	76.3	1.0293 X
3.2505	72.9	1.0462 X	4.1335	76.4	1.0288 X
3.2708	73.0	1.0457 X	4.1653	76.5	1.0284 X
3.2914	73.1	1.0451 X	4.1976	76.6	1.0280 X
3.3121	73.2	1.0466 X	4.2303	76.7	1.0276 X
3.3332	73.3	1.0440 X	4.2635	76.8	1.0271 X
3.3544	73.4	1.0435 X	4.2972	76.9	1.0267 X
3.3759	73 5	1.0429 X	4.3315	77.0	1.0263 X
3.3977	73.6	1.0424 X	4.3662	77.1	1.0259 X
3.4197	73.7	1.0419 X	4.4015	77.2	1.0225 X
3.4420	73.8	1.0413 X	4.4373	77.3	1.0251 X
3.4646	73.9	1.0408 X	4.4737	77.4	1.0247 X
3.4874	74.0	1.0403 X	4.5107	77.5	1.0243 X
3.5105	74.1	1.0398 X	4.5483	77.6	1.0239 X

Table 5-3. Vector Conversion (cont'd).

$\frac{X}{R}$	θ degrees	Z	$\frac{X}{R}$	θ degrees	Z
4.5864	77.7	1.0235 X	6.4596	81.2	1.0119 X
4.6252	77.8	1.0231 X	6.5350	81.3	1.0116 X
4.6646	77.9	1.0227 X	6.6122	81.4	1.0114 X
4.7046	78.0	1.0223 X	6.6911	81.5	1.0111 X
4.7453	78.1	1.0220 X	6.7720	81.6	1.0108 X
4.7867	78.2	1.0216 X	6.8547	81.7	1.0106 X
4.8288	78.3	1.0212 X	6.9395	81.8	1.0103 X
4.8716	78.4	1.0208 X	7.0264	81.9	1.0101 X
4.9151	78.5	1.0205 X	7.1154	82.0	1.0098 X
4.9594	78.6	1.0201 X	7.2066	82.1	1.0096 X
5.0045	78.7	1.0198 X	7.3002	82.2	1.0093 X
5.0504	78.8	1.0194 X	7.3961	82.3	1.0091 X
5.0970	78.9	1.0191 X	7.4946	82.4	1.0089 X
5.1445	79.0	1.0187 X	7.5957	82.5	1.0086 X
5.1929	79.1	1.0184 X	7.6996	82.6	1.0084 X
5.2422	79.2	1.0180 X	7.8062	82.7	1.0082 X
5.2923	79.3	1.0177 X	7.9158	82.8	1.0079 X
5.3434	79.4	1.0174 X	8.0285	82.9	1.0077 X
5.3955	79.5	1.0170 X	8.1443	83.0	1.0075 X
5.4486	79.6	1.0167 X	8.2635	83.1	1.0073 X
5.5026	79.7	1.0164 X	8.3862	83.2	1.0071 X
5.5578	79.8	1.0160 X	8.5126	83.3	1.0069 X
5.6140	79.9	1.0157 X	8.6427	83.4	1.0067 X
5.6713	80.0	1.0154 X	8.7769	83.5	1.0065 X
5.7297	80.1	1.0151 X	8.9152	83.6	1.0063 X
5.7894	80.2	1.0148 X	9.0579	83.7	1.0061 X
5.8502	80.3	1.0145 X	9.2051	83.8	1.0059 X
5.9123	80.4	1.0142 X	9.3572	83.9	1.0057 X
5.9758	80.5	1.0139 X	9.5144	84.0	1.0055 X
6.0405	80.6	1.0136 X	9.6768	84.1	1.0053 X
6.1066	80.7	1.0133 X	9.8448	84.2	1.0051 X
6.1742	80.8	1.0130 X	10.019	84.3	1.0050 X
6.2432	80.9	1.0127 X	10.199	84.4	1.0048 X
6.3137	81.0	1.0125 X	10.385	84.5	1.0046 X
6.3859	81.1	1.0122 X	10.579	84.6	1.0044 X

Table 5-3. Vector Conversion (cont'd).

$\dfrac{X}{R}$	θ degrees	Z	$\dfrac{X}{R}$	θ degrees	Z
10.780	84.7	1.0043 X	22.022	87.4	1.0010 X
10.988	84.8	1.0041 X	23.904	87.5	1.0009 X
11.205	84.9	1.0040 X	23.859	87.6	1.0009 X
11.430	85.0	1.0038 X	24.898	87.7	1.0008 X
11.664	85.1	1.0037 X	26.031	87.8	1.0007 X
11.909	85.2	1.0035 X	27.271	87.9	1.0007 X
12.163	85.3	1.0034 X	28.636	88.0	1.0006 X
12.429	85.4	1.0032 X	30.145	88.1	1.0005 X
12.706	85.5	1.0031 X	31.820	88.2	1.0005 X
12.996	85.6	1.0029 X	33.693	88.3	1.0004 X
13.229	85.7	1.0028 X	35.800	88.4	1.0004 X
13.617	85.8	1.0027 X	38.188	88.5	1.0003 X
13.951	85.9	1.0026 X	40.917	88.6	1.0003 X
14.301	86.0	1.0024 X	44.066	88.7	1.0002 X
14.668	86.1	1.0023 X	47.739	88.8	1.0002 X
15.056	86.2	1.0022 X	52.081	88.9	1.0002 X
15.464	86.3	1.0021 X	57.290	89.0	1.0001 X
15.894	86.4	1.0020 X	63.657	89.1	1.0001 X
16.350	86.5	1.0019 X	71.615	89.2	1.0001 X
16.832	86.6	1.0018 X	81.847	89.3	1.0001 X
17.343	86.7	1.0017 X	95.489	89.4	1.0000 X
17.886	86.8	1.0016 X	114.59	89.5	1.0000 X
18.464	86.9	1.0015 X	143.24	89.6	1.0000 X
19.081	87.0	1.0014 X	190.98	89.7	1.0000 X
19.740	87.1	1.0013 X	286.48	89.8	1.0000 X
20.446	87.2	1.0012 X	572.96	89.9	1.0000 X
21.205	87.3	1.0011 X	R = 0	90.0	1.0000 X

□ **Example:**

The impedance of a circuit, expressed in polar form, is Z equals 3000−j4000. What is the absolute magnitude of the impedance and the phase angle?

3000 represents the resistance R; 4000 is the reactance X. The ratio X/R is 4000/3000 equals 1.3333. The closest value of X/R in Table 5-3 is 1.3319. Immediately to the right is the phase angle of 53.1 degrees. However, because the reactance is negative and is given as −j4000, the phase angle is also negative and is −53.1 degrees. The value for Z in the column directly to the right of the phase angle is 1.2505X. Since X is 4000, Z equals 1.2505X, 5002 ohms. The answer, then, is that Z is 5002 ohms and the phase angle is −53.1 degrees.

□ **Example:**

Take the example just given and work it backward. Suppose we are told that the circuit impedance is 5000 ohms (5002 rounded off to 5000) and the phase angle is −53.1 degrees. How could we express this in j-notation?

Start with the phase angle. In the third column 1.25X equals Z. But Z equals 5000 ohms. Since 1.25X equals 5000, then X equals 5000/1.25, or 4,000 ohms. Now move back to the first column. Here X/R equals 1.3319. Thus, 4000/R equals 1.3319. Solving for R we get R equals 3000 ohms. And, since we know that the phase angle is negative, we also know that our j term will also be negative. Our answer, then, is 3000−j4000.

□ **Example:**

What is the impedance of a circuit whose resistance, R, is 1000 ohms and whose reactance, X, is 775 ohms?

The ratio, X/R is 775/1000 or 0.775. Locating the nearest equivalent number in the first column of Table 5-3 gives 0.77568. To the right of this number, the phase angle is 37.8 degrees and, continuing to the right, Z equals 1.6316X. Thus, the impedance, Z equals 1.6316 × 775, or 1264.49 ohms.

IMPEDANCE AND TURNS RATIO

Transformers are conveniently used as impedance transformation devices (Fig. 5-2). The impedance of a transformer varies as the square of the turns ratio. These ratios, in steps of 1 to 100, are given in Table 5-4.

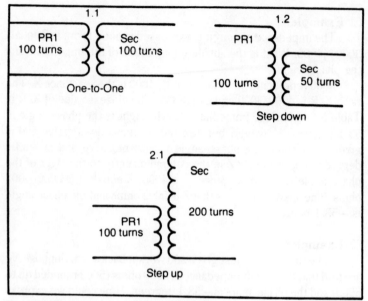

Fig. 5-2. The turns ratio is the number of turns on the secondary winding compared to the number of turns on the primary.

$$N = \frac{\sqrt{Z_s}}{\sqrt{Z_p}}$$

$$N^2 = \frac{Z_s}{Z_p}$$

$$Z_s = Z_p N^2$$

$$Z_p = \frac{Z_s}{N^2}$$

$$\frac{Z_s}{Z_p} = \frac{N_s^2}{N_p^2} = N^2$$

N = turns ratio
Z_s = secondary impedance
Z_p = primary impedance
N_s = secondary turns
N_p = primary turns

Table 5-4. Impedance Ratio and Turns Ratio of a Transformer.

Turns Ratio	Impedance Ratio	Turns Ratio	Impedance Ratio
100:1	10,000:1	65:1	4,225:1
99:1	9,801:1	64:1	4,096:1
98:1	9,604:1	63:1	3,969:1
97:1	9,409:1	62:1	3,844:1
96:1	9,216:1	61:1	3,721:1
95:1	9,025:1	60:1	3,600:1
94:1	8,836:1	59:1	3,481:1
93:1	8,649:1	58:1	3,364:1
92:1	8,464:1	57:1	3,249:1
91:1	8,281:1	56:1	3,136:1
90:1	8,100:1	55:1	3,025:1
89:1	7,921:1	54:1	2,916:1
88:1	7,744:1	53:1	2,809:1
87:1	7,569:1	52:1	2,704:1
86:1	7,396:1	51:1	2,601:1
85:1	7,225:1	50:1	2,500:1
84:1	7,056:1	49:1	2,401:1
83:1	6,889:1	48:1	2,304:1
82:1	6,724:1	47:1	2,209:1
81:1	6,571:1	46:1	2,116:1
80:1	6,400:1	45:1	2,025:1
79:1	6,241:1	44:1	1,936:1
78:1	6,084:1	43:1	1,849:1
77:1	5,929:1	42:1	1,764:1
76:1	5,776:1	41:1	1,681:1
75:1	5,625:1	40:1	1,600:1
74:1	5,476:1	39:1	1,521:1
73:1	5,329:1	38:1	1,444:1
72:1	5,184:1	37:1	1,369:1
71:1	5,041:1	36:1	1,296:1
70:1	4,900:1	35:1	1,225:1
69:1	4,761:1	34:1	1,156:1
68:1	4,624:1	33:1	1,089:1
67:1	4,489:1	32:1	1,024:1
66:1	4,356:1	31:1	961:1

Table 5-4. Impedance Ratio and Turns Ratio of a Transformer (cont'd).

Turns Ratio	Impedance Ratio	Turns Ratio	Impedance Ratio
30:1	900:1	1:100	1:10,000
29:1	841:1	1:99	1:9,801
28:1	784:1	1:98	1:9,604
27:1	729:1	1:97	1:9,409
26:1	676:1	1:96	1:9,216
25:1	625:1	1:95	1:9,025
24:1	576:1	1:94	1:8,836
23:1	529:1	1:93	1:8,649
22:1	484:1	1:92	1:8,464
21:1	441:1	1:91	1:8,281
20:1	400:1	1:90	1:8,100
19:1	361:1	1:89	1:7,921
18:1	324:1	1:88	1:7,744
17:1	289:1	1:87	1:7,569
16:1	256:1	1:86	1:7,396
15:1	225:1	1:85	1:7,225
14:1	196:1	1:84	1:7,056
13:1	169:1	1:83	1:6,889
12:1	144:1	1:82	1:6,724
11:1	121:1	1:81	1:6,571
10:1	100:1	1:80	1:6,400
9:1	81:1	1:79	1:6,241
8:1	64:1	1:78	1:6,084
7:1	49:1	1:77	1:5,929
6:1	36:1	1:76	1:5,776
5:1	25:1	1:75	1:5,625
4.1	16:1	1:74	1:5,476
3:1	9:1	1:73	1:5,329
2:1	4:1	1:72	1:5,184
1:1	1:1	1:71	1:5,041
		1:70	1:4,900

Table 5-4. Impedance Ratio and Turns Ratio of a Transformer (cont'd).

Turns Ratio	Impedance Ratio	Turns Ratio	Impedance Ratio
1:69	1:4,761	1:33	1:1,089
1:68	1:4,624	1:32	1:1,024
1:67	1:4,489	1:31	1:961
1:66	1:4,356	1:30	1:900
1:65	1:4,225	1:29	1:841
1:64	1:4,096	1:28	1:784
1:63	1:3,969	1:27	1:729
1:62	1:3,844	1:26	1:676
1:61	1:3,721	1:25	1:625
1:60	1:3,600	1:24	1:576
1:59	1:3,481		
1:58	1:3,364	1:23	1:529
1:57	1:3,249	1:22	1:484
1:56	1:3,136	1:21	1:441
1:55	1:3,025	1:20	1:400
1:54	1:2,916	1:19	1:361
1:53	1:2,809	1:18	1:324
1:52	1:2,704	1:17	1:289
1:51	1:2,601	1:16	1:256
1:50	1:2,500	1:15	1:225
1:49	1:2,401	1:14	1:196
1:48	1:2,304	1:13	1:169
1:47	1:2,209	1:12	1:144
1:46	1:2,116	1:11	1:121
1:45	1:2,025	1:10	1:100
1:44	1:1,936		
1:43	1:1,849	1:9	1:81
1:42	1:1,764	1:8	1:64
1:41	1:1,681	1:7	1:49
1:40	1:1,600	1:6	1:36
1:39	1:1,521	1:5	1:25
1:38	1:1,444	1:4	1:16
1:37	1:1,369	1:3	1:9
1:36	1:1,296	1:2	1:4
1:35	1:1,225	1:1	1:1
1:34	1:1,156		

□ **Example:**

A step-down transformer has a turns ratio of 37 to 1. What is the impedance ratio?

Locate the turns ratio, 37:1. Immediately alongside is the impedance ratio of 1369:1.

□ **Example:**

What is the impedance transformation of a step-up-transformer having a turns ratio of 1:53?

Table 5-4 shows that the impedance transformation for this turns ratio is 1:2809.

The turns ratio of a transformer is specifically that—a ratio— and gives no indication of the actual number of turns used by the transformer. A transformer having 50 primary turns and 50 secondary turns has a 1:1 ratio. So does a transformer having 5,000 primary and 5,000 secondary turns.

To get the ratio, divide the larger number of turns by the smaller. Table 5-4 will then supply the impedance ratio.

□ **Example:**

A step-down transformer has 2465 primary turns and 85 secondary turns. What is the impedance ratio?

Divide 2465 by 85. This supplies a ratio of 29:1. Table 5-4 shows the impedance ratio is 841:1. If the transformer had 85 primary turns and 2465 secondary turns, the impedance ratio would be 1:841.

ANGULAR VELOCITY

In an ac generator, the rotating armature produces a voltage sine wave (Fig. 5-3). We can represent the armature by a rotating vector r. The angle through which this vector sweeps is usually indicated by the Greek letter θ. The instantaneous values of the voltage produced are maximum when the phase angle, θ, is 90° or 270° and zero when that phase angle is 0° or 180° .

The radius vector r rotates about the origin, 0, and is taken to rotate in a counterclockwise direction. The angular velocity in radians per second of this rotating vector is the rate at which the angle, θ, is produced by its rotation. Angular velocity is conveniently represented by ω or $2\pi f$ (radians per second) in which 2π is a constant and is equal to 6.28. The frequency f is in hertz.

Angular velocity appears in formulas involving instantaneous values of sine waves of voltage or current (e equals $E_{max} \sin \omega t$) and

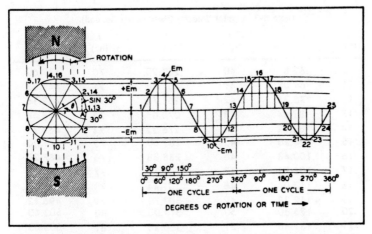

Fig. 5-3. Production of a sine wave by a coil rotating in a fixed magnetic field.

in formulas involving inductive reactance (X_L equals $2 \pi fL$) and capacitive reactance (X_C equals 1 divided by $2 \pi fC$).

The solution of problems involving angular velocity is simplified by Table 5-5. Here we have angular velocity corresponding to particular values of frequency.

Table 5-5 can be used for values of frequency other than those indicated by moving the decimal point. To find ω at a frequency of 4 kHz, find 40 in the f column. To change 40 Hz to 4 kHz multiply 40 by 100. This means we must also multiply the value of 251.20 (which is the value of ω for 40) by 100. The answer is 25,120. To get megahertz values, multiply both f and ω by 1,000,000.

□ **Example:**

What is the angular velocity of a generator, whose armature rotates at 1800 revolutions per minute?

Divide 1800 revolutions by 60 (60 seconds equals 1 minute) to obtain the number of revolutions per second. 1800/60 equals 30 revolutions per second. In 30 revolutions, the armature generates 30 complete sine wave cycles. The frequency, then, is 30. Locate 30 in the frequency (f) column of Table 5-5. The corresponding angular velocity is found to the right in the ω column. The answer is 188.4 radians per second.

Table 5-5. Angular Velocity (Radians per Second).

f (Hz)	ω	f (Hz)	ω	f (Hz)	ω
10	62.8	40	251.20	70	439.60
11	69.08	41	257.48	71	445.88
12	75.36	42	263.76	72	452.16
13	81.64	43	270.04	73	458.44
14	87.92	44	276.32	74	464.72
15	94.20	45	282.60	75	471.00
16	100.48	46	288.88	76	477.28
17	106.76	47	295.16	77	483.56
18	113.04	48	301.44	78	489.84
19	119.32	49	307.72	79	496.12
20	125.60	50	314.00	80	502.40
21	131.88	51	320.28	81	508.68
22	138.16	52	326.56	82	514.96
23	144.44	53	332.84	83	521.24
24	150.72	54	339.12	84	527.52
25	157.00	55	345.40	85	533.80
26	163.28	56	351.68	86	540.08
27	169.56	57	357.96	87	546.36
28	175.84	58	364.24	88	552.64
29	182.12	59	370.52	89	558.92
30	188.40	60	376.80	90	565.20
31	194.68	61	383.08	91	571.48
32	200.96	62	389.36	92	577.76
33	207.24	63	395.64	93	584.04
34	213.52	64	401.92	94	590.32
35	219.80	65	408.20	95	596.60
36	226.08	66	414.48	96	602.88
37	232.36	67	420.76	97	609.16
38	238.64	68	427.04	98	615.44
39	244.92	69	433.32	99	621.72
				100	628.00

Permeability

PERMEABILITY

Permeability can be considered the ability of a substance to conduct magnetic lines of force. More technically, permeability is the ratio of flux density in gausses (B) to a magnetizing force, H, in oersteds. The permeability of a vacuum, or air, is considered unity. See Table 6-1.

Table 6-1. Permeabilities of Magnetic Materials.

Cobalt	170
Iron-cobalt alloy (Co 34%)	13,000
Iron, commercial annealed	6000 to 8000
Nickel	400 to 1000
Permalloy (Ni 78.5%, Fe 21.5%)	over 80,000
Perminvar (Ni 45%, Fe 30%, Co 25%)	2000
Sendust	30,000 to 120,000
Silicon steel (Si 4%)	5000 to 10,000
Steel, cast	1500
Steel, open hearth	3000 to 7000

Power

WATTS AND HORSEPOWER

Motors are generally rated in terms of horsepower. But since a motor is an electromechanical device, it also has a rating in watts. The relationship between horsepower (hp) and power in watts is 1 hp per 745.7 watts.

Table 7-1 and Table 7-2 give the conversion between these two quantities. The tables can be extended by moving the decimal point in the same direction, for the same number of places, in both columns. For horsepower and electrical power equivalents, see Table 7-3.

☐ **Example:**
A small motor is rated at one-tenth horsepower. What is its rating in watts?

The closest value to one-tenth horsepower in Table 7-1 is 0.100575 shown in the right-hand column under the heading hp. The power corresponding to this value is 75 watts.

☐ **Example:**
A motor-generator is rated at one-half kilowatt. What is its equivalent horsepower rating?

One-half kilowatt is 500 watts. Table 7-1 does not list such a value but we can use the number 50 in place of 500. Locate the

Table 7-1. Watts vs. Horsepower.

Watts	hp	Watts	hp	Watts	hp
1	.001341	34	045594	67	.089847
2	.002682	35	.046935	68	.091188
3	.004023	36	.048278	69	.092529
4	.005364	37	.049617	70	.093870
5	.006705	38	.050958	71	.095211
6	.008046	39	.052299	72	.096552
7	.009387	40	.053640	73	.097893
8	.010728	41	.054981	74	.099234
9	.012069	42	.056322	75	.100575
10	.013410	43	.057663	76	.101916
11	.014751	44	.059004	77	.103257
12	.016092	45	.060345	78	.104598
13	.017433	46	.061686	79	.105939
14	.018774	47	.063027	80	.107280
15	.020115	48	.064368	81	.108261
16	.021456	49	.065709	82	.109962
17	.022797	50	.067050	83	.111303
18	.024138	51	.068391	84	112644
19	.025479	52	.069732	85	.113985
20	.026820	53	.071073	86	115326
21	.028161	54	.072414	87	.116667
22	.029502	55	.073755	88	.118008
23	.030843	56	.075096	89	.119349
24	.032184	57	.076437	90	.120690
25	.033525	58	.077778	91	.122031
26	.034866	59	.079119	92	.123372
27	.036207	60	.080460	93	.124713
28	.037548	61	.081801	94	.126054
29	.038889	62	.083142	95	.127395
30	.040230	63	.084483	96	.128736
31	.041571	64	.085824	97	.130077
32	.042912	65	.087165	98	.131418
33	.044253	66	.088506	99	.132759
				100	.134100

Table 7-2. Horsepower vs. Watts.

hp	Watts	hp	Watts
.01	7.457	.26	193.882
.02	14.914	.27	201.339
.03	22.371	.28	208.796
.04	29.828	.29	216.253
.05	37.285	.30	223.710
.06	44.742	.31	231.167
.07	52.199	.32	238.624
.08	59.656	.33	246.081
.09	67.113	.34	253.538
.10	74.570	.35	260.995
.11	82.027	.36	268.452
.12	89.484	.37	275.909
.13	96.941	.38	283.366
.14	104.398	.39	290.823
.15	111.855	.40	298.280
.16	119.312	.41	305.737
.17	126.769	.42	313.194
.18	134.226	.43	320.651
.19	141.683	.44	328.108
.20	149.140	.45	335.565
.21	156.597	.46	343.022
.22	164.054	.47	350.479
.23	171.511	.48	357.936
.24	178.968	.49	365.393
.25	186.425	.50	372.850
.51	380.307	.76	556.732
.52	387.764	.77	574.189
.53	395.221	.78	581.646
.54	402.678	.79	589.103
.55	410.135	.80	596.560
.56	417.592	.81	604.017
.57	425.049	.82	611.474
.58	432.506	.83	618.931
.59	439.963	.84	626.388
.60	447.420	.85	633.845
.61	454.877	.86	641.302
.62	462.334	.87	648.759
.63	469.791	.88	656.216
.64	477.248	.89	663.673
.65	484.705	.90	671.130
.66	492.162	.91	678.587
.67	499.619	.92	686.044
.68	507.076	.93	693.501
.69	514.533	.94	700.958
.70	521.990	.95	708.415
.71	529.447	.96	715.872
.72	536.904	.97	723.329
.73	544.361	.98	730.786
.74	551.818	.99	738.243
.75	559.275	1.00	745.700

Table 7-3. Horsepower and Electrical Power Equivalents.

Unit	Equivalents
1 hp	745.7 watts
1 hp	0.746 kW.
1 hp	33,000 ft.-lbs. per min.
1 hp	550 ft.-lbs. per sec.
1 hp	2,545 BTU per hr.
1 hp	0.175 lbs. carbon oxidized per hr.
1 hp	17 lbs. water per hr. heated from 62-212° F.
1 hp	2.64 lbs. water per hr. evaporated at 212° F.
1 kW	1,000 joules per sec.
1 kW	1.34 hp
1 kW	44,250 ft. lbs. per min.
1 kW	737.3 ft.-lbs. per sec.
1 kW	3,412 BTU per hr.
1 kW	0.227 lbs. carbon oxidized per hr.
1 kW	22.75 lbs. water per hr. heated from 62 to 212° F.
1 kW	3.53 lbs. water per hr. evaporated at 212° F.

number 50 in the column headed by watts. Move the decimal point one place to the right and 50 becomes 500. The corresponding value of horsepower is 0.067050 hp. Moving the decimal point of this number by one decimal place (to the right) supplies 0.6705 hp. In practice this could be rounded off to 0.7 hp.

☐ **Example:**

What is the horsepower rating of a 1-kilowatt generator?

1 kilowatt is equal to 1,000 watts. Use Table 7-1 by selecting the number 100 and moving its decimal point one place to the right, thus changing 100 to 1,000. The corresponding value of horsepower is 0.134100, but remember to move the decimal point here as well. The answer is 1.341 hp.

☐ **Example:**

A fractional hp motor is rated at 0.07 hp. What is its power rating in watts?

There is no value in Table 7-1 that corresponds exactly to 0.07. There are two values, though, which are very close. One of these is 0.069732 and the other 0.071073. Thus, this motor has a rating between 52 and 53 watts. However, if you want a more precise value than this, consult Table 7-2. There, you will see that .07 hp corresponds to 52.199 watts.

Decibels

VOLTAGE AND CURRENT RATIOS VS. POWER RATIOS AND DECIBELS

Table 8-1 supplies voltage or current ratios vs. power (watts) ratios for decibels (dB) ranging from 0.1 to 50. A decibel (one-tenth of a bel) is a ratio, a means of comparing the relative strengths of a pair of currents, voltages, or powers. In itself, a decibel is not indicative of any particular amount of power, voltage, or current.

Since the decibel is a comparison unit, and not an absolute value, some reference level must be indicated. A common reference, also called zero level, is 1 milliwatt. Other reference levels can be used, but in any event, the reference should be specified. If 1 milliwatt is the reference, the letter m is added to dB, the unit being called the dBm (m for milliwatt). Thus, if an amplifier (assuming equal input and output resistances) has an output of 1 watt, the relationship of the power output to the power input is:

$$dB = 10 \log P2/P1$$

$$dB = 10 \log 1/.001 = 10 \log 1000 = 30 \ dB$$

Instead of using the formula, we could have obtained our answer by consulting Table 8-1. The power ratio is 1000—that is, an output of 1 watt is 1000 times greater than the reference level of 1 milliwatt. Locate 1000 under the heading of power ratio in the table. Move to the left and note a gain of 30 dB.

Table 8-1. Voltage or Current Ratios vs. Power Ratios and Decibels.

Voltage or Current Ratio	Power Ratio	− dB +	Voltage or Current Ratio	Power Ratio
1.000	1.000	0	1.0000	1.0000
0.989	0.977	0.1	1.0116	1.0233
0.977	0.955	0.2	1.0233	1.0471
0.966	0.933	0.3	1.0351	1.0715
0.955	0.912	0.4	1.0471	1.0965
0.944	0.891	0.5	1.0593	1.1220
0.933	0.871	0.6	1.0715	1.1482
0.923	0.851	0.7	1.0839	1.1749
0.912	0.832	0.8	1.0965	1.2023
0.902	0.813	0.9	1.1092	1.2303
0.891	0.794	1.0	1.1220	1.2589
0.881	0.776	1.1	1.135	1.288
0.871	0.759	1.2	1.1482	1.3183
0.861	0.741	1.3	1.161	1.349
0.851	0.724	1.4	1.175	1.380
0.841	0.708	1.5	1.189	1.413
0.832	0.692	1.6	1.202	1.445
0.822	0.676	1.7	1.216	1.479
0.813	0.661	1.8	1.230	1.514
0.803	0.646	1.9	1.245	1.549
0.749	0.631	2.0	1.2589	1.5849
0.776	0.603	2.2	1.288	1.660
0.759	0.575	2.4	1.318	1.738
0.750	0.562	2.5	1.334	1.778
0.724	0.525	2.8	1.380	1.905
0.708	0.501	3.0	1.4125	1.9953
0.692	0.479	3.2	1.445	2.089
0.676	0.457	3.4	1.479	2.188
0.668	0.447	3.5	1.4962	2.2387
0.661	0.436	3.6	1.514	2.291
0.646	0.417	3.8	1.549	2.399
0.631	0.398	4.0	1.5849	2.5119
0.596	0.355	4.5	1.6788	2.8184
0.562	0.316	5.0	1.7783	3.1623
0.531	0.282	5.5	1.8836	3.5481
0.501	0.251	6.0	1.9953	3.9811
0.473	0.224	6.5	2.113	4.467
0.447	0.200	7.0	2.239	5.012
0.422	0.178	7.5	2.371	5.623
0.398	0.159	8.0	2.512	6.310
0.376	0.141	8.5	2.661	7.079
0.355	0.126	9.0	2.818	7.943
0.335	0.112	9.5	2.985	8.913
0.316	0.100	10	3.162	10.00
0.282	0.0794	11	3.55	12.6
0.251	0.0631	12	3.98	15.9
0.224	0.0501	13	4.47	20.0
0.200	0.0398	14	5.01	25.1
0.178	0.0316	15	5.62	31.6
0.159	0.0251	16	6.31	39.8

115

Table 8-1. Voltage or Current Ratios vs. Power Ratios and Decibels (cont'd).

Voltage or Current Ratio	Power Ratio	– dB +	Voltage or Current Ratio	Power Ratio
0.141	0.0200	17	7.08	50.1
0.126	0.0159	18	7.94	63.1
0.112	0.0126	19	8.91	79.4
0.10000	0.0100	20	10.00	100.0
0.08913	0.0079	21	11.22	125.9
0.07943	0.0063	22	12.59	158.5
0.07079	0.0050	23	14.13	199.5
0.06310	0.00398	24	15.85	251.2
0.05623	0.03162	25	17.78	316.2
0.05012	0.002512	26	19.95	398.1
0.04467	0.001995	27	22.39	501.2
0.03981	0.001585	28	25.12	631.0
0.03548	0.001259	29	28.18	794.3
0.03162	0.001000	30	31.62	1000
0.02818	0.000794	31	35.48	1259
0.02512	0.000631	32	39.81	1585
0.02239	0.000501	33	44.67	1995
0.01995	0.000398	34	50.12	2512
0.01778	0.000316	35	56.23	3162
0.01585	0.000251	36	63.10	3981
0.01413	0.000199	37	70.79	5012
0.01259	0.000158	38	79.43	6310
0.01122	0.000126	39	89.13	7943
0.01000	0.000100	40	100.00	10000
0.00891	0.000079	41	112.2	12590
0.00794	0.000063	42	125.9	15850
0.00708	0.000050	43	141.3	19950
0.00631	0.000040	44	158.5	25120
0.00562	0.000032	45	177.8	31620
0.00501	0.000025	46	199.5	39810
0.00447	0.000020	47	223.9	50120
0.00398	0.000016	48	251.2	63100
0.00355	0.000013	49	281.8	79430
0.00316	0.000010	50	316.2	100000

□ **Example**:

The input voltage to an amplifier is 1 volt, and the output is 20 volts. What is the voltage gain of the amplifier in dB?

The ratio of the two voltages is 20. The nearest comparable value in the Table is 19.95. The gain is 26 dB.

□ **Example**:

An amplifier is said to have a gain of 20 dB. What is its output in watts?

Locate the number 20 in the dB column in Table 8-1. Move to the right to see that this represents a power ratio of 100. If no power input is given and a reference level of 1 milliwatt is indicated, then the output is 100 milliwatts. If an input power is specified, multiply the input power by 100 to get the value of output power.

Note that the first two columns in Table 8-1 are the reciprocals of the last two columns. It makes no difference whether you divide the output power of an amplifier by the input or vice versa. If you put the larger number in the numerator of the dB power formula the answer will be a whole number and you will work with the two columns at the right in Table 8-1. If you use the smaller power value in the numerator, the answer will be a decimal, as indicated in the first two columns in the Table. In either case, the answer in decibels will be the same. If you have an electronic device with a power ratio of 1 to 1000 or, conversely, 1000 to 1, the gain or loss in dB will be 30 in either case. Some technicians and engineers prefer working with whole numbers and put a minus sign in front of their answer to indicate a loss.

It is essential to remember that the decibel has no absolute value. All it does is indicate how many times greater a power, voltage or current is than some reference level (or how many times smaller).

Still another fact is that the relationship between decibel levels isn't linear. You cannot regard 6 dB as simply being twice as much as 3 dB. Arithmetically, 6 dB is twice as much as 3 dB, but the values these numbers represent do not have a 2 to 1 relationship. For example:

10 dB is 3.1 × the reference level
20 dB is 10 × the reference level
30 dB is 31.6 × the reference level
40 dB is 100 × the reference level
50 dB is 316 × the reference level
60 dB is 1,000 × the reference level

117

POWER AND VOLTAGE GAIN

Table 8-2 shows power gain in watts and voltage gain (in volts) when expressed in terms of decibels. A gain of 3 dB means a doubling of power, or, expressed another way, a doubling of power means a 3 dB increase. As an example, locate .2 in the left hand column. This number represents 0.2 watt. Doubling the power results in 0.4 watt. But if 0.2 watt corresponds to 3 dB then 0.4 watt corresponds to 6 dB.

In terms of voltage, find 1.6 volts in the right hand column. Doubling this voltage results in 3.2 volts. In doubling the voltage we have a 6 dB increase because we go from the corresponding 24 dB to 30 dB. Hence, a doubling of voltage means a 6 dB increase.

The reverse, of course, is also valid. A decrease in voltage by 50% is a 6 dB decrease. A 50 percent decrease in power is a 3 dB decrease.

In Table 8-3, the reference level, 0 dB, is 1,000 microvolts (μV). Since the decibel is referenced to the millivolt it is written as dBmV. This is shown in the heading of the left-hand column. In the column to its right, the heading is microvolts (μV).

Any voltage below 1,000 microvolts is less than the reference and so dBmV are indicated by a minus sign placed in front of the

Table 8-2. Power and Voltage Gain in Decibels.

Power (watts)	dB Gain (reference)	Volts
.1	0	.1
.2	3	
.4	6	.2
.8	9	
1.6	12	.4
3.2	15	
6.4	18	.8
12.8	21	
25.6	24	1.6
51.2	27	
102.4	30	3.2
	33	
	36	6.4
	39	
	42	12.8
	45	
	48	25.6
	51	
	54	51.2
	57	
	60	102.4

Table 8-3. dBmV vs. μV.

dBmV	μV	dBmV	μV	dBmV	μV
-40	10	0	1,000	40	100,000
-39	11	1	1,100	41	110,000
-38	13	2	1,300	42	130,000
-37	14	3	1,400	43	140,000
-36	16	4	1,600	44	160,000
-35	18	5	1,800	45	180,000
-34	20	6	2,000	46	200,000
-33	22	7	2,200	47	220,000
-32	25	8	2,500	48	250,000
-31	28	9	2,800	49	280,000
-30	32	10	3,200	50	320,000
-29	36	11	3,600	51	360,000
-28	40	12	4,000	52	400,000
-27	45	13	4,500	53	450,000
-26	50	14	5,000	54	500,000
-25	56	15	5,600	55	560,000
-24	63	16	6,300	56	630,000
-23	70	17	7,000	57	700,000
-22	80	18	8,000	58	800,000
-21	90	19	9,000	59	900,000
-20	100	20	10,000	60	1.0 volt
-19	110	21	11,000	61	1.1
-18	130	22	13,000	62	1.3
-17	140	23	14,000	63	1.4
-16	160	24	16,000	64	1.6
-15	180	25	18,000	65	1.8
-14	200	26	20,000	66	2.0
-13	220	27	22,000	67	2.2
-12	250	28	25,000	68	2.5
-11	280	29	28,000	69	2.8
-10	320	30	32,000	70	3.2
-9	360	31	36,000	71	3.6
-8	400	32	40,000	72	4.0
-7	450	33	45,000	73	4.5
-6	500	34	50,000	74	5.0
-5	560	35	56,000	75	5.6
-4	630	36	63,000	76	6.3
-3	700	37	70,000	77	7.0
-2	800	38	80,000	78	8.0
-1	900	39	90,000	79	9.0
-0	1,000	40	100,000	80	10.0

number. For any values above the 1,000 microvolt level, dBmV are expressed as positive numbers.

This table is useful when making comparisons between voltages that are less than 1 volt. These are expressed in microvolts to avoid using decimal numbers. Thus, 60 dBmV in the chart is a comparison between 1.0 volt (shown in the right-hand column under the heading of μV) and the reference of 1,000 μV or 0 dBmV.

Immediately below 60 dBmV is 59 dBmV while the number to

its right is 900,000. This is 900,000 microvolts. This could also have been written decimally as 0.9 volt, but is avoided by using 900,000 microvolts instead.

☐ **Example:**

An amplifier has an output of 0.08 volt. What is its output when referenced to 1,000 microvolts (0 dBmV)?

0.08 volt is equal to 80,000 microvolts. Locate 80,000 in the column headed μV. To the left of this number find 38. This is 38 dBmV.

DECIBELS-NEPERS CONVERSIONS

The neper, like the decibel, is a dimensionless unit. While the decibel is derived from common logarithms (logarithms to the base 10), the neper is used to express the ratio of two power levels using the natural system of logarithms—logarithms to the base e (e equals 2.71828). The formula for finding the number of nepers is:

$$\text{nepers} = \tfrac{1}{2}\log_e P2/P1$$

The relationships between decibels and nepers are as follows:

1 decibel	= 0.1 bel	1 bel	= 10 decibels
1 decibel	= 0.1151 neper	1 neper	= 0.8686 bel
1 bel	= 1.151 nepers	1 neper	= 8.686 decibels

As in the case of decibels, nepers must be used with some reference level if just one value of power, either input or output, is specified.

Table 8-4 supplies the conversion between decibels and nepers. Table 8-5 gives nepers to decibels.

☐ **Example:**

An amplifier has a gain of 10 dB. What is its gain in nepers?

Locate the number 10 in the dB column in Table 8-4. To the right of this number you will find 1.1510 nepers.

☐ **Example:**

Assuming a zero reference level of 1 milliwatt, what is the gain in nepers of an amplifier whose output is 50 milliwatts?

The power ratio in the problem is 50 to 1. Locate the nearest number to this in Table 8-1. This is shown as 50.1. The gain in dB is 17. Now consult Table 8-4. Locate 17 dB in the left-hand column. The number of nepers corresponding to 17 dB is 1.9567.

Table 8-4. Decibels vs. Nepers.
(n, nepers; dB, decibels)

dB	n	dB	n	dB	n
1	0.1151	34	3.9134	67	7.7117
2	0.2302	35	4.0285	68	7.8268
3	0.3453	36	4.1436	69	7.9419
4	0.4604	37	4.2587	70	8.0570
5	0.5755	38	4.3738	71	8.1721
6	0.6906	39	4.4889	72	8.2872
7	0.8057	40	4.6040	73	8.4023
8	0.9208	41	4.7191	74	8.5174
9	1.0359	42	4.8342	75	8.6325
10	1.1510	43	4.9493	76	8.7476
11	1.2661	44	5.0644	77	8.8627
12	1.3812	45	5.1795	78	8.9778
13	1.4963	46	5.2946	79	9.0929
14	1.6114	47	5.4097	80	9.2080
15	1.7265	48	5.5248	81	9.3231
16	1.8416	49	5.6399	82	9.4382
17	1.9567	50	5.7550	83	9.5533
18	2.0718	51	5.8701	84	9.6684
19	2.1869	52	5.9852	85	9.7835
20	2.3020	53	6.1003	86	9.8986
21	2.4171	54	6.2154	87	10.0137
22	2.5322	55	6.3305	88	10.1288
23	2.6473	56	6.4456	89	10.2439
24	2.7624	57	6.5607	90	10.3590
25	2.8775	58	6.6758	91	10.4741
26	2.9926	59	6.7909	92	10.5892
27	3.1077	60	6.9060	93	10.7043
28	3.2228	61	7.0211	94	10.8194
29	3.3379	62	7.1362	95	10.9345
30	3.4530	63	7.2513	96	11.0496
31	3.5681	64	7.3664	97	11.1647
32	3.6832	65	7.4815	98	11.2798
33	3.7983	66	7.5966	99	11.3949
				100	11.5100

Table 8-5. Neper vs. Decibel Conversion.

n	dB	n	dB	n	dB
1	8.686	34	295.324	67	581.962
2	17.372	35	304.010	68	590.648
3	26.058	36	312.696	69	599.334
4	34.744	37	321.382	70	608.020
5	43.430	38	330.068	71	616.706
6	52.116	39	338.754	72	625.392
7	60.802	40	347.440	73	634.078
8	69.488	41	356.126	74	642.764
9	78.174	42	364.812	75	651.450
10	86.860	43	373.498	76	660.136
11	95.546	44	382.184	77	668.822
12	104.232	45	390.870	78	677.508
13	112.918	46	399.556	79	686.194
14	121.604	47	408.242	80	694.880
15	130.290	48	416.928	81	703.556
16	138.976	49	425.614	82	712.252
17	147.662	50	434.300	83	720.938
18	156.348	51	442.986	84	729.624
19	165.034	52	451.672	85	738.310
20	173.720	53	460.358	86	746.996
21	182.406	54	469.044	87	755.682
22	191.092	55	477.730	88	764.368
23	199.778	56	486.416	89	773.054
24	208.464	57	495.102	90	781.740
25	217.150	58	503.788	91	790.426
26	225.836	59	512.474	92	799.112
27	234.522	60	521.160	93	807.798
28	243.208	61	529.846	94	816.484
29	251.894	62	538.532	95	825.170
30	260.580	63	547.218	96	833.856
31	269.266	64	555.904	97	842.542
32	277.952	65	564.590	98	851.228
33	286.638	66	573.276	99	859.914
				100	868.600

Sensitivity

RECEIVER OR TUNER SENSITIVITY

Sensitivity is the ability of a receiver or tuner to respond to signals, previously specified in microvolts (μV). These specs have been changed from voltage to power ratings. Sensitivity in microvolts can be converted to dBf by:

$$dBf = 20 \log (\mu V/0.55)$$

Log is log to the base 10 and μV is microvolts based on a 300-ohm antenna input. Sensitivity is not the same for monaural and stereo so tuners and receivers should include separate dBf figures for each; f is the reference and is the femtowatt or 10^{-15} watt.

☐ **Example:**
A tuner has an input sensitivity of 2 μV when set in the stereo mode. What is its sensitivity in dBf?

$$dBf = 20 \log (\mu V/0.55)$$
$$= 20 \log 2/0.55$$
$$= 20 \log 3.636363$$
$$\log 3.636363 = 0.5599$$

$$dBf = (20)(0.5599) = 11.198 \text{ dBf}$$

123

Table 9-1. Microvolts (μV) vs. dBf.

μV	dBf	μV	dBf
1.5	8.71	3.5	16.07
1.6	9.28	3.6	16.3
1.7	9.8	3.7	16.548
1.8	10.3	3.8	16.776
1.9	10.77	3.9	17.012
2.0	11.198	4.0	17.2
2.1	11.6	4.5	18.256
2.2	12.04	5.0	19.17
2.3	12.424	6.0	20.748
2.4	12.8	7.0	22.075
2.5	13.15	8.0	22.984
2.6	13.478	9.0	24.296
2.7	13.804	10.0	25.19
2.8	14.134	30.0	34.74
2.9	14.436	32.0	35.3
3.0	14.74	40.0	37.23
3.1	15.01	50.0	39.17
3.2	15.3	55.0	40.0
3.3	15.7	100.0	45.19
3.4	15.8	1000.0	65.154

Instead of doing this work, consult Table 9-1. Locate 2 μV in the left column. Move directly across to the answer, 11.198 dBf.

☐ **Example**:
A tuner has an input sensitivity of 40 μV when set for monaural reception. What is its sensitivity in dBf?

$$dBf = 20 \log (\mu V/0.55)$$
$$= 20 \log 40/0.55$$
$$= 20 \log 72.7272$$
$$\log 72.7272 = 1.8615$$

dBf = (20) (1.8615) = 37.23 dBf

Using Table 9-1, locate 40 μV in the left column. Move across to the right for an answer of 37.23 dBf.

124

10

Sound And Acoustics

WAVELENGTHS OF SOUND

The distance between two successive positive peaks, two successive negative peaks, or between any two corresponding points of a sine wave is known as its wavelength. As mentioned earlier, this is often represented by the letter λ. This description is not only applicable to radio-frequency waves, but also to sound waves. The reference here is not to a complex sound waveform, but to a pure sine wave only.

A sound wave of constant velocity (represented by the letter μ) will travel a distance of one wavelength in a one period interval. This is more concisely stated in the formula μ equals λ/T. But the period of a wave has an inverse relationship to the frequency. Thus, T equals 1/f. By substituting in the formula μ equals λ/T, we get μ equals f λ. We can rearrange this formula to read λ equals μ/f.

The velocity of sound in air at a temperature of 20 degrees C (68 degrees F) is 1130 feet per second. Using this information, we can conveniently set up Table 10-1 which gives the relationship between the frequency of sound in air in hertz and the wavelengths of sound in feet and in inches.

Conversion from feet to meters and from inches to millimeters is explained in Chapter 23.

□ **Example:**
What is the wavelength, in feet, of a 60-hertz sine wave?

Locate the number 60 in the left-hand column of Table 10-1. The corresponding value is shown as 18.83 feet.

□ **Example:**

A sound wave has a length of 7 feet. What is its frequency in hertz?

The closest value given in Table 10-1 is 7.06 feet. The frequency of this sound wave, then, is approximately 160 Hz.

Table 10-1. Sound Wavelengths.

(1130 ft/sec, in air, at 20 degrees C; 68 degrees F)

Frequency (Hz)	Wavelength (inches)	Frequency (Hz)	Wavelength (inches)	Frequency (Hz)	Wavelength (inches)
20	56.50	220	5.14	900	1.26
25	45.20	230	4.91	950	1.19
30	37.67	240	4.71	975	1.16
35	32.29	250	4.52	990	1.14
40	28.25	260	4.35	1000	13.56
45	25.11	270	4.19	2000	6.78
50	22.60	280	4.04	3000	4.52
55	20.55	290	3.90	4000	3.39
60	18.83	300	3.77	5000	2.71
65	17.38	320	3.53	6000	2.26
70	16.14	340	3.32	7000	1.94
75	15.07	360	3.14	8000	1.70
80	14.13	380	2.97	9000	1.51
85	13.29	400	2.83	10000	1.36
90	12.56	420	2.69	11000	1.23
95	11.89	440	2.57	12000	1.13
100	11.30	460	2.46	13000	1.04
110	10.27	480	2.35	14000	0.97
120	9.42	500	2.26	15000	0.90
130	8.69	525	2.15	16000	0.85
140	8.07	550	2.05	17000	0.80
150	7.53	575	1.97	18000	0.75
160	7.06	600	1.88	19000	0.71
170	6.65	650	1.74	20000	0.68
180	6.28	700	1.61		
190	5.95	750	1.51		
200	5.65	800	1.41		
210	5.38	850	1.33		

□ **Example**:

What is the wavelength, in meters, of a wave whose frequency is 75 Hz?

First, locate the wavelength in feet corresponding to a frequency of 75 Hz. Table 10-1 shows that this is 15.07 feet. Consult Table 23-4 on page 326. The closest value to 15.07 is 15 feet. Move one column and the answer is 4.5720 meters.

RANGE OF MUSICAL INSTRUMENTS

The fundamental range of musical instruments is limited. At the low-frequency end, few musical instruments can produce tones below 50 Hz. The human voice doesn't go much lower than 70 Hz. At the high-frequency end, all musical instruments and voices are under 5 kHz in fundamental frequency. It is the fundamental frequency that determines the pitch of a tone. See Fig. 10-1.

What you can hear depends on your age, sex, the physical condition of your ears and brain, and prior musical training. While the audio spectrum is assumed to have a range of 20 Hz to 20 kHz, few of us have a hearing capability that goes down to 20 Hz and equally few can hear as high as 20 kHz. The pipe organ, contrabassoon and the harp can reach below 40 Hz. Natural sounds include hardly any low frequencies, and what you may find at such frequencies is noise See Fig. 10-2.

THE OCTAVE

An octave is a doubling of frequency. From 30 Hz to 60 Hz could be called an octave. We could regard 60 Hz to 120 Hz as another octave. We do not start with 0 Hz for this is actually dc. 32 Hz is often selected as a practical beginning but we can start with 16 Hz as a bottom limit. If we select 16 Hz as our starting point, we can have 10 octaves up to approximately 16 kHz.

The ten octaves in Table 10-2 are of particular interest because they roughly represent a substantial part of the range of human hearing capability. Sounds below 64 Hz and above 16 kHz cannot be heard by most. Consider 16 Hz and 16 kHz as the outermost hearing limits.

FREQUENCY RANGE OF THE PIANO

Middle C on the piano (Table 10-3) is indicated by the letter C

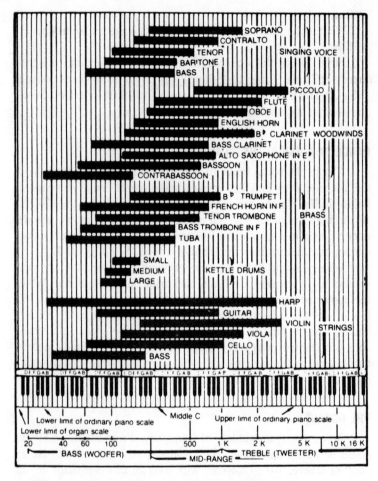

Fig. 10-1. The fundamental range of musical instruments is quite limited. Shown here are the ranges of fundamanetal components of tones for the principal musical instruments and voices. (courtesy of Ziff-Davis Publishing Company, from the January 1976 issue of Stereo Review).

(not followed by a number) corresponding to a frequency of 261.63 Hz. From key A3 to A4 is an octave; from key A3 to A2 is also an octave. Starting with key C3 below middle C there are seven octaves to key C4 above middle C. Figure 10-2 shows the frequencies of organ and piano tones.

The frequency range of other musical instruments appears in Table 10-4, and Table 10-5 supplies the frequency range of musical voices.

Table 10-2. Frequency Range in Octaves.

Frequency Range, Hz	Octave
16 to 32	first
32 to 64	second
64 to 128	third
128 to 256	fourth
256 to 512	fifth
512 to 1024	sixth
1024 to 2048	seventh
2048 to 4096	eighth
4096 to 8192	ninth
8192 to 16,384	tenth

Figure 10-3 is a comparison of the range of fundamental frequencies of various instruments and male and female voices.

VELOCITY OF SOUND IN AIR

Sound velocity can be calculated from either of the following equations:

$$V = 49\sqrt{459.4} + °F \text{ feet/second}$$

or

$$V = 20.06 \sqrt{273} + °C \text{ meters/second}$$

Table 10-3. Frequency Range of the Piano.

Key	Frequency (Hz)	Key	Frequency (Hz)	Key	Frequency (Hz)
A_4	27.50	E_1	164.81	B^1	987.77
B_4	30.87	F_1	174.61	C^2	1046.50
C_3	32.70	G_1	196.00	D^2	1174.70
D_3	36.71	A_1	220.00	E^2	1318.50
E_3	41.20	B_1	246.94	F^2	1396.90
F_3	43.65	C	261.63	G^2	1568.00
G_3	49.00	D	293.66	A^2	1760.00
A_3	55.00	E	329.63	B^2	1975.50
F_3	61.74	F	349.23	C_3	2023.00
C_2	65.41	G	392.00	D_3	2349.30
D_2	73.42	A	440.00	E_3	2637.00
E_2	82.41	B	493.88	F_3	2793.80
F_2	87.31	C^1	523.25	G_3	3136.00
G_2	98.00	D^1	587.33	A_3	3520.00
A_2	110.00	E^1	659.26	B_3	3951.10
B_2	123.47	F^1	698.46	C_4	4186.00
C_1	130.81	G^1	783.99		
D_1	146.83	A^1	880.00		

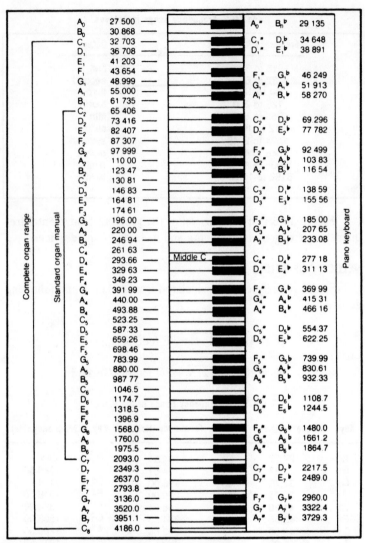

Fig. 10-2. Frequencies of organ and piano tones.

V is the velocity in feet/second or meters/second, °F is the temperature in degrees Fahrenheit, °C the temperature in degrees Celsius.

The velocity of sound isn't affected by frequency, somewhat slightly by humidity, much more so by temperature, and varies considerably depending on the material (medium) through which it moves. In solid substances, such as brick or steel, the velocity of

Table 10-4. Frequency Range of Musical Instruments.

Instrument	Low Hz	High Hz
Bass clarinet	82.41	493.88
Bass tuba	43.65	349.23
Bass viola	41.20	246.94
Bassoon	61.74	493.88
Cello	130.81	698.46
Clarinet	164.81	1567.00
Flute	261.63	3349.30
French horn	110.00	880.00
Trombone	82.41	493.88
Trumpet	164.81	987.77
Oboe	261.63	1568.00
Violin	130.81	1174.70
Violin	196.00	3136.00

Table 10-5. Frequency Range of Musical Voices.

	Low Hz	High Hz
Alto	130.81	698.46
Baritone	98.00	392.00
Bass	87.31	349.23
Soprano	246.94	1174.70
Tenor	130.81	493.88

sound is far greater than in air. In air, the velocity of sound increases by about two feet per second for each increase of one degree Celsius. Refer to Tables 10-6 and 10-7.

RELATIVE VOLUME LEVELS OF ORDINARY SOUNDS

The threshold of hearing is zero dB and the threshold of hearing pain is 130 dB. See Fig. 10-4.

Table 10-8 indicates the intensity levels of various musical instruments measured at a distance of 10 feet. These are referenced to the threshold of hearing or 0 dB.

While the decibel is generally used to indicate the relative volume level of ordinary sounds, there are two other units that are also used. These are the sone and the phon. Measured at a frequency of 1 kHz, 40 dB above the threshold of hearing produces a loudness of 1 sone.

The phon, a loudness unit, is based on the fact that doubling the intensity of a sound does not result in a doubling of a hearing sensation. The phon is a loudness unit that is an attempt to reflect the true loudness of a sound as perceived by the ear. The sone, on the other hand, is a unit of loudness that is graduated in equal steps.

131

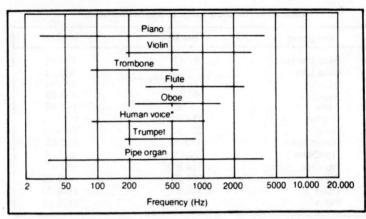

Fig. 10-3. Comparison of the range of fundamental frequencies of male and female voices.

Table 10-6. Velocity of Sound in Air at Various Temperatures.

Deg. F	Speed (ft/second)
32	1087
50	1107
59	1117
68	1127
86	1147

Deg. C	Speed (meters/second)
0	331.32
10	337.42
15	340.47
20	343.51
30	349.61

Table 10-7. Velocity of Sound in Liquids and Solids.

Material	Sound Velocity	
	Feet/second	Meters/second
Alcohol	4724	1440
Aluminum	20,407	6220
Brass	14,530	4430
Copper	15,157	4620
Glass	17,716	5400
Lead	7,972	2430
Magnesium	17,487	5330
Mercury	4,790	1460
Nickel	18,372	5600
Polystyrene	8,760	2670
Quartz	18,865	5750
Steel	20,046	6110
Water	4,757	1450

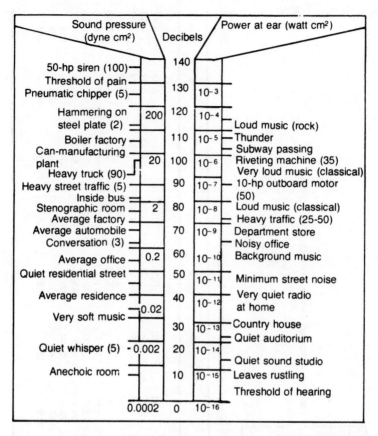

Fig. 10-4. Relative loudness levels of common sounds. (Courtesy of Ziff-Davis Publishing Company, from the January, 1976 issue of Stereo Review).

Table 10-9 shows some common sounds and their loudness in phons.

Table 10-10 can be used for the conversion of phons to sones or sones to phons.

Table 10-8. Intensity Levels of Various Musical Instruments.

Piano	60 to 100 dB
Organ	35 to 110 dB
Bass drum	35 to 115 dB
Trumpet	55 to 95 dB
Violin	42 to 95 dB
Tympani	30 to 110 dB
Cymbal	40 to 110 dB

Sound	Phons
Threshold of Hearing	0
Movement of leaves	10 - 20
Ticking of a clock. Whispering	20 - 30
Soft music	30 - 40
Street noises	50 - 60
Loud voice	60 - 80
Orchestra playing loudly	90
Motorcycle racing its engine	100
Threshold of pain	130

Table 10-9. Ordinary Sounds and Their Level in Phons.

Table 10-10. Conversion of Phons to Sones or Sones to Phons.

Phons	0	+1	+2	+3	+4	+5	+6	+7	+8	+9
20	0.25	0.27	0.29	0.31	0.33	0.35	0.38	0.41	0.44	0.47
30	0.50	0.54	0.57	0.62	0.66	0.71	0.76	0.81	0.87	0.93
40	1.0	1.07	1.15	1.23	1.32	1.41	1.52	1.62	1.74	1.87
50	2.0	2.14	2.30	2.46	2.64	2.83	3.03	3.25	3.48	3.73
60	4.0	4.29	4.59	4.92	5.28	5.66	6.06	6.50	6.96	7.46
70	8.0	8.60	9.20	9.80	10.6	11.3	12.1	13.0	13.9	14.9
80	16.0	17.1	18.4	19.7	21.1	22.6	24.3	26.0	27.9	29.9
90	32.0	34.3	36.8	39.4	42.2	45.3	48.5	52.0	55.7	59.7
100	64.0	68.6	73.5	78.8	84.4	90.5	97.0	104	111	119
110	128	137	147	158	169	181	194	208	223	239
120	256	274	294	315	338	362	388	416	446	478

Table 10-11. Sound Absorption Coefficients for 3/16" Plywood.

Frequency (Hz)	Sound Absorption Coefficient
125	0.35
250	0.25
500	0.20
1,000	0.15
2,000	0.05
4,000	0.05

SOUND ABSORPTION

The sound absorption coefficient of materials varies with frequency. Table 10-11 indicates that plywood, a common building material, has a greater sound absorption coefficient at frequencies below 500 Hz, but tends to level off in the region above 2 kHz.

The limits of sound absorption coefficients are 0 and 1. 1, or 100 percent, would indicate a substance that absorbed sound completely, and is the goal of construction materials used in anechoic chambers. 0 or 0 percent indicates that the material does not absorb sound at all. This does not necessarily mean total reflectivity. An open window or rather that portion of it that constitutes open space can be regarded as having no reflectivity. In an echo chamber, the surfaces may have extremely high values of reflectivity. How effective a material is for the absorption of sound is indicated by its sound absorption coefficient. A cement floor has an absorption coefficient of 0.015 or 1.5%. This means that sound striking such a floor will dissipate 1.5 percent of the sound as heat, or slightly less, depending on how much of the sound will pass through the material. 98.5 percent of the sound will be reflected.

Table 10-12 indicates the effect of frequency on the sound absorption coefficients of various materials. For some materials, the absorption coefficient increases with frequency; with others it decreases; and for some it remains relatively fixed.

Absorption coefficients of building materials are often supplied as an overall range without indicating the effect of frequency, as indicated in Table 10-13.

Table 10-12. Frequency vs. Sound Absorption.

Material	Frequency (Hz)					
Glass window	125	250	500	1000	2000	4000
Lightweight drapes	.35	.25	.18	.12	.07	.04
Heavy drapes	.03	.04	.11	.17	.24	.35
Wood floor	.14	.35	.55	.72	.70	.65
Carpet	.15	.11	.10	.07	.06	.07
(on concrete)	02	.06	.14	.37	.60	.65

Table 10-13. Sound Absorption Coefficients of Various Materials.

Material	Coefficient Absorption Range
Linoleum on concrete floor	0.03 to 0.08
Upholstered seats	0.05
Ventilating grilles	0.15 to 0.50
Painted brick	0.02 to 0.04
Plaster on brick	0.02 to 0.04
Plaster on lath	0.3 to 0.04
Door	0.3 to 0.05
Window glass	0.3 to 0.05
Thick carpeting	0.15 to 0.5
Heavy curtains	0.2 to 0.8

REVERBERATION TIME

Reverberation time is a function of room volume and sound absorption. It is directly proportional to the volume of an enclosed space, such as a recording studio or an in-home listening room and inversely proportional to the total amount of sound absorption in that enclosure (Fig. 10-5).

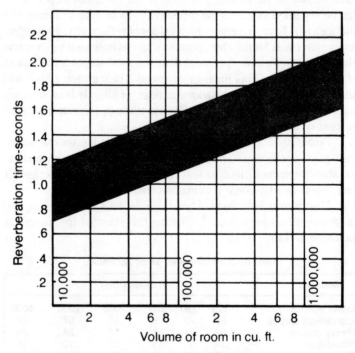

Fig. 10-5. Dark area indicates acceptable reverberation time for rooms of different volumes.

Table 10-14. Sound Absorption of Seats and Audience at 512 Hertz.

	Equivalent Absorption (in sabins)
Audience, seated, units per person, depending on character of seats, etc.	3.0-4.3
Chairs, metal or wood	0.17
Pew cushions	1.45-1.90
Theater and auditorium chairs	
Wood veneer seat and back	0.25
Upholstered in leatherette	1.6
Heavily upholstered in plush or mohair	2.6-3.0
Wood pews	0.4

Reverberation time is not the same for all frequencies. The formula for calculating reverberation time, developed by Professor Wallace C. Sabine in 1895, indicates average reverberation time and does not take frequency into consideration.

$$T_{60} = \frac{0.05v}{S_a}$$

T_{60} is the reverberation time in seconds, V is the volume of the room in cubic feet and S_a is the total equivalent sound absorption in sabins per square foot of surface material. Reverberation time is how long it takes for reverberant sound to decrease by 60 dB, one millionth of the original sound source intensity.

Table 10-14 shows the sound absorption of seats and an audience at 512 Hz. The absorption is indicated in sabins. Table 10-15 indicates the effect of an audience on reverberation time.

It takes time for sounds to reach a reflecting surface and then the ears of listeners. Table 10-16 shows the approximate time in seconds for reverberant sound to be heard.

LISSAJOUS PATTERNS

An oscilloscope can provide Lissajous figures for calibrating an

Table 10-15. Effect of an Audience on Reverberation Time.

Audience (number present)	Absorption (sabins)	Reverberation Time (seconds)
0	1,201	7.3
200	2,011	4.3
400	2,821	3.1
600	3,631	2.4
800	4,441	2.0

Table 10-16. Reverberation Time.

Total Distance Traveled (feet)	Distance from Ear to Barrier (feet)	Approximate Time Required (seconds)
11.2	5.6	0.01
22.4	11.2	0.02
33.6	16.8	0.03
44.8	22.4	0.04
56.0	28.0	0.05
63.2	31.6	0.06
78.4	39.2	0.07
89.6	44.8	0.08
100.8	50.4	0.09
112.0	56.0	0.1
224.0	112.0	0.2
336.0	168.0	0.3
448.0	224.0	0.4
560.0	280.0	0.5
632.0	316.0	0.6
784.0	392.0	0.7
896.0	448.0	0.8
1008.0	504.0	0.9
1120.0	560.0	1.0

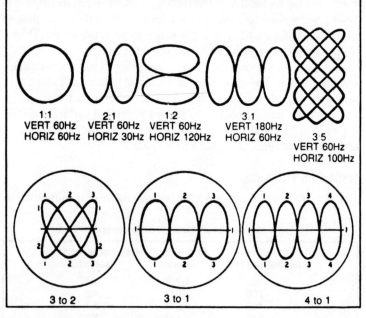

Fig. 10-6. Lissajous figures.

Table 10-17. Wire Gauge for Speaker Connections.

Length of Wire from Amplifier to Speaker	Gauge of Stranded Wire
up to 25 feet	18 AWG
25 to 40 feet	16 AWG
40 to 60 feet	14 AWG
above 60 feet	12 AWG

unknown against a known frequency. Figure 10-6 shows patterns indicating the frequency ratio of horizontal to vertical input sine waves. Count the number of loops across the top and down the side.

SPEAKER CONNECTIONS

The size of the wire to be used in connecting loudspeakers to the output terminals of a power amplifier depends on the distance of the speakers from the amplifier. Table 10-17 indicates the wire gauge that should be used. The smaller the gauge number the thicker the wire. Thus, No. 12 wire is thicker than No. 18. There is no harm in using a thicker wire than that suggested in the table. However, the thicker the wire, the more expensive it is.

For the reproduction of strong bass tones whose momentary current can be quite high, so-called *monster* cable is preferable. Such cable is stranded, and has twice the thickness, or more, of the wire gauges shown in Table 10-17.

Recording

PLAYBACK EQUALIZATION FOR RECORDING TAPES

The point at which treble boost begins is referred to as a time constant. CrO_2 and cobalt-treated ferric tapes have a time constant of 70 microseconds. Ferric oxide tapes have a 120-microsecond time constant. See Table 11-1.

Table 11-1. Time Constants of Magnetic Tapes and Their Equalization.

(Time Constants in Microseconds)

Formulation	Bias	Equalization
Ferric Oxide (FeO2)	normal	120
Chromium Dioxide (CrO2)	high	70
Beridox (Fuji)	high	70
Ferric Oxide/Cobalt	high	70
Ferrichrome (FeCr)	normal	70
Pure Metal	high	70

TREBLE BOOST

159,154.9431 divided by the time constant in microseconds supplies the treble boost frequency. See Table 11-2.

Table 11-2. Time Constants of Magnetic Tape vs. Starting Point of Treble Boost.

Time Constant	Treble Boost Frequency (Hz)
10	15,915.4943
15	10,610.3295
20	7,957.7471

140

Time Constant	Treble Boost Frequency (Hz)
25	6,366.1977
30	5,305.1467
35	4,547.284
40	3,978.8735
45	3,536.7765
50	3,183.0988
55	2,893.7262
60	2,652.5823
65	2,448.5375
70	2,273.462
75	2,122.0659
80	1,989.4367
85	1,872.411
90	1,768.3882
95	1,675.3151
100	1,591.5494

OPERATING TIME OF MAGNETIC TAPE

The amount of time required by a tape for recording or playback can be calculated with the following formula:

$$T \text{ (seconds)} = \frac{\text{tape length (inches}}{\text{operating speed (inches per second)}}$$

☐**Example:**

What is the operating time of a single track tape that is 150 feet long, running at 15 ips (inches per second)

Convert 150 feet to inches by multiplying by 12, so $150 \times 12 = 1800$ inches.

$$T = \frac{1800}{15} = 120 \text{ seconds} = 2 \text{ minutes}$$

For two tracks, forward and reverse, operating time would be doubled or 4 minutes. See Table 11-3.

Table 11-3. Recording or Playback Time of Open Reel Tape.

No. of feet	No. of tracks	15 (ips) (38 cm/sec.)	7½ (ips) (19 cm/sec.)	Speed 3¾ (ips) (9.5 cm/sec.)	1⅞ (ips) (4.75 cm/sec.)	15/16 (ips) (2.375cm/sec.)
4800	1 Track	1 hr. 4 min.	2 hrs. 8 min.	4 hrs. 16 min.	8 hrs. 32 min.	17 hrs. 4 min.
	2 Tracks	2 hrs. 8 min.	4 hrs. 16 min.	8 hrs. 32 min.	17 hrs. 4 min.	34 hrs. 8 min.
	4 Tracks	4 hrs. 16 min.	8 hrs. 32 min.	17 hrs. 4 min.	34 hrs. 8 min.	68 hrs. 16 min.
3600	1 Track	48 min.	1 hr. 36 min.	3 hrs. 12 min.	6 hrs. 24 min.	12 hrs. 48 min.
	2 Tracks	1 hr. 36 min.	3 hrs. 12 min.	6 hrs. 24 min.	12 hrs. 48 min.	25 hrs. 36 min.
	4 Tracks	3 hrs. 12 min.	6 hrs. 24 min.	12 hrs. 48 min.	25 hrs. 36 min.	51 hrs. 12 min.

Table 11-3. Recording or Playback Time of Open Reel Tape (Cont'd).

NO. OF FEET	NO. OF TRACKS	SPEED				
		15 (IPS) (38 cm/sec.)v	7½ (IPS) (19 cm/sec.)	3¾ (IPS) (9.5 cm/sec)	1⅞ (IPS) (4.75 cm/sec.)	15/16 (IPS) (2.375 cm/sec.)
2500	1 Track	33 min.	1 hr. 6 min.	2 hrs. 13 min.	4 hrs. 26 min.	8 hrs. 52 min.
	2 Tracks	1 hr. 6 min.	2 hrs. 12 min.	4 hrs. 26 min.	8 hrs. 52 min.	17 hrs. 44 min.
	4 Tracks	2 hrs. 12 min.	4 hrs. 24 min.	8 hrs. 52 min.	17 hrs. 44 min.	35 hrs. 28 min.
2400	1 Track	32 min.	1 hr. 4 min.	2 hrs. 8 min.	4 hrs. 16 min.	8 hrs. 32 min.
	2 Tracks	1 hr. 4 min.	2 hrs. 8 min.	4 hrs. 16 min.	8 hrs. 32 min.	17 hrs. 4 min.
	4 Tracks	2 hrs. 8 min.	4 hrs. 16 min.	8 hrs. 32 min.	17 hrs. 4 min.	34 hrs. 8 min.
1800	1 Track	24 min.	48 min.	1 hr. 36 min.	3 hrs. 12 min.	6 hrs. 24 min.
	2 Tracks	48 min.	1 hr. 36 min.	3 hrs. 12 min.	6 hrs. 24 min.	12 hrs. 48 min.
	4 Tracks	1 hr. 36 min.	3 hrs. 12 min.	6 hrs. 24 min.	12 hrs. 48 min.	25 hrs. 36 min.
1200	1 Track	16 min.	32 min.	1 hr. 4 min.	2 hrs. 8 min.	4 hrs. 16 min.
	2 Tracks	32 min.	1 hr. 4 min.	2 hrs. 8 min.	4 hrs. 16 min.	8 hrs. 32 min.
	4 Tracks	1 hr. 4 min.	2 hrs. 8 min.	4 hrs. 16 min.	8 hrs. 32 min.	17 hrs. 4 min.
900	1 Track	12 min.	24 min.	48 min.	1 hr. 36 min.	3 hrs. 12 min.
	2 Tracks	24 min.	48 min.	1 hr. 36 min.	3 hrs. 12 min.	6 hrs. 24 min.
	4 Tracks	48 min.	1 hr. 36 min.	3 hrs. 12 min.	6 hrs. 24 min.	12 hrs. 48 min.
600	1 Track	8 min.	16 min.	32 min.	1 hr. 4 min.	2 hrs. 8 min.
	2 Tracks	16 min.	32 min.	1 hr. 4 min.	2 hrs. 8 min.	4 hrs. 16 min.
	4 Tracks	32 min.	1 hr. 4 min.	2 hrs. 8 min.	4 hrs. 16 min.	8 hrs. 32 min.
300	1 Track	4 min.	8 min.	16 min.	32 min.	1 hr. 4 min.
	2 Tracks	8 min.	16 min.	32 min.	1 hr. 4 min.	2 hrs. 8 min.
	4 Tracks	16 min.	32 min.	1 hr. 4 min.	2 hrs. 8 min.	4 hrs. 16 min.
225	1 Track	3 min.	6 min.	12 min.	24 min.	48 min.
	2 Tracks	6 min.	12 min.	24 min.	48 min.	1 hr. 36 min.
	4 Tracks	12 min.	24 min.	48 min.	1 hr. 36 min.	3 hrs. 12 min.
150	1 Track	2 min.	4 min.	8 min.	16 min.	32 min.
	2 Tracks	4 min.	8 min.	16 min.	32 min.	1 hr. 4 min.
	4 Tracks	8 min.	16 min.	32 min.	1 hr. 4 min.	2 hrs. 8 min.

RECORDING AND PLAYBACK SPEEDS

While recording and playback speeds are commonly given in ips (inches per second), these are sometimes supplied in cm/sec (centimeters per second). Table 11-4 shows the relationship between ips and cm/sec.

Table 11-4. Tape Speeds in ips and cm/sec.

ips	30	15	7 ½	3 ¾	1 ⅞	15/16
cm/sec	76.2	38.1	19.05	9.525	4.7625	2.38125

TAPE LENGTHS IN FEET AND METERS

Tape lengths are now being supplied in feet and meters. Table

Table 11-5. Tape Lengths in Feet and Meters.

Tape Length, Feet	Tape Length, Meters
100	30.48
150	45.72
200	60.96
210	64.008
300	91.44
400	121.92
420	128.016
450	137.16
500	152.40
600	182.88
700	213.36
800	243.84
900	274.32
1000	304.8
1200	365.76
1500	457.20
1800	548.64
2000	609.60
2400	731.52
3000	914.40
3600	1097.28

11-5 supplies data on various lengths of tape in the English and metric systems.

OPEN REEL SIZES

The diameters of the reels used in open reel decks are now

Table 11-6. Open Reel Diameters in Inches and Centimeters.

Inches	Centimeters	Actual Diameter (cm)
3	8	7.62
4	10	10.16
5	13	12.70
5 ¾	15	14.605
7	18	17.78
10 ½	26	26.67

Table 11-7. Recording and Playback Time of Audio Cassette Tape.

Type	Length		Recording or Playback Time (minutes)	
	(meters)	(feet)	One Way	Both Ways
C30	45	150	15	30
C60	90	300	30	60
C90	135	450	45	90
C120	180	600	60	120

measured in inches and centimeters. Table 11-6 supplies diameters of the most commonly used reels in both types of measurements. The center column shows the diameters as specified. The column at the right shows the actual diameters.

Because cassette tape is confined to a plastic shell, longer playing times can be achieved only by making the tape thinner. There is a practical limit to this method since a tape that is excessively thin can be stretched by sudden starts. Table 11-7 supplies the lengths, recording and playback times of the most commonly used cassette tapes. While a C30 and C60 have approximately the same tape thickness, a C120 tape has half the thickness of a C60.

12

Video

TV CHANNELS AND FREQUENCIES

Table 12-1 supplies the channel width, and the frequencies in megahertz (MHz) of the picture and sound carriers of tv channels. The table covers VHF channels 2 to 13 (channel 1 is not used for tv) and UHF channels 14 to 83.

The bandwidth of each channel is 6 megahertz regardless of frequency. In each instance, the video carrier is 1.25 megahertz above the lower edge of the band while the sound carrier is 0.25 megahertz lower than the high-frequency end of the channel. The separation between carriers, video and sound, is 4.5 megahertz. Worldwide standards are listed in Table 12-2.

☐**Example**:

What is the sound carrier frequency of channel 16?

Locate channel 16 in the left-hand column. Move to the right and under the heading of sound carrier find 487.75 MHz.

The frequencies between 806 and 890 MHz are now allocated to the land mobile services. These frequencies are designated as UHF channels 70 to 83 inclusive. Hence, channel 69 is the highest frequency UHF channel available for tv broadcasting and reception.

Television channels, 2 through 13, aren't continuous bands. Channels 2 through 6, known as the low band tv channels, extend from 54 MHz, the low-frequency end of VHF channel 2, to 88 MHz,

Table 12-1. TV Channels and Frequencies.

VHF Television Frequencies

Channel No.	Frequency Band (MHz)	Video Carrier (MHz)	Sound Carrier (MHz)
(1 not assigned)			
2	54-60	55.25	59.75
3	60-66	61.25	65.75
4	66-72	67.25	71.75
5	76-82	77.25	81.75
6	82-88	83.25	87.75
7	174-180	175.25	179.75
8	180-186	181.25	185.75
9	186-192	187.25	191.75
10	192-198	193.25	197.75
11	198 204	199.25	203.75
12	204-210	205.25	209.75
13	210-216	211.25	215.75

UHF Television Frequencies

Channel No.	Frequency Band (MHz)	Video Carrier (MHz)	Sound Carrier (MHz)
14	470 476	471.25	475.75
15	476-482	477.25	481.75
16	482-488	483.25	487.75
17	488-494	489.25	493.75
18	494-500	495.25	499.75
19	500-506	501.25	505.75
20	506-512	507.25	511.75
21	512-518	513.25	517.75
22	518-524	519.25	523.75
23	524-530	525.25	529.75
24	530-536	531.25	535.75
25	536-542	537.25	541.75
26	542-548	543.25	547.75
27	548-554	549.25	553.75
28	554-560	555.25	559.75
29	560-566	561.25	565.75
30	566-572	567.25	571.75
31	572-578	573.25	577.75
32	578-584	579.25	583.75
33	584-590	585.25	589.75
34	590-596	591.25	595.75
35	596-602	597.25	601.75

Table 12-1. TV Channels and Frequencies (cont'd).

Channel No.	Frequency Band (MHz)	Video Carrier (MHz)	Sound Carrier (MHz)
36	602-608	603.25	607.75
37	608-614	609.25	613.75
38	614-620	615.25	619.75
39	620-626	621.25	625.75
40	626-632	627.25	631.75
41	632-638	633.25	637.75
42	638-644	639.25	643.75
43	644-650	645.25	649.75
44	650-656	651.25	655.75
45	656-662	657.25	661.75
46	662-668	663.25	667.75
47	668-674	669.25	673.75
48	674-680	675.25	679.75
49	680-686	681.25	685.75
50	686-692	687.25	691.75
51	692-698	693.25	697.75
52	698-704	699.25	703.75
53	704-710	705.25	709.75
54	710-716	711.25	715.75
55	716-722	717.25	721.75
56	722-728	723.25	727.75
57	728-734	729.25	733.75
58	734-740	735.25	739.75
59	740-746	741.25	745.75
60	746-752	747.25	751.75
61	752-758	753.25	757.75
62	758-764	759.25	763.75
63	764-770	765.25	769.75
64	770-776	771.25	775.75
65	776-782	777.25	781.75
66	782-788	783.25	787.75
67	788-794	789.25	793.75
68	794-800	795.25	799.75
69	800-806	801.25	805.75
70	806-812	807.25	811.75

Table 12-1. TV Channels and Frequencies (cont'd).

Channel No.	Frequency Band (MHz)	Video Carrier (MHz)	Sound Carrier (MHz)
71	812-818	813.25	817.75
72	818-824	819.25	823.75
73	824-830	825.25	829.75
74	830-836	831.25	835.75
75	836-842	837.25	841.75
76	842-848	843.25	847.75
77	848-854	849.25	853.75
78	854-860	855.25	859.75
79	860-866	861.25	865.75
80	866-872	867.25	871.75
81	872-878	873.25	877.75
82	878-884	879.25	883.75
83	884-890	885.25	889.75

the high-frequency end of VHF channel 6. Following this is the FM band from 88 to 108 MHz. As shown in Table 12-3, this is followed by the mid band, a group of frequencies from 120 MHz to 174 MHz. Then the regular VHF channels continue from 174 MHz to 216 MHz followed by the super band channels from 216 MHz to 300 MHz.

The low and high tv bands are identified by numbers and are channels 2 through 6 and 7 through 13. The mid band channels begin with the letter A and extend through the letter I. The super band channels start with the letter J and continue through the letter W.

In all instances, broadcast tv follows the requirements of the NTSC signal. Operators of cable tv, however, can make changes. For example, they might invert the sound and picture carrier frequencies as a means of encoding their programs.

STANDARD NTSC SIGNAL FOR COLOR TV TRANSMISSION

The standard NTSC (National Television Standards Committee) color television signal waveform, shown in Fig. 12-1, isn't an international standard. Other nations, as listed in Table 12-2, use systems, known as PAL and SECAM.

H, as indicated in the illustration, represents the time from the start of one horizontal line to the start of the next line. Since the horizontal sweep frequency is 15,750 Hz, the time duration of 1 line is 1/15,750 or 63.5 microseconds.

Table 12-2. Worldwide Television Standards.

Country	Lines/Fields	System	Voltage (V) (Nominal)	Frequency (Hz)
AFGHANISTAN	625/50	PAL	220	50
ALBANIA	625/50	SECAM	220	50
ALGERIA	625/50	PAL	127-220	50
ANDORRA	625/50		220	50
ANGOLA	625/50		220	50
ARGENTINA	625/50	PAL	220	50
AUSTRALIA	625/50	PAL	240	50
AUSTRIA	625/50	PAL	220	50
AZORES	525/60	PAL	220	50
BAHAMAS	525/60	NTSC	120	60
BAHRAIN	625/50	PAL	220	50
BANGLADESH	625/50	PAL		
BARBADOS	625/50	NTSC	120	50
BELGIUM	625/50	PAL	127-220	50
BERMUDA	525/60	NTSC	120	60
BOLIVIA	625/50	PAL	115-230	50
BRAZIL	525/60	PAL	220	60
BULGARIA	625/50	SECAM	220	50
BURUNDI	625/50		220	50
CAMEROON	625/50		127-220	50
CANADA	525/60	NTSC	120-240	60
CANARY IS.	625/50	PAL	127	50
CENTRAL AFRICAN REP.	625/50		220	50
CEYLON	625/50		230	50
CHAD	625/50		220	50
CHILE	525/60	NTSC	220	50
CHINA (PEOPLES REP.)	625/50	PAL	220	50
COLOMBIA	525/60	NTSC	150-240	60
CONGO (PEOPLES REP.)	625/50	SECAM	220	50
COSTA RICA	525/60	NTSC	110	60
CUBA	525/60	NTSC	120	60
CURACAO	525/60	NTSC	120	60
CYPRUS	625/50	PAL	220	50
CZECHOSLOVAKIA	625/50	SECAM	220	50
DAHOMEY	625/50		220	50
DENMARK	625/50	PAL	220	50
DOMINICAN REP.	525/60	NTSC	110	60
ECUADOR	525/60	NTSC	120	60
EGYPT	625/50	SECAM	220	50
EL SALVADOR	525/60	NTSC	110	60
ETHIOPIA	625/50		127	50
FIJI	625/50		240	50
FINLAND	625/50	PAL	220	50
FRANCE	625/50	SECAM	115-230	50
GABON	625/50	SECAM	127-220	50
GAMBIA	625/50			
GERMANY (DEM. REP.)	625/50	SECAM	220	50
GERMANY (FED. REP.)	625/50	PAL	220	50
GHANA	625/50		230	50
GIBRALTAR	625/50		230	50
GREAT BRITAIN	625/50	PAL	240	50

Table 12-2. Worldwide Television Standards (cont'd).

Country	Lines/Fields	System	Voltage (V) (Nominal)	Frequency (Hz)
GREECE	625/50		110-220	50
GREENLAND	525/60		220	50
GUAM	525/60	NTSC	110	60
GUATEMALA	525/60	NTSC	110-220	60
GUINEA	625/50		127-220	50
GUYANA	625/50	SECAM	127	50
HAITI	625/50	SECAM	115-220	50
HAWAII	525/60	NTSC	117	60
HONDURAS	525/60	NTSC	110-220	60
HONG KONG	625/50	PAL	220	50
HUNGARY	625/50	SECAM	220	50
ICELAND	625/50	PAL	220	50
INDIA	625/50		230	50
INDONESIA	625/50	PAL	110	50
IRAN	625/50	SECAM	220	50
IRAQ	625/50	SECAM	220	50
IRELAND	625/50	PAL	220	50
ISRAEL	625/50	PAL	230	50
ITALY	625/50	PAL	127-220	50
IVORY COAST	625/50	SECAM	220	50
JAMAICA	625/50	PAL	110	50,60
JAPAN	525/60	NTSC	100-200	50,60
JORDAN	625/50	PAL	220	50
KENYA	625/50	PAL	240	50
KOREA (NORTH)	625/50			
KOREA (SOUTH)	525/60	NTSC	100	60
KUWAIT	625/50	PAL	240	50
LEBANON	625/50	SECAM	110-190	50
LIBERIA	625/50	PAL	120	60
LIBYA	625/50	SECAM	120	50
LUXEMBOURG	625/50	SECAM	120-208	50
MALAGASY REP.	625/50		127-220	50
MALAWI	625/50		220	50
MALAYSI	625/50	PAL	240	50
MALI	625/50		125	50
MALTA	625/50		240	50
MARTINIQUE	625/50	SECAM	125	50
MAURETANIA	625/50		220	50
MAURITIUS	625/50	SECAM	220	50
MEXICO	525/60	NTSC	127-220	50,60
MONACO	625/50	SECAM	125	50
MONGOLIA	625/50			
MOROCCO	625/50	SECAM	115	50
MOZAMBIQUE	625/50	PAL	220	50
NETHERLANDS	625/50	PAL	220	50
NETHERLANDS ANTILLES	525/60	NTSC	120-220	50,60
NEW CALEDONIA	625/50	SECAM	220	50
NEW ZEALAND	625/50	PAL	230	50
NICARAGUA	525/60	NTSC	117	60

Table 12-2. Worldwide Television Standards (cont'd).

Country	Lines/Fields	System	Voltage (V) (Nominal)	Frequency (Hz)
NIGER (REP.)	625/50		220	50
NIGERIA	625/50	PAL	220	50
NORWAY	625/50	PAL	230	50
OMAN	625/50	PAL	220	50
PAKISTAN	625/50	PAL	220	50
PANAMA	525/60	NTSC	110	60
PARAGUAY	625/50	PAL	220	50
PERU	525/60	NTSC	220	60
PHILIPPINES	525/60	NTSC	115	60
POLAND	625/50	SECAM	220	50
PORTUGAL	625/50		110-220	50
PUERTO RICO	525/60	NTSC	120	60
RHODESIA	625/50	PAL	220	50
RUMANIA	625/50		220	50
RWANDA	625/50		220	50
SAMOA	525/60	NTSC	120	60
SAUDI ARABIA	625/50	SECAM	120-230	50,60
SENEGAL	625/50	SECAM	125	50
SIERRA LEONE	625/50	PAL	230	50
SINGAPORE	625/50	PAL	220	50
SOMALIA (REP.OF)	625/50		220	50
SOUTH AFRICA	625/50	PAL	220	50
SPAIN	625/50	PAL	127-220	50
SPANISH SAHARA	625/50			
ST. KITTS	525/60	NTSC	220	60
SUDAN	625/50	PAL	220	50
SURINAM	525/60	NTSC	115-127	50,60
SWAZILAND	625/50	PAL		
SWEDEN	625/50	PAL	220	50
SWITZERLAND	625/50	PAL	220	50
SYRIA	625/50	SECAM	115-220	50
TAHITI	625/50	SECAM		
TAIWAN	525/60	NTSC	100	60
TANZANIA	625/50	PAL	230	50
THAILAND	625/50	PAL	220	50
TOGOLESE REP.	625/50		127-220	50
TRINIDAD & TOBAGO	525/60	NTSC	117	60
TUNISIA	625/50	SECAM	117-220	50
TURKEY	625/50	PAL	110-220	50
UGANDA	625/50	PAL	220	50
UPPER VOLTA	625/50		220	50
URUGUAY	625/50	PAL	220	50
U.S.A.	525/60	NTSC	120	60
U.S.S.R.	625/50	SECAM	220	50
VENEZUELA	525/60	NTSC	110-220	60
VIETNAM	525/60		120	50
VIRGIN IS.	525/60	NTSC	115	60
YEMEN	625/50		220	50
YUGOSLAVIA	625/50	PAL	220	50
ZAIRE	625/50	SECAM		
ZAMBIA	625/50		230	50

Table 12-3. Low Band, Mid Band, High Band, and Super Band Channels.

	Channel Number	Frequency Band MHz	Picture Carrier MHz
Low Band	2	54-60	55.25
	3	60-66	61.25
	4	66-72	67.25
	5	76-82	77.25
	6	82-88	83.25
	FM	88-108	(100.00)
Mid Band	A	120-126	121.25
	B	126-132	127.25
	C	132-138	133.25
	D	138-144	139.25
	E	144-150	145.25
	F	150-156	151.25
	G	156-162	157.25
	H	162-168	163.25
	I	168-174	169.25
High Band	7	174-180	175.25
	8	180-186	181.25
	9	186-192	187.25
	10	192-198	193.25
	11	198-204	199.25
	12	204-210	205.25
	13	210-216	211.25
Super Band	J	216-222	217.25
	K	222-228	223.25
	L	228-234	229.25
	M	234-240	235.25
	N	240-246	241.25
	O	246-252	247.25
	P	252-258	253.25
	Q	258-264	259.25
	R	264-270	265.25
	S	270-276	271.25
	T	276-282	277.25
	U	282-288	283.25
	V	288-294	289.25
	W	294-300	295.25

V represents the time from the start of one field to the start of the next field. Each field (Fig. 12-2) is 262½ lines, and two fields, interlaced, form a frame, or a single complete picture. Since a field requires 60 seconds for its completion, the time duration of a single field is 1/60 or 16,667 microseconds.

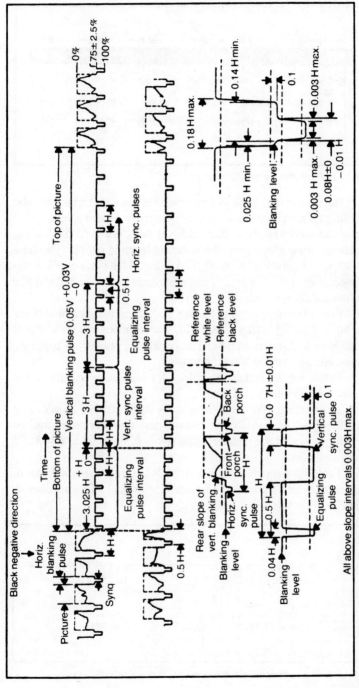

Fig. 12-1. Standard NTSC television signal.

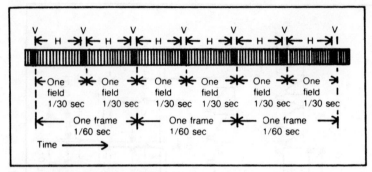

Fig. 12-2. Field and frame relationship.

During the time the electron scanning beam in the picture tube retraces its path from the bottom of the screen (end of a field) to the top of the screen (start of the next field) there is a vertical blanking interval during which time the scanning beam is cut off. The time duration is equivalent to 0.05 V or about 14H (14.5 × 63.5 microseconds) or 0.05 × 16,667 microseconds (between 820 and 920 microseconds approximately).

Following the sweep of the last horizontal line in a field there are six equalizing pulses (Fig. 12-3). These pulses last for a total of 3.025H or 192 microseconds. The equalizing pulses are superimposed on the vertical blanking pulse.

Following the equalizing pulses are vertical sync pulses, and their time is equal to 3H.

The vertical sync pulse interval is then followed by six more equalizing pulses, and like the first six, have a similar time duration. The frequency of the equalizing pulses is 0.5H or 63.5/2 = 31.75 microseconds.

Table 12-4 supplied details of the NTSC television waveform. Figure 12-5 furnishes data about the video signal following demodulation in the television receiver.

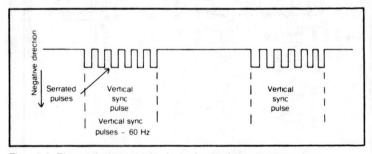

Fig. 12-3. The vertical sync pulse is made up of six serrated pulses.

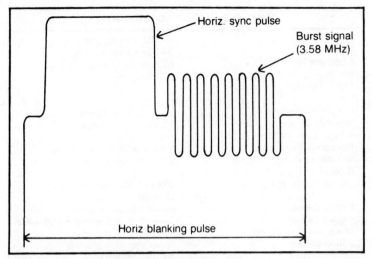

Fig. 12-4. The color burst is transmitted for a few microseconds after each horizontal sync pulse.

TELEVISION INTERFERENCE

The fundamental frequencies and the harmonics of other broadcast services can interfere with television signal reception. Table 12-5 indicates these frequencies, their harmonics, and the tv channels that are subject to interference.

Table 12-6 lists tv channels and amateur and citizens band harmonics. Figure 12-6 illustrates the relationship of amateur band harmonics to VHF tv channels. Table 12-7 lists the harmonic relationship of ham bands to UHF tv channels.

PICTURE TUBE DESIGNATIONS

Picture tube designations are alphanumerical, using nomenclature such as 3AP1. The first number indicates the screen size, measured diagonally. The letter P is the type of phosphor used and is listed in Table 12-8. A 5AP7 would be a picture tube whose screen size is 5 inches and uses a blue phosphor with a long persistence. The letter A indicates some internal tube change to distinguish it from a 5P7.

PLAYING AND RECORDING TIME OF VIDEOTAPE

The amount of recording and playback time of a videotape in minutes and hours is supplied in Table 12-9. The amount of time for VHS tapes is generally indicated directly on the package and is for

155

Table 12-4. NTSC Television Standards.

Field	16.667 microseconds
Frame	33.334 microseconds
Complete horizontal line from start of one line to beginning of next line	63.5 microseconds (1/15.750 second)
Horizontal blanking	10.16 to 11.4 microseconds
Horizontal trace without blanking time	53.34 microseconds
Horizontal sync pulse	5.08 to 5.68 microseconds
Vertical sync pulse interval (total of six)	190.5 microseconds
Vertical blanking interval	833 to 1300 microseconds per field
Picture carrier	sound carrier - 4.5 MHz
Picture carrier	low end of tv channel + 1.25 MHz
Lower video sideband	0.75 MHz
Sound carrier	upper end of tv channel minus 0.25 MHz
Sound carrier	minus picture carrier = 4.5 MHz
Channel width	6 MHz
Color burst frequency	3.579545 MHz (Fig. 12-4)
Horizontal sync pulse frequency	15.750 Hz
Vertical sync pulse frequency	60 Hz
Type of scanning	interlaced
Single frame	odd field + even field
H	time of 1 line of picture information
Picture information, horizontal blanking, and horizontal sync	93% of one field time
Vertical blanking, sync, and retrace period	7% of one field or 1270 microseconds
One field	1/30 second
One frame	1/60 second
Blanking or black reference level	75% of carrier
Equalizing pulses	0.5H or 31.75 microseconds
Front porch	1.27 microseconds
Back porch	3.8 microseconds

standard play (SP). Thus, a T-60 videocassette for VHS use can play and record for 1 hour. If the machine has an EP (extended play and an SLP (super long play) capability, that same tape can be used for 2 hours and 3 hours respectively.

Beta format videocassettes do not indicate playing time on the package but rather the length of the tape.

TYPES OF VIDEOTAPES

The most commonly used videotapes (other than industrial types) are VHS and Beta format. Both have a ½-inch width. However, the physical dimensions of the tape shell and different

156

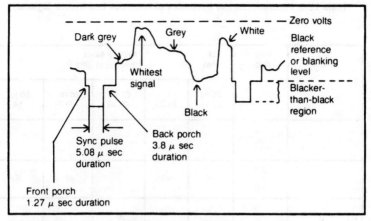

Fig. 12-5. The video signal following demodulation.

methods of loading the videotapes into a video cassette recorder make these tapes incompatible.

Table 12-10 indicates the tapes that are available, the type of scanning, and tape usage.

VHS TAPE SPEEDS

Tape speed is controlled by the tape speed selector in a VCR, assuming the component has such an adjustment. The three commonly used speeds are SP (standard play), LP (long play), and EP (extended play, also known as SLP or super long play). Table 12-11 lists the speed of VHS tapes in inches per second and millimeters per second.

TAPE CONSUMPTION

The amount of information to be played back or recorded on videotape depends on the amount of surface area presented to the head gap. This is determined by the tape width and the tape speed. The greater the width of the tape and the greater its speed, the more tape surface area that is scanned by a head.

Table 12-12 supplies data for various tapes. Some of the more recent tapes have ¼-inch width but these aren't widely used and aren't included. Tapes for consumer use are either VHS or Beta and these are both ½-inch wide. U-Matic and EIAJ #1 are used in industry while the 2-inch quad is a standard broadcast tape.

TAPE LENGTHS

The designations on the package for VHS videocassettes indi-

Table 12-5. Fundamentals and Harmonics of Television Interference Signals.

Channel	Freq. MHz	CB MHz 26.96 27.41	Amateur Bands Frequency in MHz 14 to 14.35	21 to 21.45	28 to 29.7	50 to 54
tv IF	40 to 47					
tv CH.2	54 to 60	2F	4F		2F	F
CH. 3	60 to 66			3F		
CH. 4	66 to 72					
CH. 5	76 to 82	3F				
CH. 6	82 to 88	3F		4F	3F	
FM	88–108	4F				
tv CH. 7	174 to 180					
CH. 8	180 to 186					
CH. 9	186 to 192	7F				
CH. 10	192 to 198	7F				
CH. 11	198 to 204					4F
CH. 12	204 to 210					4F
CH. 13	210 to 216					4F

	AM	FM	FM Osc.	Color Osc.
	.55–1.6 MHz	88-108 MHz	98 to 118 MHz	3.53 MHz
				20F
				21F
		F		23F
	ALL T.V. CHANNELS			
				49F
			2F 99 to 102	57F
			2F 102 to 105	58F
			2F 105 to 108	60F

Table 12-6. TV Channels and Amateur and Citizens Band Harmonics.

Channel	Freq. Range	Picture Carrier Freq.	CB	Harmonics			
				40 meters	20 meters	15 meters	10 meters
TV I-F	41-47	42	—	—	42-43	42-43	—
2	54-60	55.25	53.9-54.8 (2nd)	56-58.4 (8th)	56-57.3 (4th)	—	56-59.4 (2nd)
3	60-66	61.25	—	63-65.7 (9th)	—	63-64.35 (3rd)	—
4	66-72	67.25	—	70-73 (10th)	70-72 (5th)	—	—
5	76-82	77.25	80.9-82.2 (3rd)	—	—	—	—
6	82-88	83.25	82-82.2 (3rd)	—	84-86.4 (6th)	84-85.8 (4th)	84-89.1 (3rd)

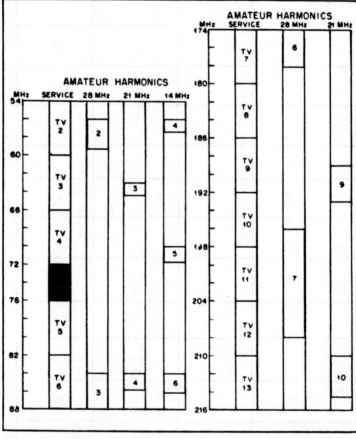

Fig. 12-6. The relationship of amateur-band harmonics to the VHF TV channels. (Courtesy of American Radio Relay League. Inc.)

Table 12-7. Harmonic Relationship—Amateur VHF Bands and UHF TV Channels.

Amateur Band	Harmonic	Fundamental Freq. Range	Channel Affected	Amateur Band	Harmonic	Fundamental Freq. Range	Channel Affected
144 MHz	4th	144.0-144.5	31	220 MHz	3rd	220.00-220.67	45
		144.5-146.0	32			220.67-222.87	46
		146.0-147.5	33			222.67-224.67	47
		147.5-148.0	34			224.67-225.00	48
	5th	144.0-144.4	55		4th	220-221	82
		144.4-145.6	56			221.0-222.5	83
		145.6-146.8	57	420 MHz	2nd	420-421	75
		146.8-148.0	58			421-424	76
	6th	144.00-144.33	79			424-427	77
		144.33-145.33	80			427-430	78
		145.33-147.33	81			430-433	79
		147.33-148.00	82			433-436	80

Table 12-8. Picture Tube Phosphor Designations.

Type	Fluorescent Color	Persistence
P1	green	medium
P2	blue-green	long
P3	yellow-green	medium
P4	white	medium
P5	blue	very short
P6	white	medium
P7	blue	long
P11	blue	very short

Table 12-9. Maximum Recording and Playback Times for VHS and Beta Format Videocassettes.

	Minutes			Hours		
VHS	SP	EP	SLP	SP	EP	SLP
T-20	20	40	60	1/3	2/3	1
T-30	30	60	90	1/2	1	1 1/2
T-40	40	80	120	2/3	1 1/3	2
T-60	60	120	180	1	2	3
T-80	80	160	240	1 1/3	2	4
T-90	90	180	270	1 1/2	3	4 1/2
T-100	100	200	300	1 2/3	3 1/3	5
T-120	120	240	360	2	4	6
T-160	160	320	480	2 2/3	5 1/3	8
T-180	180	360	540	3	6	9
BETA	**B1**	**B11**	**B111**	**B1**	**B11**	**B111**
L-125	15	30	45	1/4	1/2	3/4
L-250	30	60	90	1/2	1	1 1/2
L-370	45	90	135	3/4	1 1/2	2 1/4
L-500	60	120	180	1	2	3
L-750	90	180	270	1 1/2	3	4 1/2
L-830	100	200	300	1 2/3	3 1/3	5

Tape Format (width)	Scan Method	Mode	Where used	How used	Why used
2-inch	Quadruplex	Hi-band color Lo-band color	Broadcast tv studios—NBC CBS ABC PBS local TV stations large universities	1 Over-the-air broadcasting 2 To make multiple copies and go many generations	Excellent color quality Very high picture definition Easily edited
2-inch	Helical	Monochrome and color	Schools, univer-sities, industry	Large closed-circuit systems Master recording	Good to excellent quality picture Good copies
1-inch	Helical	Monochrome and color	Schools, industry professions, cable tv broadcast tv	For anything recordable— For copies of all tape	Moderate to excel-lent quality pic-ture good editing
¾-inch U-Matic standard	Helical	Monochrome and color	Schools, industry professions, cable tv home use	Video cassettes	Interchangeability Good quality color 2 sound tracks (stereo sound) Easy to use
½-inch EIAJ standard	Helical	Monochrome and color	Education, industry professions, cable tv home use	Video cartridge Teaching, experimen-tal closed circuit, surveillance	Low cost Interchangeability Easy maintenance Simplicity fair editing
¼-inch	Helical	Monochrome and color	Education, schools cable tv	Education, home use on-location recording, distribution limited by nonstandardization and lack of ¼-inch equipment	Low cost Small size Light weight

Table 12-10. Videotapes and Usage.

Table 12-11. VHS Tape Speeds.

Mode	ips (decimals)	ips (fractions)	mms
Standard Play (SP)	1.31	1-5/16	33.27
Long Play (LP)	0.87	87/100	22.10
Extended Play (super long play SP)	0.44	7/16	11.18

mms = millimeters per second

Table 12-12. Tape Area Used per Hour.

Format	Tape Width (inches)	Tape Speed (in./sec.)	Square Feet/Hour
VHS standard play	0.5	1.32	16.37
long play	0.5	0.66	8.19
extended play	0.5	0.45	5.59
BETA I	0.5	1.54	20.15
II	0.5	0.79	10.3
III	0.5	0.53	6.95
U-MATIC	0.75	3.75	70.03
EIAJ #1	0.5	7.5	93.8
QUAD (tv broadcast)	2.0	15.0	750.0

cates possible operating times; those on Beta show tape lengths. The chart in Table 12-13 lists the tape length in feet for both types of tape.

VCR CHARACTERISTICS

The various types of videocassettes differ not only in physical characteristics, but also in electrical characteristics as well, as shown in the listing in Table 12-14.

S/N FOR VIDEO DECKS

The signal-to-noise ratios for video decks not only depends on

Table 12-13. Lengths of VHS and Beta Tapes.

VHS	Length (feet)	Beta	Length (feet)
T-15	115	L-125	125
T-30	225	L-250	250
T-45	335	L-370	370
T-60	420	L-500	500
T-90	645	L-750	750
T-120	815	L-830	830
T-180	1260		

the tape that is used, but also on the speed at which the tape is operated. As tape speed is decreased to supply longer playing or recording time, the signal-to-noise (S/N) ratio becomes poorer, as shown in Table 12-15.

VHS CONNECTORS

Table 12-16 lists the connectors used for VHS video cassette decks.

VIDEODISC SYSTEMS

While a number of videodisc systems have been suggested, there are three currently available, as indicated in Table 12-17.

SATCOM 1 RECEPTION

Certain data must be available so as to adjust a dish for the reception of a particular satellite. This information includes the look angle, the tangent of that angle, the maximum height of any obstruction at a measured distance (in feet) from the dish, and the azimuth angle.

The look angle is important for it determines the maximum obstruction height. For Satcom 1, or for any other satellite, obtain the value of the look angle. A table of trigonometric functions will supply the value of the tangent of this angle. Move the decimal point two places to the right to obtain the maximum obstruction height at a distance of 100 feet from the dish. Table 12-18 supplies information for all the states for Satcom 1 reception.

While the chart in Table 12-18 supplies maximum obstruction height figures for a distance of 100 feet, it is possible to determine this number for any distance. Multiply the measured distance by the tangent of the look angle for the particular state.

Table 12-14. Video Tape Characteristics.

Recorder types	Video track width µm	Audio track width mm	Drum diameter mm	Drum speed rpm	Luminance frequency MHz	Chroma frequency kHz	Cassette dimensions mm	Cassette volume cm³
Consumer VCR format:								
Betamax standard-play	58.5	1.05	74.5	1800	3.5-4.8	688	156×96×25	374
Betamax long-play	29.2	1.05	74.5	1800	---	---	156×96×25	374
VHS standard-play	58	1.0	62	1800	3.4-4.4	629	188×104×25	489
VHS long-play	35	1.0	62	1800	3.4-4.4	629	188×104×25	489
VR-1000 (VX-2000)	48	0.4	48	3600	3.1-4.6	688	213×146×44	1368
Institutional & industrial:								
V-Cord II	60	1.0	81.3	---	3.1-4.3	688	156×108×25	421
V-Cord (skip-frame mode)	---	1.0	81.3	---	---	---	156×108×25	421
U-Matic	85	0.8	110	1800	3.8-5.4	688	222×140×32	995
EIAJ open reel	110	1.0	115.8	---	3.1-4.5	767	---	---

Table 12-15. Typical Values of S/N for Video Decks.

Tape Speed	S/N in dB
SP	46
LP	43
EP	40

Table 12-16. VHS Connectors.

Connection	Connector
Earphone or Microphone	3.5 mm mini jack
Power Supply Terminal	7-pin DIN
Camera Terminal	Round 10-pin

☐ **Example:**

What is the maximum permitted obstruction height in the state of Pennsylvania for objects located at a distance of 165 feet from the dish?

For Pennsylvania, the look angle in degrees is 16. The tangent of this angle is given as 0.29. $165 \times 0.29 = 47.85$ feet. The farther the object is from the dish, the greater the allowable obstruction size, in feet.

SIGNAL LOSS

The download from an antenna to the antenna input terminals of a tv receiver is either 75-ohm coaxial cable or 300-ohm transmission line. Table 12-19 shows the losses that can occur in 300-ohm line. Losses increase when the line is wet and also increase with frequency.

SIGNAL RANGE

Area designations for tv reception for VHF and UHF reception are shown in Table 12-20. These distances can be used to determine the type of tv antenna to select for satisfactory reception.

ANTENNA DIMENSIONS

Television antennas are broadly tuned resonant circuits. Table 12-21 supplies the dimensions of the dipole, the lengths, and spacing of directors and reflectors.

ROTOR CABLES

In some areas, tv antennas are operated by a rotor since tv

Table 12-17. Types of Videodisc Systems.

	VHD	VLP	CED
Playback method	Grooveless disc with stylus contact	No disc contact	Grooved disc with stylus contact
Type of pickup	Capacitance	Laser beam	Capacitance
Tracking method	Electronic	Optical	Groove
Speed of rotation	900 rpm	1800 rpm	450 rpm
Stylus	Sapphire or diamond	Laser beam	Diamond
Playing time	60 minutes each side	30 minutes each side or 60 minutes each side	60 minutes each side
Disc diameter	10"	12"	12"

VHD = Video High Density
VLP = Video Long Play
CED = Capacitive Electronic Disc

167

Table 12-18. Dish Adjustments for Satcom 1 Reception.

State	Look Angle In Degrees	Tangent Of Angle	Max. Ht. Of Obstruction at 100 ft.	Azimuth Angle In Degrees
Alabama	25	.47	47	244
Alaska	16	.29	29	162
Arizona	43	.93	93	217
Arkansas	29	.55	55	239
California	43	.93	93	203
Colorado	36	.73	73	222
Connecticut	11	.19	19	252
Delaware	14	.25	25	250
Florida	23	.42	42	251
Georgia	24	.45	45	148
Hawaii	54	.99	99	129
Idaho	33	.65	65	208
Illinois	24	.45	45	238
Indiana	22	.40	40	241
Iowa	27	.51	51	233
Kansas	31	.60	60	229
Kentucky	22	.40	40	242
Louisiana	30	.58	58	241
Maine	10	.18	18	251
Maryland	15	.27	27	250
Massachusetts	11	.19	19	252
Michigan	20	.36	36	240
Minnesota	24	.45	45	229
Mississippi	29	.55	55	241
Missouri	27	.51	51	236
Montana	31	.60	60	213
Nebraska	30	.58	58	227
Nevada	41	.87	87	208
New Hampshire	11	.19	19	253
New Jersey	14	.25	25	250
New Mexico	39	.81	81	224
North Carolina	19	.34	34	248
North Dakota	26	.49	49	222
Ohio	18	.33	33	244
Oklahoma	31	.60	60	234
Oregon	38	.78	78	201
Pennsylvania	16	.29	29	246
Rhode Island	14	.25	25	250
South Carolina	21	.38	38	248
South Dakota	28	.53	53	225
Tennessee	24	.45	45	243
Texas	38	.78	78	243
Utah	38	.78	78	214
Vermont	11	.19	19	252
Virginia	18	.33	33	247
West Virginia	18	.33	33	247
Washington	33	.65	65	200
Washington, DC	15	.27	27	250
Wisconsin	22	.40	40	235
Wyoming	34	.68	68	216

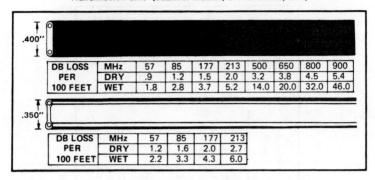

DB LOSS PER 100 FEET	MHz	57	85	177	213	500	650	800	900
	DRY	.9	1.2	1.5	2.0	3.2	3.8	4.5	5.4
	WET	1.8	2.8	3.7	5.2	14.0	20.0	32.0	46.0

DB LOSS PER 100 FEET	MHz	57	85	177	213
	DRY	1.2	1.6	2.0	2.7
	WET	2.2	3.3	4.3	6.0

Table 12-20. Area Designations for TV Reception.

Area Designation	For VHF	For UHF
Deepest Fringe	100+ Miles	60+ Miles
Deep Fringe	100 Miles	55 Miles
Fringe	80 Miles	45 Miles
Near Fringe	60 Miles	35 Miles
Far Suburban	45 Miles	25 Miles
Suburban	30 Miles	15 Miles
Urban	20 Miles	10 Miles

stations can be in different directions. Rotor cables are either three
or four conductor types of 20 gauge. 7 strand wire covered with a
black vinyl jacket. Representative rotor wires and cable dimensions
are shown in Fig. 12-7.

MICROWAVE BANDS

The microwave bands extend from 1 to 300 GHz and are
identified by letters except the band from 40 to 300 GHz. This is
known as the millimeter band, a reference to the wavelengths.
Table 12-22 is a listing of the microwave bands.

Table 12-21. Dimensions of Dipole Elements for VHF.

Channel Number	Channel Freq. MHz	Dipole Length	Reflector Length	Spacing of Reflector	Director Length	Spacing of Director
2	54-60	8 ft. 5⅝ in.	8 ft. 10⅞ in.	2 ft. 8⅛ in.	8 ft. 1¾ in.	1 ft. 9⅜ in.
3	60-66	7 ft. 7¾ in.	8 Ft. ⅜ in.	2 ft. 3¾ in.	7 ft. 4¼ in.	1 ft. 7¼ in.
4	66-72	6 ft. 11½ in.	7 ft. 3⅞ in.	2 ft. 2⅝ in.	6 ft. 8⅜ in.	1 ft. 5½ in.
5	76-82	6 ft. ¾ in.	6 ft. 4½ in.	1 ft. 11 in.	5 ft. 10 in.	1 ft. 1½ in.
6	82-88	5 ft. 7½ in.	5 ft. 10⅝ in.	1 ft. 9¼ in.	5 ft. 4⅞ in.	1 ft. 2⅛ in.
7	174-180	2 ft. 8 in.	2 ft. 9⅝ in.	10⅛ in.	4 ft. 6¾ in.	6¾ in.
8	180-186	2 ft. 6 in.	2 ft. 8½ in.	9¾ in.	2 ft. 5¾ in.	6½ in.
9	186-192	2 ft. 9¾ in.	2 ft. 7½ in.	9½ in.	2 ft. 4¾ in.	6¼ in.
10	192-198	2 ft. 5 in.	2 ft. 6½ in.	9¼ in.	2 ft. 4 in.	6⅛ in.
11	198-204	2 ft. 4⅛ in.	2 ft. 5⅝ in.	8⅞ in.	2 ft. 3⅛ in.	5⅞ in.
12	204-210	2 ft. 2¾ in.	2 ft. 4¾ in.	8⅝ in.	2 ft. 2¼ in.	5¾ in.
13	210-216	2 ft. 2⅝ in.	2 ft. 4 in.	8⅜ in.	2 ft. 1½ in.	5⅝ in.

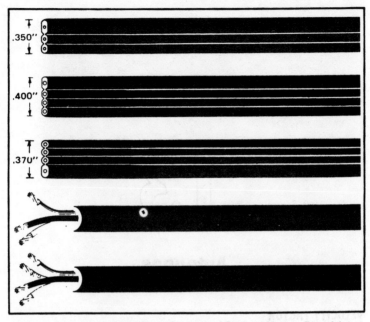

Fig. 12-7. Rotor wire dimensions. (Courtesy of Channel Master, Div. of Avnet, Inc.)

Table 12-22. Microwave Bands.

Designation	Frequency (GHz)
L-band	1.0- 2.0
S-band	2.0- 4.0
C-band	4.0- 8.0
X-band	8.0- 12.0
Ku-band	12.0- 18.0
K-band	18.0- 27.0
Ka-band	27.0- 40.0
Millimeter	40.0-300.0

171

13

Antennas

VELOCITY FACTOR

The velocity of a wave along a conductor, such as a transmission line, is not the same as the velocity of that wave in free space. The ratio of the two (actual velocity vs. velocity in space) is known as the velocity factor. Obviously, velocity factor must always be less than 1, and, in typical lines varies from 0.6 to 0.97. See Table 13-1.

Table 13-1. Velocity Factors of Transmission Lines.

Type of Line	Velocity factor (V)
Two-wire open line (wire with air dielectric)	0.975
Parallel tubing (air dielectric)	0.95
Coaxial line (air dielectric)	0.85
Coaxial line (solid plastic dielectric)	0.66
Two-wire line (wire with plastic dielectric)	0.68-0.82
Twisted-pair line (rubber dielectric)	0.56-0.65

Table 13-2 lists the various types of antennas, supplying a description and applications.

COAXIAL CABLES

Coaxial cable is a type of transmission line used to connect the CB transceiver to its antenna. The inner conductor of the cables

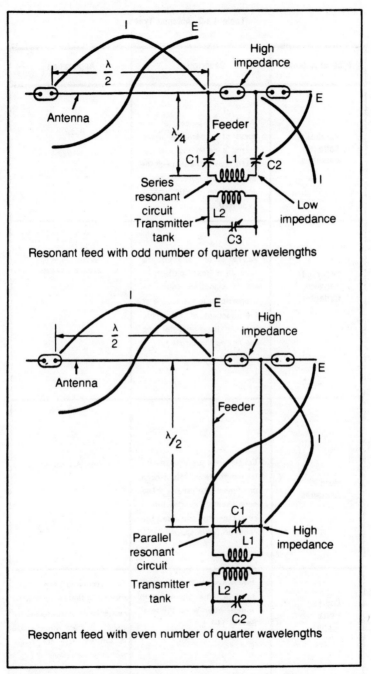

Fig. 13-1. Methods of feeding an end-fed half-wave Hertz antenna.

Table 13-2. Antenna Types.

Type of Antenna	Description	Application
Parabolic Reflector Antennas	A radiator placed at the focus of a parabola which forms a reflecting surface. Variations in the shape of the parabola provide changes in the shape of the beam produced.	Radar
Cosecant-Squared Reflector	A reflector shaped to produce a beam pattern in which signal strength is proportional to the square of the cosecant of the angle between the horizontal and the line to the target.	Surface search by airborne radar sets
Horn Antennas	Consists of a waveguide with its mouth flared into a horn or funnel-like shape. The horn usually radiates into a reflector to provide the required beam shape.	Radar Applications
End-Fed Hertz (Zepp)	Half wavelength voltage fed radiator fed at one end with tuned, open-wire feeders See Fig. 13-1.	For receiving and transmitting in the 1.6 to 30 MHz range. Most useful for multi-band operation where space is limited. Use for fixed station installations.

Table 13-2. Antenna Types (cont'd).

Type of Antenna	Description	Application
Center-Fed Hertz (tuned doublet or center-fed Zepp)	A center-fed, half-wave doublet usually employing spaced feeders. Current fed on fundamental and voltage fed on all even harmonics. See Fig. 13-2.	For receiving and transmitting in the 1.6- to 30-MHz range. Can be used on any frequency if the system as a whole can be tuned to that frequency.
Fuchs Antenna	Long-wire, voltage-fed radiator an even number of quarter waves long. One end of radiator brought directly to the transmitter or tuning unit without using a transmission line.	For transmitting and receiving on any frequency where simplicity and convenience are desired.
Corner Reflector	A half-wave radiator with two large metal sheets arranged so their surfaces meet at an angle whose apex lies behind the radiator	Used in the VHF and UHF ranges to provide directivity in the plane which bisects the angle formed by the reflector.
Marconi	A vertical radiator approximately one-quarter wavelength long at operating frequency. One end is grounded or worked against ground. May be fed at or near base with low-impedance line. Electrical length may be increased by using loading coil in series with base or near center of radiator or by using capacitive loading at the top. The length, L, in feet can be computed by: $L = \frac{234}{f}$	Widely used for medium- and low-frequency receiving and transmitting where vertical polarization is desirable.

175

Table 13-2. Antenna Types (cont'd).

Type of Antenna	Description	Application
	f is in megaHertz. L is the overall length. in feet. from the top of the antenna to the point where it connects to ground or counterpoise The total power dissipated in and radiated from a Marconi antenna can be calculated by $I^2 (R_r + R_g)$ I is the antenna current measured at the antenna base. R_r is the radiation resistance and R_g is the ground resistance The useful radiated power is the difference between the total power consumed and the power lost in the ground resistance. See Fig. 13-3.	
Parasitic Array	Consists of a radiator with a reflector behind and/or one or more directors in front Produces a unidirectional radiation pattern. May be either vertically or horizontally polarized.	Used to develop high gain in one direction with little or no radiation or pickup in other directions Used on all frequencies where these characteristics are desired and space is available
Rhombic Antenna	A system consisting of four long wire radiators arranged in the form of a diamond and fed at one end. If the corner opposite the feed point is open. response is bidirectional in a line running through these two corners. If the open end is terminated	Widely used where high gain and directivity are required Can be used over a wide range of frequencies and is particularly useful when each leg is two or more wavelengths long on lowest frequency. Angle of radiation is lowered and vertical

Table 13-2. Antenna Types (cont'd).

Type of Antenna	Description	Application
	with the proper resistance, response is unidirectional in the direction of the terminated end. Gain may vary from 20 to 40 times that of a dipole, depending on the number of wavelengths in each leg.	directivity narrowed by increasing length of legs and/or increasing operating frequency
Vertical J	A one-half wavelength vertical radiator fed at the bottom through a quarter-wave matching stub. It is omnidirectional, produces vertical polarization, and can be fed conveniently from a wide range of feed-line impedances.	Practical for use at frequencies above about 7 MHz. Normally used for fixed frequency applications because of its extreme sensitivity to frequency changes. Efficiency falls off as frequency is raised
Coaxial Antenna (sleeve antenna)	Vertical radiator one-half wavelength long. Upper half consists of a relatively thin radiator and the bottom half a large diameter cylinder. Fed at the center from coaxial cable of 70 to 120 ohms.	Practical for frequencies above about 7 MHz. Normally used for fixed frequency applications. Changes in frequency require that the antenna be retuned by varying length of the two halves of the radiator. Practical for operation up to about 100 MHz.
Ground-Plane Antenna	Omnidirectional quarter-wave vertical radiator mounted above a horizontal reflecting surface. Its impedance is approximately 36 ohms or less.	Practical for producing vertically polarized waves at frequencies above about 7 MHz and frequently used at frequencies as high as 300 MHz.

Table 13-2. Antenna Types (cont'd).

Type of Antenna	Description	Application
Half Rhombic (inverted V or tilted wire)	A two-wire antenna with the legs in a vertical plane and in the shape of an inverted V. Directivity is in the plane of the legs. Feeding one end and leaving the other open results in bidirectivity. Terminating the free end with a suitable resistor produces unidirectional radiation in the direction of the termination. Gain and angle between legs depend on frequency and the number of wavelengths in each leg.	Used to provide high gain. Used where low angle of radiation is desirable. Usable over a wide frequency range. Bandwidth is greatest for terminated type. Angle of radiation is lowered as leg length and/or operating frequency is increased.
Beverage Antenna	A directional long-wire horizontal antenna, two or more wavelengths long. The end nearest the distant receiving station is terminated with a 500-ohm resistor connected to a good counterpoise. The antenna, generally suspended 10 to 20 feet above ground, is non-resonant.	Used for transmitting and receiving vertically polarized waves. Often used for long-wave transoceanic broadcasts. Its input impedance is fairly constant so it can be used over a wide frequency range. Useful for frequencies between 300 kHz and 3 MHz. Highly suitable for use over dry, rocky soil. Never use over salt marshes or water.
Folded Dipole	A simple center-fed dipole with a second half-wave conductor connected across its ends. Spacing between the conductors is a very small fraction of a wavelength.	Its impedance is higher than that of a simple dipole. Applications same as simple dipole. Often used in parasitic arrays to raise the feedpoint impedance to a value which can be conveniently matched to transmission line.

Table 13-2. Antenna Types (cont'd).

Type of Antenna	Description	Application
	The upper antenna is a V-shaped dipole radiator with a parasitic element. Marker beacon antennas may consist of colinear dipoles or arrays.	
Omni-Range (VOR)	Consists of two pairs of square-loop radiators surrounding a single square-loop radiator.	Use to provide navigation signals for aircraft in all directions from the range station.
Adcock Antenna	Consists of vertical radiators which produce bidirectional vertically polarized radiation.	Used in low-frequency radio ranges and for direction finding.
Loop Antennas	A loop of wire consisting of one or more turns arranged in the shape of a square, circle, or other convenient form. It produces a bidirectional pattern along the plane of the loop.	Normally used for direction-finding applications, particularly in ships and air craft.
Stub Mast	A quarter wave vertical radiator consisting of a metal sheath over a hard-wood supporting mast. Fed with 50-ohm line with the outer conductor connected to a large metal ground sur face.	Used for wide band reception and transmission of frequencies above 100 MHz Normally used in aircraft installations.

Table 13-2. Antenna Types (cont'd).

Type of Antenna	Description	Application
Crow-Foot Antenna	A low-frequency antenna consisting of comparatively short vertical radiator with a 3-wire V-shaped flat top and a counterpoise having the same shape and size as the flat top See Fig. 13-4.	Normally used where it is impractical to erect a quarter-wave vertical radiator. Used most frequently for reception and transmission in the 200 to 500 kHz range.
Turnstile Antenna	An omni-directional, horizontally polarized antenna consisting of two half-wave radiators mounted at right angles to each other in the same horizontal plane. They are fed with equal currents 90 degrees out of phase. Gain is increased by stacking. Dipoles may be single, folded, or special broadband types.	Normally used for transmission and reception of FM and television broadcast signals.
Skin Antennas	Usually consist of an insulated section of the skin of an aircraft. Its radiation pattern varies with frequency, size of the radiating section, and position of the radiator on the aircraft.	Used for VHF and UHF reception and transmission in high-speed aircraft. Often used to replace fixed-wire antennas used in the 2 to 2.5 MHz range.
ILAS Antennas	Localizer antennas are of several different types. One type consists of two or more square loops. Glide path is usually produced by two stacked antennas. The lower antenna is usually a horizontal loop bisected by a metal screen and supported about 6 feet off the ground.	Used to enable pilots to locate the airport and to land the plane on the desired runway when weather conditions would prohibit a landing under visual flight reference.

Table 13-2. Antenna Types (cont'd).

Type of Antenna	Description	Application
V antenna	Bidirectional antenna made of two long-wire antennas in the form of a V and fed 180 degrees out of phase at its apex. The V antenna is a combination of two long-wire antennas. As the length is increased, more power is concentrated near the axis of the wire. The length of each leg of a V antenna can be found by: $$L = \frac{492 \ (N - 0.05)}{f}$$ N is the number of half wavelengths in each leg, and f is the frequency in megaHertz More gain can be obtained by stacking a second V, one-half wavelength above the first. See Fig. 13-5.	Military and commercial applications.

listed in Table 13-3 is copper and so is the outside shield braid. The characteristic impedance is 50 ohms. The cables in this table use a polyethylene dielectric.

CB ANTENNAS

Table 13-4 lists the gain, VSWR, length, and typical weight of base station antennas.

CB MOBILE ANTENNAS

Antennas for mobile CB transceivers can be trunk mounted, bumper mounted, put on a fender, on the car or truck roof. Table 13-5 supplies comparative data on a number of antennas.

MARINE ANTENNAS

Marine antennas are described in Table 13-6.

COAXIAL CABLES FOR VIDEO

Various types of coaxial cable are used in video for connecting an antenna to the antenna terminals of a television receiver or for interconnecting video components. Coaxial cable is also used for

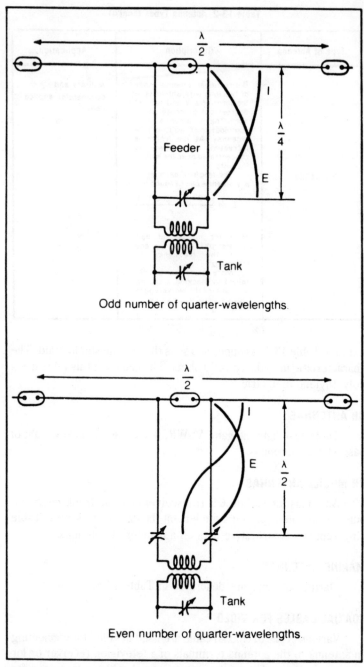

Fig. 13-2. Methods of feeding a center-fed half-wave Hertz antenna.

Table 13-3. Coaxial Cables for CB.

RG/U CABLE TYPE	NOMINAL ATTENUATION dB / 100 FT, MHz			
	100	1000	5000	10,000
8/U	1.9	8.0	27.0	100.0
8A/U	1.9	8.0	27.0	100.0
58/U	4.6	17.5	60.0	100.0
58A/U	4.9	24.0	83.0	100.0
59/U	3.4	12.0	42.0	100.0
59A/U	3.4	12.0	42.0	100.0

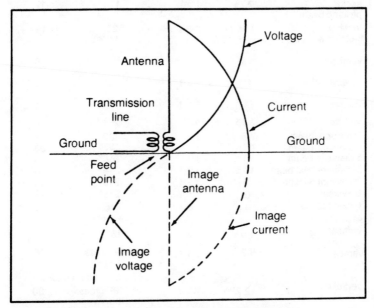

Fig. 13-3. Current and voltage distribution of a Marconi antenna.

connecting a dish antenna to a satellite receiver. The cable impedance is 75 ohms.

Table 13-7 and Table 13-8 indicate the nominal attenuation in dB per hundred feet of representative types of cables. Note that the attenuation, actually loss of signal along the line, increases with frequency (that is, in going from channel 2 to channel 83).

STANDING WAVE RATIO

The maximum current in a transmission line, compared to the minimum current, or the maximum voltage compared to the minimum voltage, is known as the standing-wave ratio (SWR). For voltage this is sometimes shown as VSWR. SWR is an indication of

Table 13-4. Base Station Antennas for CB.

Type	Gain (dB)	VSWR	Length	Average Wgt. (lbs.)
hor/vert 3-element	hor: 8 vert: 9.75	1.5:1	108" boom	26
half-wave shunt-fed	3.75	1.5:1	216"	10
sector phased beam	7.75	1.5:1	17.5'	17
vert/hor 5-element	hor:11 vert: 12.5	1.5:1	22' boom	19
vertical	-	-	234"	9
vertical	4	1.4:1	12'	3.5
vertical	11	1.2:1	11.9'	13.5
vertical	3.75	1:1	214"	5
3-element beam	7.5	1:1	120"	15
4-element beam	9	1:1	192"	20
5-element beam	10	1:1	288"	25
6-element dual beam	10.5	1.2:1	216"	30
8-element dual beam	12	1.2:1	216"	45
Universal	0	1.5:1	216"	2
Ground plane	0	1.2:1	216"	5
vertical	-	-	17'3"	-
vertical	1.5	1.5:1	18'	-
vertical	9.2	1.1:1	9' cross boom 3'1" beam boom	16
vertical	12.7	1.1:1	14' cross boom 12'2" beam boom	30
vertical	13.9	1.1:1	cross boom 20' beam boom	55
vertical	8	1.4:1	12' boom	7
vertical	9.5	1.1:1	18' boom	18
vertical	unity	1.1:1	9' ht	3
vertical	3	1.1:1	17' ht	6
vertical	3.4	1.1:1	19' 10" ht	8
vertical	3.4	1.1:1	19'10" ht	14
attic	-	2.1:0	18"	3
dipole	-	2.1:0	36"	6
vertical	-	2.1:0	18"	3
ground plane	-	2.1:0	18"	3
vertical	-	2.1:0	16"	10
beam	12.3	1.5:1	18'8-¾" each element	15
				4

Table 13-4. Base Station Antennas for CB (cont'd).

Type	Gain (dB)	VSWR	Length	Average Wgt. (lbs.)
vertical	unity	1.5:1	9'	4
vertical	5	1.5:1	9'	7
vertical	6	1.2:1	19'8"	
vertical	3.75	1.17:1	210"	10
vertical	10	1.5 1	204" boom	15
vertical	8	1.6:1	198" elements	11
vertical	-	1.8:1	108" element 108" radials	4
vertical	-	-	19'	12
vertical	-	-	19'	11
beam	-	-	19'	23
vertical	-	-	17'	8
vertical	-	-	20'	23
vertical	3.7	1.5:1	17'3"	-
beam	8	1.5:1	224-¾"	12.5
beam	8.7	1.5:1	224-¾"	15
beam	9.5	1.5:1	224-¾'	16.5
beam	8	1.5:1	216-½"	14
beam	9.5	1.5:1	224-¾"	20.5
beam and stacking kit	11	1.5:1	224-¾"	29
beam and stacking kit	12	1.5:1	224-¾"	38
stacking kit	13	1.5:1	224-¾"	47
conversion kit	-	-	-	10
stacking kit	-	-	-	18
stacking kit	-	-	-	8
vertical	-	1.5:1	235.5"	7.5
vertical	-	1.5:1	245"	7.6
vertical	-	-	19'8"	16
vertical	-	-	19'8"	15
vertical	-	-	19'10-¾"	8.75
beam	-	-	18'10" boom	8.5
beam	-	-	14'1" boom	10.75
gutter lamp	-	-	25"	-
vertical	7.5	1.1:1	-	11
vertical	9	1.1:1	-	16
vertical	10.4	1.1:1	-	22
vertical	-	1.5:1	18'6"	7.5
vertical	3.4	1.1:1	234"	9
vertical	unity	1.4:1	216"	7

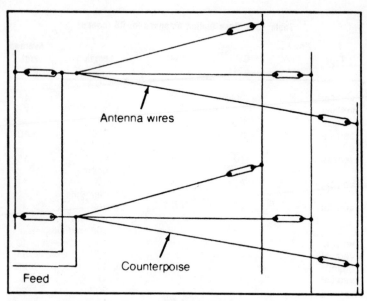

Fig. 13-4. Arrangement of a crow-foot antenna.

Table 13-5. CB Mobile Antennas.

Type	Gain (dB)	VSWR	Length	Average Wgt. (lbs.)
whip. base spring	-	1.5:1	108"	6
trunk mount	-	1.5:1	48"	1.5
roof mount	-	1.5:1	44"	1.5
trunk mount	-	1.5:1	48"	2
fender mount	-	1.5:1	44"	1.5
combination CB/AM	-	1.5:1	46"	1.5
roof mount	-	1.5:1	44"	1.5
roof mount	-	1.5:1	46"	1.5
roof mount	-	1.5:1	18"	1.25
gutter mount	-	1.5:1	18"	1.25
magnetic mount	-	1.5:1	-	1.25
trunk mount	-	1.5:1	46"	1.5
gutter mount	-	1.5:1	48"	1.5
vertical	-	-	96'	1.5
vertical	-	-	49"	1.5
vertical	-	-	103"	1.5
CB/AM	-	-	48"	2
vertical	-	-	96"	-
vertical	-	-	30"	-
vertical	-	-	48"	-
vertical	1.5	1.5:1	86"	-
vertical	-	1.2:1	-	-
vertical	-	1.1:1	26"	1
vertical	unity	1.1:1	46"	2

Table 13-5. CB Mobile Antennas (cont'd).

Type	Gain (dB)	VSWR	Length	Average Wgt. (lbs.)
vertical	unity	1.1:1	50'	2
vertical	unity	1.1:1	46"	2
vertical	unity	1.1:1	32"	1
vertical	unity	1.1:1	59'	3
vertical	unity	1.1:1	50"	2
vertical	unity	1.1:1	28"	2
vertical	unity	1.1:1	46"	1
vertical	unity	1.1:1	19"	2
vertical	unity	1.1:1	46"	3
vertical	unity	1.1:1	102"	3
vertical	unity	1.1:1	47"	3
vertical	unity	1.1:1	23 7/8"	1.7
vertical	unity	1.1:1	34 5/8"	3.2
vertical	unity	1.1:1	47"	3.1
vertical	-	1.1:1	47"	3.1
vertical	-	2.1:0	18"	3
vertical	-	2.1:0	30"	3
vertical	-	-	30"	1
AM-FM/CB	-	-	50"	1
vertical	-	-	102"	7
vertical	-	1.7:1	39"	2
vertical	-	1.7:1	39"	5
vertical	-	1.8:1	46"	4
vertical	-	1.6:1	102"	3
vertical	-	1:1	20"	3
Gutter vertical	-	-	6'	1.5
vertical		-	4'	.5
vertical		-	5'	1
vertical		-	4'	1
vertical	-	-	6'	1.25
vertical	unity	1.5:1	63"	-
vertical	unity	1.5:1	42"	-
vertical	unity	1.5:1	44 1/2"	-
vertical	unity	1.5:1	45"	-
vertical	-	1.5:1	8'7"	1.75
vertical	-	1.5:1	45"	2
vertical	-	1.5:1	43 3/4"	1
vertical	-	1.5:1	17"	1
vertical	-	1.5:1	36"	1
vertical	-	1.5:1	100"	3
vertical	-	-	83'	2.75
mobile res.			29"	.5
mast	•	-	54"	2.25
mast	•	-	54"	2.5
vertical	•	-	-	1
vertical	•	-	-	.33
vertical	•	-	26 1/2"	1.5
			60 1/2"	

Table 13-5. CB Mobile Antennas (cont'd).

Type	Gain (dB)	VSWR	Length	Average Wgt. (lbs.)
vertical	•	•	26 ½'	.75
vertical	•	•	60 ½"	
vertica	•	•	-	.75
vertical	•	•	-	.75
vertical	•	•	-	1
AM/FM/CB		•	-	1.5
vertical	•	•	-	
roof mount	•	•	30"	.5
gutter clamp	•	•	25"	1
trunk groove	•	•	-	1.25
base loaded	•	•	45 ¾"	3
trans. ant.	•	•	-	-
vertical	•	•	108"	4.5
bumper mount	•	•	108"	5
vertical	•	•	102"	1.5
vertical	•	•	108"	1.5
vertical	•	1.5:1	8'	1
vertical	•	1.5:1	4'	.5
vertical	•	1.5:1	9'	1
vertical	•	1.5:1	36"	1
magnetic mount	4.1	1.1:25	32"	-
vertical	5.2	1.1:25	32"	-
mtch. network	6.4	1.1:0	10"	-
vertical	•	1.1:1	35"	.75
vertical	•	1.1:1	36"	1
vertical				
vertical	•	•	48"	-
vertical	•	•	96"	-
vertical	•		-	-
vertical	•	•	38"	
vertical	•	•	20"	.
vertical	•	•	37"	
vertical	•	•	19"	-
vertical	•	•	38"	-
vertical	•	•	20"	-
vertical	•	•	3"	4 oz.
vertical	•	•	18"	6 oz.
vertical	4	1.1:1	108"	5
vertical	unity	1.1:1	46"	1.5
vertical	unity	1.1:1	45"	1.75
vertical	unity	1.1:1	18'	1.5
vertical	unity	1.1:1	45"	1.5
vertical	unity	1.1:1	46"	1.5
vertical	unity	1.1:1	48"	2
vertical	unity	1.1:1	50"	2
vertical	unity	1.2:1	93"	4
vertical	unity	1.2:1	77"	4

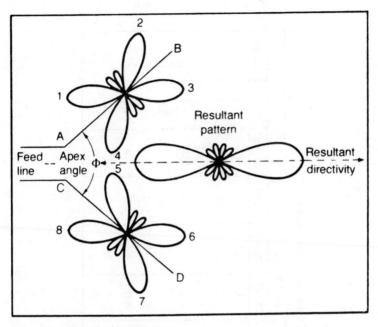

Fig. 13-5. Lobes of a V antenna.

the impedance mismatch between a transmission line and its load. Maximum delivery of energy takes place when the load impedance matches the impedance of the transmission line, and under these circumstances, SWR has a value of 1.

SWR is expressed by the following formula:

$$SWR = \frac{Z1}{Z2} \quad or \quad SWR = \frac{Z2}{Z1}$$

Table 13-6. Marine Antennas.

Type	Gain (dB)	VSWR	Length	Average Wgt. (lbs.)
vertical	-	1.5:1	97"	3
vertical	-	-	54"	2
vertical	1.5	1.1:1	19'11"	-
vertical	-	-	6'	4
vertical	1.5	1.5:1	79"	-
vertical	-	1.5:1	96 ⅜"	2
vertical	-	1.5:1	12'	7.5
vertical	-	1.5:1	18'6"	9
vertical	4.1	1.1.25	30"	-
vertical	1.4	1.51	60"	2

Table 13-7. Nominal Attenuation of Coaxial Cables at Selected TV Channels.

Cable	Nominal Attenuation, dB per 100 Feet											
	Ch. 2	Ch. 6	Ch. 7	Ch. 13	Ch. 20	Ch. 30	Ch. 40	Ch. 50	Ch. 60	Ch. 70	Ch. 83
Color Duct	2.3	2.7	3.8	4.2	6.5	7.0	7.5	7.8	8.0	8.4	9.0
Foam Color Duct	2.1	2.5	3.3	3.8	5.9	6.3	6.7	7.0	7.3	7.7	8.0
RG 59/U	2.6	3.5	4.9	5.4	8.3	8.8	9.2	9.7	10.3	11.0	11.9
RG 59/U Foam	2.3	2.7	3.8	4.2	6.2	6.6	6.8	7.1	7.3	7.7	8.0
RG-6 Foam	1.7	1.9	2.8	3.0	4.8	5.2	5.6	5.9	6.2	6.5	6.8
RG 11/U	1.4	1.7	2.2	3.2	5.1	5.3	5.5	5.7	6.1	6.2	6.8
RG 11/U Foam	1.1	1.4	1.6	2.3	3.9	4.0	4.1	4.2	4.4	4.6	4.9
.412 Cable	.74	1.0	1.4	1.5	2.5	2.6	2.7	2.9	3.1	3.3	3.5
.500 Cable	.52	.67	.72	1.1	1.5	1.8	2.1	2.4	2.7	3.0	3.1

Table 13-8. Nominal Attenuation of Coaxial Cable at Selected Frequencies.

Cable	Impedance	V_p	Attenuation in dB/100 ft.			
			50 MHz	144 MHz	220 MHz	432 MHz
RG-58C/U	52.5	.66	3.0	6.0	8.0	15.0
RG-58(F)	50	.79	2.2	4.1	5.0	7.1
RG-59B/U	73	.66	2.3	4.2	5.5	8.0
RG-59(F)	75	.79	2.0	3.4	4.6	6.1
RG-8A/U RG-213/U	52	.66	1.5	2.5	3.5	5.0
RG-8(F)	50	.80	1.2	2.2	2.7	3.9
RG-11A/U	75	.66	1.55	2.8	3.7	5.5
RG-17A/U RG-218/U	52	.66	0.5	1.0	1.3	2.3

Table 13-9. Return Loss/VSWR.

Return Loss	VSWR	Match Ratio	Reflection Coefficient	Percentage Reflection
2 dB	8.71	1.26 1	.79	79%
4 dB	4.42	1.59 1	.63	63%
6 dB	3.01	1.99 1	.50	50%
8 dB	2.32	2.52 1	.40	40%
10 dB	1.92	3.16 1	.32	32%
12 dB	1.67	3.98 1	.25	25%
14 dB	1.50	5.01 1	.20	20%
16 dB	1.37	6.31 1	.16	16%
18 dB	1.28	7.94 1	.13	13%
20 dB	1.22	10.0 1	.10	10%
22 dB	1.17	12.6 1	.079	7.9%
24 dB	1.13	15.9 1	.063	6.3%
26 dB	1.11	19.9 1	.050	5.0%
28 dB	1.08	25.1 1	.040	4.0%
30 dB	1.07	31.6 1	.032	3.2%
32 dB	1.05	39.8 1	.025	2.5%
34 dB	1.04	50.1 1	.020	2.0%
36 dB	1.032	63.1 1	.016	1.6%
38 dB	1.026	79.4 1	.013	1.3%
40 dB	1.020	100 1	.010	1.0%
46 dB	1.010	199 1	.005	0.5%
50 dB	1.006	316 1	.003	0.3%
54 dB	1.004	501 1	.002	0.2%
60 dB	1.002	1000 1	.001	0.1%

(courtesy: Channel Master, Div. of Avnet, Inc.)

Table 13-10. Master Antenna Television Distribution Symbols.

HEADEND EQUIPMENT		DISTRIBUTION SYSTEM EQUIPMENT	
Symbol	Description	Symbol	Description
	Antenna		2 Way Splitter
	Balun		3 Way Splitter
	Antenna Joiner		4 Way Splitter
	Preamplifier		Variable Isolation Wall Tap
	Power Supply		One Way Line Tap
	Hi-Lo Joiner		Two Way Line Tap
	U/V Joiner		4 Way Line Tap
	Variable Attenuator		"0" dB Wall Outlet
	Fixed Attenuator		Terminator
	Lo Band Sound Carrier Reducer		UHF/VHF Band Separator
	Hi Band Sound Carrier Reducer		Matching Transformer
	Lo Band Hi "Q" Trap		Cable Adapter RG-6
	Hi Band Hi "Q" Trap		Cable Adapter RG-11, .412, .500 to "F"
	Lo Band Mixing Unit		Cable Adapter .412, .500 Entry Mount with Pin
	Hi Band Mixing Unit		Cable Adapter 5/8" Entry to Female "F"
	Converter		Auxiliary Power Supply
	Broadband Distribution Amplifier		Line Amplifier
	Single Channel Bandpass Filter		Power Adder
	Single Channel Amplifier		Voltage Block

Z1 is the impedance of the transmission line; Z2 is the terminating impedance. The larger number is always put in the numerator so as to produce a whole number.

As the SWR approaches 1, the reflection coefficient decreases, as shown in Table 13-9. The reflection coefficient, a decimal, is multiplied by 100 to produce reflection as a percentage, as indicated in columns 4 and 5 of Table 13-9.

MATV DISTRIBUTION SYMBOLS

Special symbols are used in master antenna television distribution (MATV), and these are illustrated in Table 13-10.

Electronics
Abbreviations and Symbols

Table 14-1. Electronic Units, Abbreviations and Symbols.

Capacitance	F= farad	**Symbol C**
	μF = microfarad	
	pF = picofarad	
Frequency		**Symbol f**
	Hz = cycles per second	
	kHz = kiloHertz	
	MHz = megaHertz	
Inductance		**Symbol L**
	h = henry	
	mh = millihenry	
	μh = microhenry	
Resistance		**Symbol R**
	Ω or ω = ohm	
	Kω = kilohms	
	Mω = megohms	
Time		**Symbol t**
	sec = second	
	msec= millisecond	
	μsec = microsecond	
	nsec = nanosecond	
Current		**Symbol I or i**
	a = ampere	
	ma = milliampere	
	μ a = microampere	
Voltage		**Symbol E or v**
	v = = volt	
	mv = millivolt	
	μv = microvolt	
	kv = kilovolt	
Power		**Symbol W**
	W = watt	
	mW = milliwatt	
	μw = microwatt	
	kW = kilowatt	
	MW = megawatt	

SYMBOLS

The symbols shown in Table 14-1 are those that are commonly used, although there are some variations.

ABBREVIATIONS

Technical abbreviations Table 14-2 appear in electronics text

Table 14-2. Electronic Abbreviations.

Abbreviation	Term
A	
a	alpha, ammeter, ampere, area
A	ammeter, ampere (s)
Å	Angstrom unit
abc	automatic base compensation
ac	alternating current or voltage
acc	automatic chroma control
adf	automatic direction finder
adj	adjacent; adjustment
Aesop-1	Addressable Extension System with over-night program
AF	audio frequencies
afc	automatic frequency control
aft	audio-frequency transformer; automatic fine tuning
agc	automatic gain control
AH	ampere-hour
alc	automatic level control
AM	amplitude modulation
amp	ampere (s); amplifier
amp-hr	ampere hour
AND	gate
anl	automatic noise limiter
anr	automatic noise reduction
ant	antenna
antilog	antilogarithm
apc	automatic phase control
apld	Automatic program locating device
arm	armature
ASA	American Standards Association
att	attenuator; attenuation
atto	10^{-18}
autotrans	autotransformer
aux	auxiliary
av	average
avc	automatic volume control
ave	automatic volume expansion
AWG	American wire gauge
B	
b	bel; beta, magnetic flux density
b or base	base (transistor)
B	magnetic flux density; susceptance
bal	balance

195

Table 14-2. Electronic Abbreviations (cont'd).

Abbreviation	Term
bal mod	balanced modulator
balun	balanced-to-unbalanced transformer
batt	battery
bcd	binary coded decimal
bci	broadcast interference
bcl	broadcast listener
bfo	beat-frequency oscillator
bin	binary
bit (s)	binary digit (s)
bo	Barkhausen oscillation
bp	bandpass
bto	blocking-tube oscillator
Btu	British thermal unit

C

Abbreviation	Term
c	centi—(one-hundredth; 10^{-2}); collector (transistor); Celsius (degrees)
c, cap	capacitor (or capacitance)
C	Celsius
calib	calibrate
cath (K on diagram)	cathode
cath foll	cathode follower
catv	community antenna television
CB	citizens' band
cc	cotton covered (wire)
cctv	closed circuit television
ccw	counterclockwise
cemf	counter electromotive force
cent	centering
centi (or c)	10^{-2}
C_{gk}	grid-cathode capacitance (tube)
C_{gp}	grid-plate capacitance (tube)
cgs	centimeter-gram-second
ch	choke
chan	channel
chg	charge
chrome	chromium
cir mil (s)	circular mil(s)
ckt brkr	circuit breaker
cm	centimeters
coax	coaxial cable
colog	cologarithm
com	common
cond	conductor
conn	connection
cont	control
conv	convergence; converter
cos	cosine
cosh	hyperbolic cosine
cot	cotangent
counter emf; cemf	counter electromotive force
cps	cycles per second (Hertz)

Table 14-2. Electronic Abbreviations (cont'd).

Abbreviation	Term
c-r	cathode-ray
cro	cathode-ray oscilloscope
CrO_2	chromium dioxide
crt	cathode-ray tube
csc; cosc	cosecant
ct	center tap
cw	clockwise; continuous wave

D

d	diode; deci—(one-tenth; 10^{-1}); diameter
dB	decibel
dBf	signal reference to power
dblr	doubler
dbm	decibels referred to 1 milliwatt across 600 ohms
DBS	direct broadcast satellite
dc	direct current or voltage
dc rest	dc restorer
dcc	double cotton covered (wire)
deci	10^{-1}
defl	deflection
deg	degrees (angle); degrees (temp.)
deka	ten
demod	demodulator
det	detector
dielec	dielectric
diff	differentiator
DIN	Deutsche Industrie Normen (German Industrial Standards)
disch	discharge
discrim	discriminator
dk	deka—(ten)
dkm	decameters
dm	decimeters
dpdt	double-pole, double-throw (switch)
dpst	double-pole, single-throw (switch)
dsc	double silk covered (wire)
dx	distance
dyn	dynamic

E

e	emitter (transistor); voltage; electronic charge
ec	enamel covered
eco	electron-coupled oscillator
eff	effective (rms)
ehf	extremely high frequencies (30 gHz to 300 gHz)
EDP	Electronic Data Processing
EIA	Electronic Industries Association
EIAJ	Electronic Industries Association

197

Table 14-2. Electronic Abbreviations (cont'd).

Abbreviation	Term
	of Japan
elec	electric; electrolytic
elect	electrode
ELF	Extremely Low Frequencies
emf	electromotive force (voltage)
emu	electromagnetic unit (s)
enam	enameled (wire)
encl	enclosure
EP	extended play
eq	equalization
equiv	equivalent
erase hd	erase head
erf	error function
erp	effective radiated power
esu	electrostatic unit (s)
ev (or eV)	electron volt(s)
ext	external or extension
F	
f	femto (10^{-15})
f or freq	frequency
F	Fahrenheit (degrees)
F	farad(s)
fax	facsimile
fa	femtoampere 10^{-15} ampere
FCC	Federal Communications Commission
FeCr	ferrichrome
femto	10^{-15}
FET	field effect transistor
ff	fast forward
FHR	Fixed Head Recorder
fil (f in diagrams)	filament
fm	frequency modulation
foll	follower (-ing)
fone	headphones; earphones (see also phone)
freq	frequency
FS	Fourier series
G	
G	conductance
g	gram; grid (in diagrams); conductance
gb	gilberts
gca	ground-controlled approach
gdo	grid-dip oscillator
gen	generator
gHz	gigahertz (kilomegahertz)
giga	10^9
Gm	mutual conductance
gnd	ground

Table 14-2. Electronic Abbreviations (cont'd).

Abbreviation	Term
H	
H	magnetic intensity
h	henry (s)
ham	**radio amateur operator**
hd	head
hecto	10^2 (one hundred)
hex	hexadecimal
hf	high frequency (3,000 to 30,000 kHz); high filter
hi fi	high fidelity
hi pot	high potential
hm	hectometers
hor or horiz	horizontal
hp	horsepower
hr	hour
htr (in diagrams)	heater
hv	high voltage
hvr	home video recorder
hy	henry
Hz	hertz (cycles per second)
I	
i	current (instantaneous value)
ic	internal connection (in tubes); integrated circuit
icw	interrupted continuous waves
i-f or i.f.	intermediate frequency
i-f or i.f.t.	intermediate-frequency transformer
IHF	Institute of High Fidelity
ils	instrument landing system
im	intermodulation; intermodulation distortion
in	inch; input
int	integrator
I_p	plate current
ips	inches per second
J	
j;J	joule; imaginary number
jb	junction box
jct	junction
K	
K	numerical constant; dielectric constant; coupling coefficient
k	kilo; thousand; 10^3 cathode (tube)
kg	kilogram
kHz	kilohertz
kilo	10^3
km	kilometers

Table 14-2. Electronic Abbreviations (cont'd).

Abbreviation	Term
kv	kilovolt (s)
kva	kilovolt-ampere
kvar	reactive kilovolt amperes
kw	kilowatt
kwhr	kilowatt-hour
L	
L	coil; inductor; inductance; load
LC	inductance-capacitance
lcd	liquid crystal diode
led	light emitting diode
lf	low frequency (30 to 300 kHz)
lin	linearity; linear
lm	limiter
ln;log$_e$	Napierian logarithm
lna	low-noise amplifier
log; log$_{10}$	common logarithm
log^{-1}	antilogarithm
lp	long play
ls	limit switch
lsd	least significant digit
lsi	large scale integration
LVR	Longitudinal Video Recording
M	
m	milli (10^{-3}); one-thousandth meter; mutual inductance
μ	micro (10^{-6})—one-millionth amplication factor; permeability
μa	microampere
ma	milliampere
mag	magnetic
matv	master antenna television
max	moving coil; maximum
md	mean deviation
mds	multipoint distribution service
meg	megohm
mega	million; (10^6)
mem	memory
mev	million electron volts
μf	microfarad
mf	medium frequencies (300kHz to 3,000 kHz)
mfb	motional feedback
mh	millihenry (s)
MHz	megahertz
μh	microhenry (s)
micro	one-millionth; (10^{-6})
micromicro	one-millionth of a millionth; (10^{-12}) (see also pico)
mike (or mic)	microphone
milli	10^{-3} (one-thousandth)

Table 14-2. Electronic Abbreviations (cont'd).

Abbreviation	Term
min	minimum
mks	meter-kilogram-second
$\mu\mu$	micromicro (10^{-12}) (same as pico)
$\mu\mu$f	micromicrofarad (same as picofarad)
mm	millimeters; moving magnet
mmf	magnetomotive force
mod	modulation; modulator;modulus
modem	modulator-demodulator
mol	maximum output level
mon	monitor
mono	monophonic; monochrome; monaural
mos	metal oxide semiconductor
mosfet	metal oxide semiconductor field effect transistor
most	metal oxide semiconductor transistor
mpx	multiplex
msd	most significant digit
μsec	microsecond
msec	millisecond
mpx	multiplex
mtv	music television
mult	multiplier
μv	microvolt
mv	multivibrator; millivolt
mva	megavolt-ampere
mvb	multivibrator
mvc	manual volume control
μv/m	microvolts per meter
μw	microwatt
Mw	megawatt
mw	milliwatt
Mwh	megawatt-hour
my	myria; (ten thousand; 10^4)
mym	myriameters

N

Abbreviation	Term
n	nano; (10^{-9}) number of turns
NAB	National Assoc. of Broadcasters
NAND	not AND gate
nano(n)	10^{-9}
nbfm	narrow-band fm
nc	normally-closed (switch or relay); neutralizing capacitor; no connection
ne	neon
neg	negative; minus
net	network
nf	negative feedback
NI	ampere turns
no	normally open (switch or relay)
NOR	not OR gate

Table 14-2. Electronic Abbreviations (cont'd).

Abbreviation	Term
npn	negative-positive-negative (transistor)
NTSC	National Television System Committee
n-type	semiconductor with excess of negative carriers
nvr	no voltage release
0	
OD	outside diameter
omni	omnidirectional
OR	or gate
o/v	ohms per volt
osc	oscillator
out	output
P	
p	power; pole; plate(on diagrams); pico (10^{-12})
pa	public address; power amplifier
PAL	phase alternation line
pam	pulse amplitude modulation
pb	playback
pc	photocell
pcm	pulse code modulation
pd	potential difference
pent	pentode
perm	permanent
pf	power factor; picofarad (micromicrofarad)
phone (s) (or fone)	headphones; earphones
photo mult	photomultiplier
pico	formerly designated as micromicro (10^{-12})
pix	picture
pl	pilot lamp
pll	phase locked loop
pm	permanent magnet (speaker); phase modulation; pulse modulation
pn	diode or transistor junction
pnp	positive-negative-positive (transistor)
pos	positive; plus
pot	potentiometer;potential
pp	peak-to-peak; pushpull
ppi	plan-position indicator (radar)
ppm	parts per million
pps	pulses per second
preamp	preamplifier
prf	pulse repetition frequency
pri	primary
pt	phototube
ptm	pulse time modulation
p-type	semiconductor with excess of positive carriers
ptv	projection television

Table 14-2. Electronic Abbreviations (cont'd).

Abbreviation	Term
pwr	power
Q	
Q	reactance-resistance ratio; transistor; coulomb; Q factor; Q signal
quad	quadrature, quadraphonic (quadriphonic)
Qube	two-way cable tv
R	
r	resistance; resistor; radius
rc	resistance-capacitance
r-c	radio control
rcdg	recording
rcdr	recorder
rcvr	receiver
rec	record
rect	rectifier
reg	regulator
regen	regeneration
rev	reverse
reverb	reverberation
rf	radio frequency
rfc	radio-frequency choke
rft	radio-frequency transformer
rheo	rheostat
r-i	resistance-inductance
r-i-c	resistance-inductance-capacitance
RIAA	Record Industry Association of America
rms	root-mean-square; effective
rpm	revolutions per minute
ry	relay
ry, nc	relay, normally closed
ry, no	relay, normally open
S	
s or sw	switch
scc	single-cotton covered (wire)
sce	single cotton enameled (wire)
scope	oscilloscope
scr	silicon-controlled rectifier
sec	second; secondary; secant
SECAM	sequential with memory
sech	hyperbolic secant
sels	selsyn
sep	separator
sg	screen grid
shf	super-high frequency (3,000 to 300,000 MHz)
sig	signal
sin	sine

Table 14-2. Electronic Abbreviations (cont'd).

Abbreviation	Term
sinh	hyperbolic sine
sld	solenoid
SLP	super long play
s/n	signal-to-noise ratio
sp	single-pole
SP	standard play
spdt	single-pole, double-throw
spdtdb	single-pole, double-throw, double-break
spdtncdb	single-pole, double-throw, normally-closed, double break
spdtno	single-pole, double-throw, normally-open
spdtnodb	single-pole, double-throw, normally-open, double-break
spec(s)	specifications
spkr	loudspeaker
spl	sound pressure level
spstnc	single-pole, single-throw, normally-closed
spstno	single-pole, single-throw, normally-open
sq	square
ssb	single sideband
ssc	single silk covered (wire)
ssdd	specific signal display device
stereo	stereophonic
strobe	stroboscope
stv	subscription tv
sup	suppressor
superhet	superheterodyne
sw	switch; short wave
swl	short-wave listener
swr	standing-wave ratio
sync	synchronization; synchronous
T	
t	transformer; trimmer capacitor; tera; (10^{12})
tacho	tachometer
tan	tangent
tanh	hyperbolic tangent
teleg	telegraph; telegram
tera	10^{12}
term	terminal
thd	total harmonic distortion
TI	ampere turns
TIM	transient intermodulation distortion
Tocom	two-way cable tv
tptg	tuned-plate; tuned-grid

Table 14-2. Electronic Abbreviations (cont'd).

Abbreviation	Term
tr	transmit-receive; turns ratio; transient response
trans	transformer
trf	tuned-radio frequency
trig	trigger
tsf	telegraphie sans fil (wireless telegraphy)
tv	television
tvi	television interference

U

uhf	ultra-high frequencies (300 to 3,000 MHz)
UL	Underwriters' Laboratories, Inc.

V

v	volt(s); transistor; voltmeter
va	voltamperes (apparent power) voltage amplifier
vac;vdc	volts ac, dc
var	variable; reactive volt-amperes; varistor
vc	voice coil
vcr	video cassette recorder
vers	versed sine
vert	vertical
vfo	variable frequency oscillator
vhf	very high frequencies (30 to 300 MHz)
vid	video
vir	vertical interval reference
vlf	very low frequencies (below 30 kHz)
v/m	volts per meter
vol	volume
vom	volt-ohm-milliammeter; volt-ohmmeter
vr	voltage regulator (tube)
vt or v	tube
vtf	vertical tracking force
vtvm	vacuum-tube voltmeter
vu	volume unit

W

w	watt (s)
whr	watt-hour
wrms	weighted rms
WVDC	working volts, dc

X

x	reactance
X_c	capacitive reactance
X_L	inductive reactance
xformer	transformer
xmit	transmitter
xmission	transmission
xtal	crystal

Table 14-2. Electronic Abbreviations (cont'd).

Abbreviation	Term
Y	
Y	admittance
Z	
Z	impedance; characteristic impedance

material, in circuit diagrams, in reports, and in magazine articles. There is no industry-wide style standard. The best that can be hoped for is that a selected style will be consistent throughout a book, report, or article. Abbreviations may be in lowercase, capital letters or some combination of both. Abbreviations may or may not have points (periods). Points are desirable where the abbreviation may be mistaken for a word.

NOTATION

Components can be identified alphanumerically. Figure 14-1 shows the three different types of notation that are used. The numbers following the letters are used for identification only and should not be regarded as multipliers.

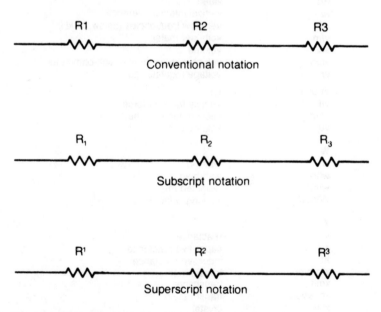

Fig. 14-1. Types of notation used in electronic circuits.

15

Solid State

TRANSISTOR ALPHA AND BETA

The current amplification factor of a transistor can be expressed in terms of either alpha (α) or beta (β). The ratio between a change in collector current for a change in emitter current is called alpha, and since alpha is less then unity, you will find it given as a decimal.

When we change emitter current we not only change collector current but base current as well. The ratio of a change in collector current to a change in base current is also used as a measure of the amplification of a transistor and is represented by beta.

The relationship between alpha and beta is shown in these formulas:

$$\beta = \alpha/1 - \alpha$$

$$\alpha = 1 - 1/[\beta + 1]$$

Because alpha is a decimal it sometimes makes a problem awkward to handle. For this reason it is sometimes more convenient to give current amplification as beta. Since beta is always in whole numbers it becomes convenient when comparing the current gains of different transistors.

Table 15-1 supplies common values of alpha and corresponding values for beta. The range of alpha is from 0.5 to 0.9964; that of beta is from 1 to 270.

□ **Example:**

The current gain (alpha) of a transistor is 0.9730. What is its current gain in terms of beta?

Table 15-1. Alpha vs. Beta.

β	α	β	α	β	α
1	0.5000	40	0.9756	79	0.9875
2	0.6666	41	0.9762	80	0.9877
3	0.7500	42	0.9767	81	0.9878
4	0.8000	43	0.9773	82	0.9880
5	0.8333	44	0.9778	83	0.9881
6	0.8571	45	0.9782	84	0.9882
7	0.8750	46	0.9786	85	0.9884
8	0.8889	47	0.9792	86	0.9885
9	0.9000	48	0.9796	87	0.9886
10	0.9091	49	0.9800	88	0.9888
11	0.9167	50	0.9804	89	0.9889
12	0.9231	51	0.9808	90	0.9890
13	0.9286	52	0.9811	91	0.9891
14	0.9333	53	0.9815	92	0.9892
15	0.9375	54	0.9818	93	0.9894
16	0.9412	55	0.9821	94	0.9895
17	0.9444	56	0.9825	95	0.9896
18	0.9474	57	0.9828	96	0.9897
19	0.9500	58	0.9831	97	0.9898
20	0.9524	59	0.9833	98	0.9899
21	0.9545	60	0.9836	99	0.9900
22	0.9565	61	0.9839	100	0.9901
23	0.9583	62	0.9841	110	0.9909
24	0.9600	63	0.9844	120	0.9917
25	0.9615	64	0.9846	125	0.9921
26	0.9630	65	0.9848	130	0.9931
27	0.9643	66	0.9851	140	0.9932
28	0.9655	67	0.9853	150	0.9933
29	0.9667	68	0.9855	160	0.9938
30	0.9677	69	0.9857	170	0.9942
31	0.9688	70	0.9859	180	0.9945
32	0.9697	71	0.9861	190	0.9948
33	0.9706	72	0.9863	200	0.9952
34	0.9714	73	0.9865	210	0.9954
35	0.9722	74	0.9867	220	0.9956
36	0.9730	75	0.9868	230	0.9958
37	0.9737	76	0.9870	240	0.9960
38	0.9744	77	0.9872	250	0.9962
39	0.9750	78	0.9873	260	0.9963
				270	0.9964

Table 15-2. Method for Testing Transistors.

Meter	Connection	(polarity)	Resistance (×100 scale)	
Emitter	Base	Collector	pnp type	npn type
positive negative	negative positive	(none) (none)	low high	high low
(none) (none)	positive negative	negative positive	high low	low high
negative positive	(none) (none)	positive negative	high mid	mid high
positive positive	to emitter to collector	negative negative	high low	
negative negative	to emitter to collector	positive positive		high low

Table 15-1 shows that an alpha of 0.9730 corresponds to a beta of 36.

☐ **Example**:

What is the alpha of a transistor if its beta value is 76?

Locate 76 in the column marked beta. To the right find the value of beta in terms of alpha, 0.9870.

TRANSISTOR TESTING

Transistors can be checked with ohmmeters that have low voltage and low current. Some meters use voltages of 10 volts or more on the high resistance scales and have no current limiting resistor on the low scale, hence such instruments can damage transistors. The method used in Table 15-2 does not indicate the quality of the transistor, simply whether it is good or bad. If the transistor is connected into the circuit, at least two of its three leads must be disconnected. The base is shorted to either of the other elements for the last four tests.

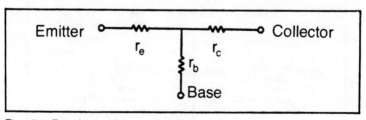

Fig. 15-1. Transistor resistance diagram.

209

Table 15-3. Transistor Resistances.

Circuit	Input resistance	Output resistance
Common base	$r_e + r_b$	$r_c + r_b$
Common emitter	$r_b + r_e$	$r_c + r_e$
Common collector	$r_b + r_c$	$r_e + r_c$

Table 15-4. Basic Transistor Circuit Characteristics.

Circuit Schematic	Common Base
	Power Gain — Yes
	Voltage Gain — Yes (approx. same CE)
	Current Gain — No (less than unity)
	Input Impedance — Lowest (approx. 50Ω)
	Output Impedance — Highest (approx. 1 MΩ)
	Phase Inversion — No

Common Emitter	Common Collector
Yes (highest)	Yes
Yes	No (less than unity)
Yes	Yes
Intermediate (approx. 1 kΩ)	Highest (approx. 300 kΩ)
Intermediate (approx. 50 kΩ)	Lowest (approx. 300 Ω)
Yes	No

210

INTERNAL INPUT AND OUTPUT RESISTANCES OF A TRANSISTOR

There is no isolation between the input and output circuits of a transistor (Fig. 15-1). These are related through their respective resistances, as indicated in Table 15-3.

Table 15-4 shows the characteristics for three basic transistor circuits: common base, common emitter, and common collector.

16

Digital Logic

LOGIC GATES

Gates are used in a wide variety of electronic equipment. In logic circuits, 1 can represent a closed circuit condition; 0 an open circuit (Fig. 16-1). A closed circuit, or 1, is sometimes called a **true** condition; an open circuit is called false. Figures 16-2 through 16-8 show basic diagrams, logic symbols and truth tables. Truth tables indicate operating conditions.

While the illustration in Fig. 16-2 shows an AND gate with two inputs, it is possible for such a gate to have three (or more) inputs.

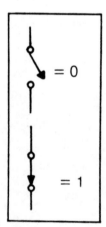

Fig. 16-1. An open switch is considered 0; a closed switch is 1.

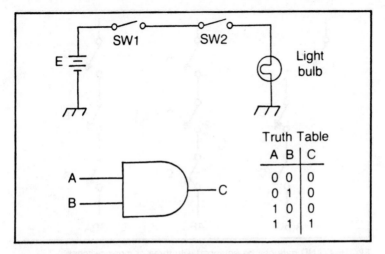

Truth Table

A	B	C
0	0	0
0	1	0
1	0	0
1	1	1

Fig. 16-2. AND gate and its truth table. SW1 and SW2 must be closed for the light bulb to be on.

Figure 16-3 illustrates an AND gate with three inputs. Each of the three switches (Fig. 16-4) must be closed for the circuit to have an output. Although switches are shown, these could represent operating conditions. Thus, each switch could be an operating voltage connected in series aiding. The AND gate would be conductive, in this case, only if all three voltages were present. The truth table for this AND arrangement shows that we would have

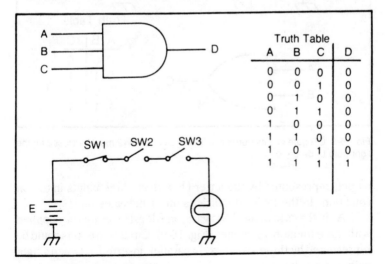

Truth Table

A	B	C	D
0	0	0	0
0	0	1	0
0	1	0	0
0	1	1	0
1	0	0	0
1	1	0	0
1	0	1	0
1	1	1	1

Fig. 16-3. AND gate and three inputs.

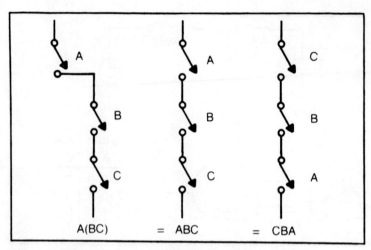

A(BC)   = ABC   = CBA

Fig. 16-4. AND gate can be equipped with three or more switches.

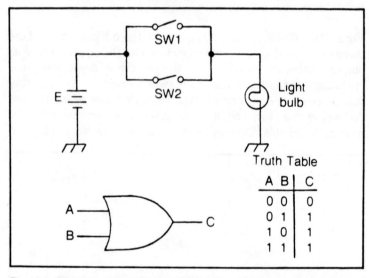

Truth Table

A	B	C
0	0	0
0	1	1
1	0	1
1	1	1

Fig. 16-5. OR gate and its truth table. Either SW1 or SW2 must be closed for the light bulb to be on.

output, represented by the letter O, with all three inputs in the on condition. In the table, 0 indicates off; 1 indicates on.

As in the case of the AND gate, an OR gate can also be supplied with three (or more) switches (Fig. 16-9). Circuitwise, this could be represented by three switches, in parallel, inserted between a light bulb and a voltage source.

214

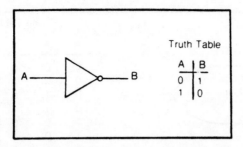

Fig. 16-6. Inverter circuit is often used with gates. This circuit changes output from plus to minus, or from minus to plus. The small circle indicates the inversion function. Without it the diagram is that of an ordinary noninverting amplifier.

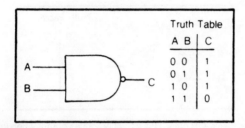

Fig. 16-7. NAND gate and its truth table. Same as AND gate but note small circle to indicate inversion of signal.

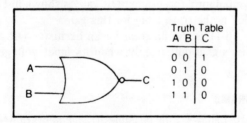

Fig. 16-8. NOR gate and its truth table. This is a not OR gate.

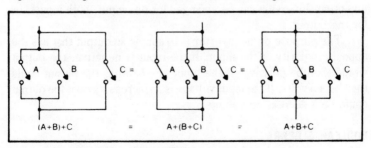

Fig. 16-9. OR gate can be equipped with three or more switches.

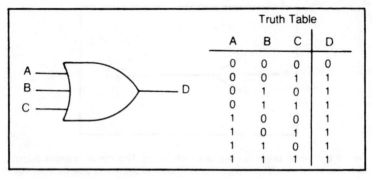

Fig. 16-10. OR gate with three inputs.

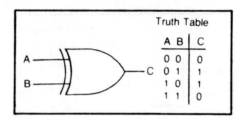

Fig. 16-11. Exclusive-OR gate. This circuit produces output when inputs are not identical.

The symbol for a three-input OR gate is shown in Fig. 16-10. Also included is the truth table for this gate.

Figure 16-11 is the diagram for an exclusive-OR gate. This gate supplies signal output only when its input voltages are not identical.

LOGIC DIAGRAMS

Table 16-1 shows some possible arrangements of AND and OR gates. The small circle attached to the symbol indicates signal inversion. Inversion can be arranged for one input, both inputs, or for the output.

The purpose of the inverter is to supply an output that has the opposite polarity of the input. If the input is negative, the output would then be positive. If the input is −5 volts, the output could then be +5 volts. If the input voltage is an increasing one, the output could be a decreasing voltage.

BOOLEAN ALGEBRA

In Boolean algebra, letters are used to represent switches.

Table 16-1. Logic Diagrams.

Gates		Truth Tables
AND	OR	A B C
A, B → C (AND gate)	A, B → C (OR gate)	0 0 0 0 1 0 1 0 0 1 1 1
A, B → C (AND gate)	A, B → C (OR gate)	0 0 1 0 1 1 1 0 1 1 1 0
A, B → C (AND gate)	A, B → C (OR gate)	0 0 0 0 1 1 1 0 1 1 1 1
A, B → C (AND gate)	A, B → C (OR gate)	0 0 1 0 1 0 1 0 0 1 1 0
A, B → C (AND gate)	A, B → C (OR gate)	0 0 0 0 1 1 1 0 0 1 1 0
A, B → C (AND gate)	A, B → C (OR gate)	0 0 0 0 1 0 1 0 1 1 1 0
A, B → C (AND gate)	A, B → C (OR gate)	0 0 1 0 1 0 1 0 1 1 1 1
A, B → C (AND gate)	A, B → C (OR gate)	0 0 1 0 1 1 1 0 0 1 1 1

Table 16-2. Boolean Algebra Definitions.

Definitions

A, B, C, etc	Symbols used in symbolic logic
A · B or AB	Read as: A AND B
A + B	Read as: A OR B
A' or $\overline{A}$	Read as: not A
1	"True" or "on"
0	"False" or "off"
$\overline{\overline{A}}$	Read as "not not" A
$\overline{AB}$	Read as "not A AND not B"
$\overline{A + B}$	Read as "not A OR not B"

217

Table 16-3. Theorems in Boolean Algebra.

$A + A = A$	$A + AB = A$
$A \cdot A = A$	$A(A+B) = A$
$A + 1 = 1$	$\overline{A+B} = \overline{A}\,\overline{B}$
$A \cdot 1 = A$	$\overline{AB} = \overline{A}+\overline{B}$
$A + 0 = A$	$A(\overline{A}+B) = AB$
$A \cdot 0 = 0$	$A + \overline{A}B = A+B$
$A + \overline{A} = 1$	$\overline{A} + AB = \overline{A}+B$
$A \cdot \overline{A} = 0$	$\overline{A} + A\overline{B} = \overline{A}+\overline{B}$
$\overline{\overline{A}} = A$	

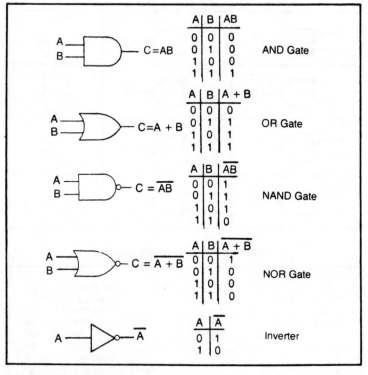

Fig. 16-12. Boolean logic.

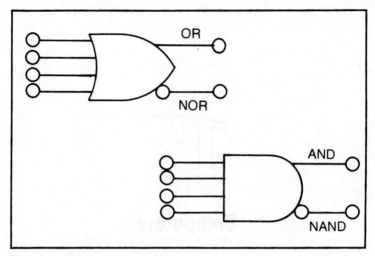

Fig. 16-13. Combined gates.

Boolean algebra is used for the design of switching circuits or logic gates. A plus symbol is used in place of the word OR. The AND symbol is a dot placed between two letters (A • B). The dot is moved up a bit so as not to confuse it with a decimal point. The dot symbol for AND is sometimes omitted and the two letters are written immediately adjacent to each other (A • B is the same as AB). The NOT function is shown by a line above a letter or by a prime mark as in $\overline{A}$ or A′. See Tables 16-2 and 16-3.

BOOLEAN EXPRESSIONS FOR GATING CIRCUITS

Logic circuits can be described in Boolean functions. Figure 16-12 shows various gates and their equivalent Boolean expressions and truth tables.

OR and NOR and AND and NAND gates can be combined in single gates as shown in Fig. 16-13.

17

Computers

BINARY NUMBER SYSTEM

The decimal system using digits ranging from 0 through 9 for all number applications is just one of a large array of number systems. Other number systems (including the decimal) such as the binary and hexadecimal, find applications in computer technology.

The binary system uses two digits: 0 and 1. As in the case of the decimal system, the value of the number depends on its horizontal position. Thus, in a decimal number such as 875, the digit 8 has a true value of 800, 7 has a value of 70 and 5 has a value of 5; 800 + 70 + 5 equals 875. The binary system is based on powers of two (Table 17-1) similar to powers of 10.

Table 17-1 supplies powers of two up to 2^{60} and 2^{-60}. To find the value of 2^{15}, as an example, move down the center column headed by the letter n, stopping at the number 15. Move to the left and the value of 2^{15} is given as 32,768. To find the value of 2^{-6}, move down the center or n column and stop at the number 6. Look to the right and the value of 2^{-6} is supplied as 0.015625.

Binary numbers, as in the case of decimal numbers, are generally arranged horizontally. The first binary number at the right has a value of 2^0, the adjacent binary a value of 2^1, etc. Thus, a number such as binary 1011 would have an equivalent decimal value of:

$$\begin{array}{cccc} 1 & 0 & 1 & 1 \\ (1 \times 2^3) + & (0 \times 2^2) + & (1 \times 2^1) + & (1 \times 2^0) \end{array}$$

$$8 \quad + \quad 0 \quad + \quad 2 \quad + \quad 1 \quad = 11$$

Table 17-1. Powers of Two.

2^n	n	2^{-n}
1	0	1 0
2	1	0 5
4	2	0 25
8	3	0 125
16	4	0 062 5
32	5	0 031 25
64	6	0 015 625
128	7	0 007 812 5
256	8	0 003 906 25
512	9	0 001 953 125
1 024	10	0.000 976 562 5
2 048	11	0.000 488 281 25
4 096	12	0.000 244 140 625
8 192	13	0.000 122 070 312 5
16 384	14	0.000 061 035 156 25
32 768	15	0.000 030 517 578 125
65 536	16	0.000 015 258 789 062 5
131 072	17	0.000 007 629 394 531 25
262 144	18	0.000 003 814 697 265 625
524 288	19	0.000 001 907 348 632 812 5
1 048 576	20	0.000 000 953 674 316 406 25
2 097 152	21	0.000 000 476 837 158 203 125
4 194 304	22	0.000 000 238 418 579 101 562 5
8 388 608	23	0.000 000 119 209 289 550 781 25
16 777 216	24	0.000 000 059 604 644 775 390 625
33 554 432	25	0.000 000 029 802 322 387 695 312 5
67 108 864	26	0.000 000 014 901 161 193 847 656 25
134 217 728	27	0.000 000 007 450 580 596 923 828 125
268 435 456	28	0.000 000 003 725 290 298 461 914 062 5
536 870 912	29	0.000 000 001 862 645 149 230 957 031 25
1 073 741 824	30	0.000 000 000 931 322 574 615 478 515 625
2 147 483 648	31	0.000 000 000 465 661 287 307 739 257 812 5
4 294 967 296	32	0.000 000 000 232 830 643 653 869 628 906 25
8 589 934 592	33	0.000 000 000 116 415 321 826 934 814 453 125
17 179 869 184	34	0.000 000 000 058 207 660 913 467 407 226 562 5
34 359 738 368	35	0.000 000 000 029 103 830 456 733 703 613 281 25
68 719 476 736	36	0.000 000 000 014 551 915 228 366 851 806 640 625
137 438 953 472	37	0.000 000 000 007 275 957 614 183 425 903 320 312 5
274 877 906 944	38	0.000 000 000 003 637 978 807 091 712 951 660 156 25
549 755 813 888	39	0.000 000 000 001 818 989 403 545 856 475 830 078 125
1 099 511 627 776	40	0.000 000 000 000 909 494 701 772 928 237 915 039 062 5
2 199 023 255 552	41	0.000 000 000 000 454 747 350 886 464 118 957 519 531 25
4 398 046 511 104	42	0.000 000 000 000 227 373 675 443 232 059 478 759 765 625
8 796 093 022 208	43	0.000 000 000 000 113 686 837 721 616 029 739 379 882 812
17 592 186 044 416	44	0.000 000 000 000 056 843 418 860 808 014 869 689 941 406
35 184 372 088 832	45	0.000 000 000 000 028 421 709 430 404 007 434 844 970 703
70 368 744 177 664	46	0.000 000 000 000 014 210 854 715 202 003 717 422 485 351
140 737 488 355 328	47	0.000 000 000 000 007 105 427 357 601 001 858 711 242 675
281 474 976 710 656	48	0.000 000 000 000 003 552 713 678 800 500 929 355 621 337
562 949 953 421 312	49	0.000 000 000 000 001 776 356 839 400 250 464 677 810 668
1 125 899 906 842 624	50	0.000 000 000 000 000 888 178 419 700 125 232 338 905 334
2 251 799 813 685 248	51	0.000 000 000 000 000 444 089 209 850 062 616 169 452 667
4 503 599 627 370 496	52	0.000 000 000 000 000 222 044 604 925 031 308 084 726 333
9 007 199 254 740 992	53	0.000 000 000 000 000 111 022 302 462 515 654 042 363 166
18 014 398 509 481 984	54	0.000 000 000 000 000 055 511 151 231 257 827 021 181 583
36 028 797 018 963 968	55	0.000 000 000 000 000 027 755 575 615 628 913 510 590 791
72 057 594 037 927 936	56	0.000 000 000 000 000 013 877 787 807 814 456 755 295 395
144 115 188 075 855 872	57	0.000 000 000 000 000 006 938 893 903 907 228 377 647 697
288 230 376 151 711 744	58	0.000 000 000 000 000 003 469 446 951 953 614 188 823 848
576 460 752 303 423 488	59	0.000 000 000 000 000 001 734 723 475 976 807 094 411 924
152 921 504 606 846 976	60	0.000 000 000 000 000 000 867 361 737 988 403 547 205 962

Hence, binary 1011 is equivalent to decimal 11. To emphasize the number system, the subscript 2 is sometimes used to identify binary numbers while 10 is used as a subscript for decimal numbers. In the example given: 1011_2 equals 11_{10}.

Table 17-2 lists the binary equivalents of decimal numbers ranging from 0 to 100.

Table 17-2. Decimal Integers to Pure Binaries.

Dec. Integer	Binary	Dec. Integer	Binary	Dec. Integer	Binary
00	00000000	33	00100001	67	01000011
01	00000001	34	00100010	68	01000100
02	00000010	35	00100011	69	01000101
03	00000011	36	00100100	70	01000110
04	00000100	37	00100101	71	01000111
05	00000101	38	00100110	72	01001000
06	00000110	39	00100111	73	01001001
07	00000111	40	00101000	74	01001010
08	00001000	41	00101001	75	01001011
09	00001001	42	00101010	76	01001100
10	00001010	43	00101011	77	01001101
11	00001011	44	00101100	78	01001110
12	00001100	45	00101101	79	01001111
13	00001101	46	00101110	80	01010000
14	00001110	47	00100111	81	01010001
15	00001111	48	00110000	82	01010010
16	00010000	49	00110001	83	01010011
17	00010001	50	00110010	84	01010100
18	00010010	51	00110011	85	01010101
19	00010011	52	00110100	86	01010110
20	00010100	53	00110101	87	01010111
21	00010101	54	00110110	88	01011000
22	00010110	55	00110111	89	01011001
23	00010111	56	00111000	90	01011010
24	00011000	57	00111001	91	01011011
25	00011001	58	00111010	92	01011100
26	00011010	59	00111011	93	01011101
27	00011011	60	00111100	94	01011110
28	00011100	61	00111101	95	01011111
29	00011101	62	00111110	96	01100000
30	00011110	63	00111111	97	01100001
31	00011111	64	01000000	98	01100010
32	00100000	65	01000001	99	01100011
		66	01000010	100	01100100

□ **Example**:

What is the binary equivalent of 27?

In Table 17-2, locate 27 in the left-hand column. Move directly across to the right and the binary equivalent is 00011011. The three zeros at the left of the binary can be omitted since they add nothing to the value of the number. Accordingly, decimal 27 equals binary 11011. Binary 11011 can be set up as:

$$
\begin{array}{ccccccccc}
1 & & 1 & & 0 & & 1 & & 1 \\
2^4 & + & 2^3 & + & 2^2 & + & 2^1 & + & 2^0 \\
16 & + & 8 & + & 0 & + & 2 & + & 1 & = 27
\end{array}
$$

11011_2 equals 27_{10}.

DECIMAL TO BINARY CONVERSION RULES

1. Write number n + 0 if even or (n − 1) + 1 if odd.
2. Divide even number obtained in (1) by 2.
   Write answer (m) below in same form:

   $$m + 0 \text{ if even, } (m - 1) + 1 \text{ if odd.}$$

3. Continue until m or (m − 1) becomes zero.
4. Column of ones and zeros so obtained is binary equivalent of n with least significant digit at the top.

□ **Example:** n = 327

$$
\begin{array}{r}
326 + 1 \\
162 + 1 \\
80 + 1 \\
40 + 0 \\
20 + 0 \\
10 + 0 \\
4 + 1 \\
2 + 0 \\
0 + 1
\end{array}
$$

Therefore, the binary equivalent of 327 is 101000111.

BINARY TO DECIMAL CONVERSION RULES

1. Start at left with first significant digit—double it if the next digit

is a zero or "dibble" it (double and add one) if the next digit is a one.
2. If the 3rd digit is a zero, double value obtained in(1), if it is a one "dibble" value obtained in (1).
3. Continue until operation indicated by least significant digit has been performed.

BINARY-CODED DECIMALS

Binary numbers can be arranged in groups of four to correspond to decimal digits. A decimal number such as 5, for example, can be represented by binary 0101. Decimal 55 would then be 0101 0101. A setup of this kind is known as a binary-coded decimal, abbreviated as BCD. Note the difference between binary-coded decimals and pure binary numbers shown earlier in Table 17-2. A decimal number such as 82 in pure binary form would be 1010010, while decimal 82 in BCD notation would be 1000 0010.

Table 17-3 lists BCD equivalents of decimal numbers ranging from 0 to 100.

Table 17-3. Decimal to Binary-Coded Decimal (BCD) Notation.

Decimal	BCD		Decimal	BCD	
00	0000	0000	22	0010	0010
01	0000	0001	23	0010	0011
02	0000	0010	24	0010	0100
03	0000	0011	25	0010	0101
04	0000	0100	26	0010	0110
05	0000	0101	27	0010	0111
06	0000	0110	28	0010	1000
07	0000	0111	29	0010	1001
08	0000	1000	30	0011	0000
09	0000	1001	31	0011	0001
10	0001	0000	32	0011	0010
11	0001	0001	33	0011	0011
12	0001	0010	34	0011	0100
13	0001	0011	35	0011	0101
14	0001	0100	36	0011	0110
15	0001	0101	37	0011	0111
16	0001	0110	38	0011	1000
17	0001	0111	39	0011	1001
18	0001	1000	40	0100	0000
19	0001	1001	41	0100	0001
20	0010	0000	42	0100	0010
21	0010	0001			

Table 17-3. Decimal to Binary-Coded Decimal (BCD) Notation (cont'd).

Decimal	BCD		Decimal	BCD	
43	0100	0011	72	0111	0010
44	0100	0100	73	0111	0011
45	0100	0101	74	0111	0100
46	0100	0110	75	0111	0101
47	0100	0111	76	0111	0110
48	0100	1000	77	0111	0111
49	0100	1001	78	0111	1000
50	0101	0000	79	0111	1001
51	0101	0001	80	1000	0000
52	0101	0010	81	1000	0001
53	0101	0011	82	1000	0010
54	0101	0100	83	1000	0011
55	0101	0101	84	1000	0100
56	0101	0110	85	1000	0101
57	0101	0111	86	1000	0110
58	0101	1000	87	1000	0111
59	0101	1001	88	1000	1000
60	0110	0000	89	1000	1001
61	0110	0001	90	1001	0000
62	0110	0010	91	1001	0001
63	0110	0011	92	1001	0010
64	0110	0100	93	1001	0011
65	0110	0101	94	1001	0100
66	0110	0110	95	1001	0101
67	0110	0111	96	1001	0110
68	0110	1000	97	1001	0111
69	0110	1001	98	1001	1000
70	0111	0000	99	1001	1001
71	0111	0001	100	0000	0000

BITS

The words "binary digit" can be contracted to form a new word, *bit*. Unlike decimal notation, the amount of bits is sometimes used to describe a particular binary number, as shown in Table 17-4.

The greater the number of bits the larger the equivalent decimal number, assuming the leftmost binary digit to be 1. Table 17-5 shows maximum binary values and their decimal equivalents.

VOLTAGE REPRESENTATION IN DIGITAL FORM

Voltages are generally analog representations as indicated in

0	One Bit
01	Two Bits
101	Three Bits
1010	Four Bits
10101	Five Bits
100110	Six Bits
1010101	Seven Bits
10101011	Eight Bits

Table 17-4. Binary Digit Designations.

Table 17-5. Maximum Binary Values and Decimal Equivalents.

Number of Bits	Maximum Binary Value	Maximum Equivalent Decimal Value
1	1	1
2	11	3
3	111	7
4	1111	15
5	11111	31
6	111111	63
7	1111111	127
8	11111111	255
9	111111111	511
10	1111111111	1023
11	11111111111	2047
12	111111111111	4095
13	1111111111111	8191
14	11111111111111	16383
15	111111111111111	32767
16	1111111111111111	65535

Fig. 17-1. However, by taking a large succession of instantaneous values and converting these from decimal to binary form, we can have a digital representation of the same waveform. Table 17-6 shows how the waveform of Fig. 17-1 can be indicated digitally.

HEXADECIMAL NUMBER SYSTEM

This number system uses digits 0 through 9 and letters A to F for its number applications. As in the case of the binary and decimal systems, the value of a hexadecimal number depends on its horizontal position. Hexadecimal numbers are based on powers of 16, as shown in Table 17-7.

Hexadecimal numbers, as in the case of binary and decimal numbers, are usually arranged horizontally. The first hexadecimal number at the right has a value of 16^0, the adjacent hexadecimal number a value of 16^1, etc.

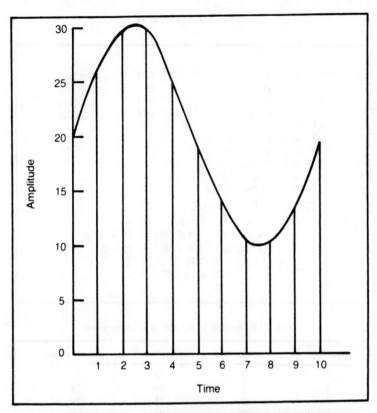

Fig. 17-1. Conversion from analog to digital form involves taking a large succession of instantaneous values in decimal form and then converting these to binary.

Table 17-6. Digital Values of Voltage Wave—Five-Bit Arrangement.

Time	Voltage in Decimal Form	Voltage in Binary Form
0	20	10100
1	25	11001
2	28	11100
3	28	11100
4	26	11010
5	20	10100
6	14	01110
7	10	01010
8	10	01010
9	14	01110
10	20	10100

Table 17-7. Powers of 16.

$$
\begin{array}{ll}
16^0 = 1 & 16^{-1} = 0.0625 \\
16^1 = 16 & 16^{-2} = 0.003906 \\
16^2 = 256 & 16^{-3} = 0.000244 \\
16^3 = 4,096 & 16^{-4} = 0.000015 \\
16^4 = 65,536 & \\
16^5 = 1,048,576 & \\
16^6 = 16,777,216 & \\
16^7 = 268,435,456 & \\
16^8 = 4,294,967,296 & \\
16^9 = 68,719,476,736 & \\
16^{10} = 1,099,511,627,776 & \\
\end{array}
$$

$$
\begin{array}{ccc}
4 & F & 3 \\
= (4 \times 16^2) + & (15 \times 16^1) + & (3 \times 16^0) \\
= (4 \times 256) + & (15 \times 16) + & (3 \times 1) \\
= 1,024 + & 240 + & 3 \\
& = 1,267 &
\end{array}
$$

Table 17-8. Decimal to Hexadecimal Conversions.

Decimal	Hexadecimal	Decimal	Hexadecimal
0	0	23	17
1	1	24	18
2	2	25	19
3	3	26	1A
4	4	27	1B
5	5	28	1C
6	6	29	1D
7	7	30	1E
8	8	31	1F
9	9	32	20
10	A	33	21
11	B	34	22
12	C	35	23
13	D	36	24
14	E	37	25
15	F	38	26
16	10	39	27
17	11	40	28
18	12	41	29
19	13	42	2A
20	14	43	2B
21	15	44	2C
22	16		

Table 17-8. Decimal to Hexadecimal Conversions (cont'd).

Decimal	Hexadecimal	Decimal	Hexadecimal
45	2D	74	4A
46	2E	75	4B
47	2F	76	4C
48	30	77	4D
49	31	78	4E
50	32	79	4F
51	33	80	50
52	34	81	51
53	35		
54	46	82	52
55	37	83	53
56	38	84	54
57	39	85	55
58	3A		
59	3B	86	56
60	3C	87	57
61	3D	88	58
62	3E	89	59
63	3F	90	5A
64	40	91	5B
65	41	92	5C
66	42	93	5D
67	43	94	5E
68	44	95	5F
69	45	96	60
70	46	97	61
71	47	98	62
72	48	99	63
73	49	100	64

In the hexadecimal system, the letter A corresponds to decimal 10, B to decimal 11, C to 12, D to 13, E to 14, and F to 15.

$$
\begin{array}{lllll}
& 4 & F & 3 & \\
= & (4 \times 16^2) & + & (15 \times 16^1) & + & (3 \times 16^0) \\
= & (4 \times 256) & + & (15 \times 16) & + & (3 \times 1) \\
= & 1{,}024 & + & 240 & + & 3 \\
= & 1{,}267 & & & &
\end{array}
$$

Hexadecimal 4F3 equals decimal 1,267.

Table 17-8 lists hexadecimal equivalents for decimal numbers from 0 to 100.

HEXADECIMAL TO DECIMAL CONVERSIONS

Every hexadecimal number consists of two parts: a coefficient or multiplier and some value of 16 raised to a power. Thus, hexadecimal 5A is the same as $(5 \times 16^1) + (10 \times 16^0)$. In this hexadecimal number, 5 is a coefficient, and so is A. (A equals decimal 10.) In a hexadecimal number such as 234, these digits have a decimal value of $(2 \times 16^2) + (3 \times 16^1) + (4 \times 16^0)$. Each of the numbers, 2, 3, and 4, are coefficients which are used to multiply various powers of 16. Since 234 looks like a decimal number, it can be identified as a hexadecimal number by using the subscript 16. Hence, 234 is more correctly written as 234_{16}.

Table 17-9 is a convenient way of obtaining decimal equivalents of hexadecimal numbers, but some addition is involved.

□ **Example:**

What is the decimal equivalent of 5FA4?

In Table 17-9, the first column at the left represents all possible coefficients or multipliers.

In converting 5FA4 to a decimal number, work with each part of this hexadecimal a digit at a time, starting at the right-hand side. The first right-hand digit of 5FA4 is 4. In Table 17-9, locate the number 4 in the left-hand column. Move completely across to 16^0 and the value of 4 is indicated as 4.

The next number of 5FA4 to consider is A. Find A in the left-hand column of Table 17-9. Move horizontally, but stop at the second or 16^1 column for you are now working with the second digit of the hex number. The value of A is 160.

The third digit of 5FA4 is F in the left-hand column of Table 17-9 and move to the third column, or the 16^2 column. The decimal value is 3840. Finally, find the decimal equivalent of the fourth digit in 5FA4. This is the number 5. Locate 5 in the left-hand column. Move across to the right and stop in the fourth column—the 16^3 column. The decimal equivalent here is 20,480.

$$
\begin{array}{r}
4 \\
160 \\
3,840 \\
\underline{20,480} \\
24,484
\end{array}
$$

Hexadecimal 5FA4 equals decimal 24484. This can also be written as $5FA4_{16}$ equals $24,484_{10}$.

Table 17-9. Hexadecimal to Decimal Integer Conversions.

	16^7 X0000000	16^6 X000000	16^5 X00000	16^4 X0000	16^3 X000	16^2 X00	16^1 X0	16^0 X
1	268,435,456	16,777,216	1,048,576	65,536	4,096	256	16	1
2	536,870,912	33,554,432	2,097,152	131,072	8,192	512	32	2
3	805,306,368	50,331,648	3,145,728	196,608	12,288	768	48	3
4	1,073,741,824	67,108,864	4,194,304	262,144	16,384	1024	64	4
5	1,342,177,280	83,886,080	5,242,880	327,680	20,480	1280	80	5
6	1,610,612,736	100,663,296	6,291,456	393,216	24,576	1536	96	6
7	1,879,048,192	117,440,512	7,340,032	458,752	28,672	1792	112	7
8	2,147,483,648	134,217,728	8,388,608	524,288	32,768	2048	128	8
9	2,415,919,104	150,994,944	9,437,184	589,824	36,864	2304	144	9
A	2,684,354,560	167,772,160	10,485,760	655,360	40,960	2560	160	10
B	2,952,790,016	184,549,376	11,534,336	720,896	45,056	2816	176	11
C	3,221,225,472	201,326,592	12,582,912	786,432	49,152	3072	192	12
D	3,489,660,928	218,103,808	13,631,488	851,968	53,248	3328	208	13
E	3,758,096,384	234,881,024	14,680,064	917,504	57,344	3584	224	14
F	4,026,531,840	251,658,240	15,728,640	983,040	61,440	3840	240	15

Table 17-10. Hexadecimal to Decimal Integer Conversions.

	0	1	2	3	4	5	6	7	8	9	A	B	C	D	E	F
000	0000	0001	0002	0003	0004	0005	0006	0007	0008	0009	0010	0011	0012	0013	0014	0015
010	0016	0017	0018	0019	0020	0021	0022	0023	0024	0025	0026	0027	0028	0029	0030	0031
020	0032	0033	0034	0035	0036	0037	0038	0039	0040	0041	0042	0043	0044	0045	0046	0047
030	0048	0049	0050	0051	0052	0053	0054	0055	0056	0057	0058	0059	0060	0061	0062	0063
040	0064	0065	0066	0067	0068	0069	0070	0071	0072	0073	0074	0075	0076	0077	0078	0079
050	0080	0081	0082	0083	0084	0085	0086	0087	0088	0089	0090	0091	0092	0093	0094	0095
060	0096	0097	0098	0099	0100	0101	0102	0103	0104	0105	0106	0107	0108	0109	0110	0111
070	0112	0113	0114	0115	0116	0117	0118	0119	0120	0121	0122	0123	0124	0125	0126	0127
080	0128	0129	0130	0131	0132	0133	0134	0135	0136	0137	0138	0139	0140	0141	0142	0143
090	0144	0145	0146	0147	0148	0149	0150	0151	0152	0153	0154	0155	0156	0157	0158	0159
0A0	0160	0161	0162	0163	0164	0165	0166	0167	0168	0169	0170	0171	0172	0173	0174	0175
0B0	0176	0177	0178	0179	0180	0181	0182	0183	0184	0185	0186	0187	0188	0189	0190	0191
0C0	0192	0193	0194	0195	0196	0197	0198	0199	0200	0201	0202	0203	0204	0205	0206	0207
0D0	0208	0209	0210	0211	0212	0213	0214	0215	0216	0217	0218	0219	0220	0221	0222	0223
0E0	0224	0225	0226	0227	0228	0229	0230	0231	0232	0233	0234	0235	0236	0237	0238	0239
0F0	0240	0241	0242	0243	0244	0245	0246	0247	0248	0249	0250	0251	0252	0253	0254	0255
100	0256	0257	0258	0259	0260	0261	0262	0263	0264	0265	0266	0267	0268	0269	0270	0271
110	0272	0273	0274	0275	0276	0277	0278	0279	0280	0281	0282	0283	0284	0285	0286	0287
120	0288	0289	0290	0291	0292	0293	0294	0295	0296	0297	0298	0299	0300	0301	0302	0303
130	0304	0305	0306	0307	0308	0309	0310	0311	0312	0313	0314	0315	0316	0317	0318	0319

140	0320	0321	0322	0323	0324	0325	0326	0327	0328	0329	0330	0331	0332	0333	0334	0335
150	0336	0337	0338	0339	0340	0341	0342	0343	0344	0345	0346	0347	0348	0349	0350	0351
160	0352	0353	0354	0355	0356	0357	0358	0359	0360	0361	0362	0363	0364	0365	0366	0367
170	0368	0369	0370	0371	0372	0373	0374	0375	0376	0377	0378	0379	0380	0381	0382	0383
180	0384	0385	0386	0387	0388	0389	0390	0391	0392	0393	0394	0395	0396	0397	0398	0399
190	0400	0401	0402	0403	0404	0405	0406	0407	0408	0409	0410	0411	0412	0413	0414	0415
1A0	0416	0417	0418	0419	0420	0421	0422	0423	0424	0425	0426	0427	0428	0429	0430	0431
1B0	0432	0433	0434	0435	0436	0437	0438	0439	0440	0441	0442	0443	0444	0445	0446	0447
1C0	0448	0449	0450	0451	0452	0453	0454	0455	0456	0457	0458	0459	0460	0461	0462	0463
1D0	0464	0465	0466	0467	0468	0469	0470	0471	0472	0473	0474	0475	0476	0477	0478	0479
1E0	0480	0481	0482	0483	0484	0485	0486	0487	0488	0489	0490	0491	0492	0493	0494	0495
1F0	0496	0497	0498	0499	0500	0501	0502	0503	0504	0505	0506	0507	0508	0509	0510	0511
200	0512	0513	0514	0515	0516	0517	0518	0519	0520	0521	0522	0523	0524	0525	0526	0527
210	0528	0529	0530	0531	0532	0533	0534	0535	0536	0537	0538	0539	0540	0541	0542	0543
220	0544	0545	0546	0547	0548	0549	0550	0551	0552	0553	0554	0555	0556	0557	0558	0559
230	0560	0561	0562	0563	0564	0565	0566	0567	0568	0569	0570	0571	0572	0573	0574	0575
240	0576	0577	0578	0579	0580	0581	0582	0583	0584	0585	0586	0587	0588	0589	0590	0591
250	0592	0593	0594	0595	0596	0597	0598	0599	0600	0601	0602	0603	0604	0605	0606	0607
260	0608	0609	0610	0611	0612	0613	0614	0615	0616	0617	0618	0619	0620	0621	0622	0623
270	0624	0625	0626	0627	0628	0629	0630	0631	0632	0633	0634	0635	0636	0637	0638	0639
280	0640	0641	0642	0643	0644	0645	0646	0647	0648	0649	0650	0651	0652	0653	0654	0655
290	0656	0657	0658	0659	0660	0661	0662	0663	0664	0665	0666	0667	0668	0669	0670	0671

Table 17-10. Hexadecimal to Decimal Integer Conversions (cont'd).

	0	1	2	3	4	5	6	7	8	9	A	B	C	D	E	F
2A0	0672	0673	0674	0675	0676	0677	0678	0679	0680	0681	0682	0683	0684	0685	0686	0687
2B0	0688	0689	0690	0691	0692	0693	0694	0695	0696	0697	0698	0699	0700	0701	0702	0703
2C0	0704	0705	0706	0707	0708	0709	0710	0711	0712	0713	0714	0715	0716	0717	0718	0719
2D0	0720	0721	0722	0723	0724	0725	0726	0727	0728	0729	0730	0731	0732	0733	0734	0735
2E0	0736	0737	0738	0739	0740	0741	0742	0743	0744	0745	0746	0747	0748	0749	0750	0751
2F0	0752	0753	0754	0755	0756	0757	0758	0759	0760	0761	0762	0763	0764	0765	0766	0767
300	0768	0769	0770	0771	0772	0773	0774	0775	0776	0777	0778	0779	0780	0781	0782	0783
310	0784	0785	0786	0787	0788	0789	0790	0791	0792	0793	0794	0795	0796	0797	0798	0799
320	0800	0801	0802	0803	0804	0805	0806	0807	0808	0809	0810	0811	0812	0813	0814	0815
330	0816	0817	0818	0819	0820	0821	0822	0823	0824	0825	0826	0827	0828	0829	0830	0831
340	0832	0833	0834	0835	0836	0837	0838	0839	0840	0841	0842	0843	0844	0845	0846	0847
350	0848	0849	0850	0851	0852	0853	0854	0855	0856	0857	0858	0859	0860	0861	0862	0863
360	0864	0865	0866	0867	0868	0869	0870	0871	0872	0873	0874	0875	0876	0877	0878	0879
370	0880	0881	0882	0883	0884	0885	0886	0887	0888	0889	0890	0891	0892	0893	0894	0895
380	0896	0897	0898	0899	0900	0901	0902	0903	0904	0905	0906	0907	0908	0909	0910	0911
390	0912	0913	0914	0915	0916	0917	0918	0919	0920	0921	0922	0923	0924	0925	0926	0927
3A0	0928	0929	0930	0931	0932	0933	0934	0935	0936	0937	0938	0939	0940	0941	0942	0943
3B0	0944	0945	0946	0947	0948	0949	0950	0951	0952	0953	0954	0955	0956	0957	0958	0959
3C0	0960	0961	0962	0963	0964	0965	0966	0967	0968	0969	0970	0971	0972	0973	0974	0975
3D0	0976	0977	0978	0979	0980	0981	0982	0983	0984	0985	0986	0987	0988	0989	0990	0991
3E0	0992	0993	0994	0995	0996	0997	0998	0999	1000	1001	1002	1003	1004	1005	1006	1007
3F0	1008	1009	1010	1011	1012	1013	1014	1015	1016	1017	1018	1019	1020	1021	1022	1023
400	1024	1025	1026	1027	1028	1029	1030	1031	1032	1033	1034	1035	1036	1037	1038	1039

	1040	1041	1042	1043	1044	1045	1046	1047	1048	1049	1050	1051	1052	1053	1054	1055
410	1040	1041	1042	1043	1044	1045	1046	1047	1048	1049	1050	1051	1052	1053	1054	1055
420	1056	1057	1058	1059	1060	1061	1062	1063	1064	1065	1066	1067	1068	1069	1070	1071
430	1072	1073	1074	1075	1076	1077	1078	1079	1080	1081	1082	1083	1084	1085	1086	1087
440	1088	1089	1090	1091	1092	1093	1094	1095	1096	1097	1098	1099	1100	1101	1102	1103
450	1104	1105	1106	1107	1108	1109	1110	1111	1112	1113	1114	1115	1116	1117	1118	1119
460	1120	1121	1122	1123	1124	1125	1126	1127	1128	1129	1130	1131	1132	1133	1134	1135
470	1136	1137	1138	1139	1140	1141	1142	1143	1144	1145	1146	1147	1148	1149	1150	1151
480	1152	1153	1154	1155	1156	1157	1158	1159	1160	1161	1162	1163	1164	1165	1166	1167
490	1168	1169	1170	1171	1172	1173	1174	1175	1176	1177	1178	1179	1180	1181	1182	1183
4A0	1184	1185	1186	1187	1188	1189	1190	1191	1192	1193	1194	1195	1196	1197	1198	1199
4B0	1200	1201	1202	1203	1204	1205	1206	1207	1208	1209	1210	1211	1212	1213	1214	1215
4C0	1216	1217	1218	1219	1220	1221	1222	1223	1224	1225	1226	1227	1228	1229	1230	1231
4D0	1232	1233	1234	1235	1236	1237	1238	1239	1240	1241	1242	1243	1244	1245	1246	1247
4E0	1248	1249	1250	1251	1252	1253	1254	1255	1256	1257	1258	1259	1260	1261	1262	1263
4F0	1264	1265	1266	1267	1268	1269	1270	1271	1272	1273	1274	1275	1276	1277	1278	1279
500	1280	1281	1282	1283	1284	1285	1286	1287	1288	1289	1290	1291	1292	1293	1294	1295
510	1296	1297	1298	1299	1300	1301	1302	1303	1304	1305	1306	1307	1308	1309	1310	1311
520	1312	1313	1314	1315	1316	1317	1318	1319	1320	1321	1322	1323	1324	1325	1326	1327
530	1328	1329	1330	1331	1332	1333	1334	1335	1336	1337	1338	1339	1340	1341	1342	1343
540	1344	1345	1346	1347	1348	1349	1350	1351	1352	1353	1354	1355	1356	1357	1358	1359
550	1360	1361	1362	1363	1364	1365	1366	1367	1368	1369	1370	1371	1372	1373	1374	1375
560	1376	1377	1378	1379	1380	1381	1382	1383	1384	1385	1386	1387	1388	1389	1390	1391
570	1392	1393	1394	1395	1396	1397	1398	1399	1400	1401	1402	1403	1404	1405	1406	1407
580	1408	1409	1410	1411	1412	1413	1414	1415	1416	1417	1418	1419	1420	1421	1422	1423
590	1424	1425	1426	1427	1428	1429	1430	1431	1432	1433	1434	1435	1436	1437	1438	1439
5A0	1440	1441	1442	1443	1444	1445	1446	1447	1448	1449	1450	1451	1452	1453	1454	1455

Table 17-10. Hexadecimal to Decimal Integer Conversions (cont'd).

	0	1	2	3	4	5	6	7	8	9	A	B	C	D	E	F
5B0	1456	1457	1458	1459	1460	1461	1462	1463	1464	1465	1466	1467	1468	1469	1470	1471
5C0	1472	1473	1474	1475	1476	1477	1478	1479	1480	1481	1482	1483	1484	1485	1486	1487
5D0	1488	1489	1490	1491	1492	1493	1494	1495	1496	1497	1498	1499	1500	1501	1502	1503
5E0	1504	1505	1506	1507	1508	1509	1510	1511	1512	1513	1514	1515	1516	1517	1518	1519
5F0	1520	1521	1522	1523	1524	1525	1526	1527	1528	1529	1530	1531	1532	1533	1534	1535
600	1536	1537	1538	1539	1540	1541	1542	1543	1544	1545	1546	1547	1548	1549	1550	1551
610	1552	1553	1554	1555	1556	1557	1558	1559	1560	1561	1562	1563	1564	1565	1566	1567
620	1568	1569	1570	1571	1572	1573	1574	1575	1576	1577	1578	1579	1580	1581	1582	1583
630	1584	1585	1586	1587	1588	1589	1590	1591	1592	1593	1594	1595	1596	1597	1598	1599
640	1600	1601	1602	1603	1604	1605	1606	1607	1608	1609	1610	1611	1612	1613	1614	1615
650	1616	1617	1618	1619	1620	1621	1622	1623	1624	1625	1626	1627	1628	1629	1630	1631
660	1632	1633	1634	1635	1636	1637	1638	1639	1640	1641	1642	1643	1644	1645	1646	1647
670	1648	1649	1650	1651	1652	1653	1654	1655	1656	1657	1658	1659	1660	1661	1662	1663
680	1664	1665	1666	1667	1668	1669	1670	1671	1672	1673	1674	1675	1676	1677	1678	1679
690	1680	1681	1682	1683	1684	1685	1686	1687	1688	1689	1690	1691	1692	1693	1694	1695
6A0	1696	1697	1698	1699	1700	1701	1702	1703	1704	1705	1706	1707	1708	1709	1710	1711
6B0	1712	1713	1714	1715	1716	1717	1718	1719	1720	1721	1722	1723	1724	1725	1726	1727
6C0	1728	1729	1730	1731	1732	1733	1734	1735	1736	1737	1738	1739	1740	1741	1742	1743
6D0	1744	1745	1746	1747	1748	1749	1750	1751	1752	1753	1754	1755	1756	1757	1758	1759
6E0	1760	1761	1762	1763	1764	1765	1766	1767	1768	1769	1770	1771	1772	1773	1774	1775
6F0	1776	1777	1778	1779	1780	1781	1782	1783	1784	1785	1786	1787	1788	1789	1790	1791
700	1792	1793	1794	1795	1796	1797	1798	1799	1800	1801	1802	1803	1804	1805	1806	1807
710	1808	1809	1810	1811	1812	1813	1814	1815	1816	1817	1818	1819	1820	1821	1822	1823

	1824	1825	1826	1827	1828	1829	1830	1831	1832	1833	1834	1835	1836	1837	1838	1839
720	1824	1825	1826	1827	1828	1829	1830	1831	1832	1833	1834	1835	1836	1837	1838	1839
730	1840	1841	1842	1843	1844	1845	1846	1847	1848	1849	1850	1851	1852	1853	1854	1855
740	1856	1857	1858	1859	1860	1861	1862	1863	1864	1865	1866	1867	1868	1869	1870	1871
750	1872	1873	1874	1875	1876	1877	1878	1879	1880	1881	1882	1883	1884	1885	1886	1887
760	1888	1889	1890	1891	1892	1893	1894	1895	1896	1897	1898	1899	1900	1901	1902	1903
770	1904	1905	1906	1907	1908	1909	1910	1911	1912	1913	1914	1915	1916	1917	1918	1919
780	1920	1921	1922	1923	1924	1925	1926	1927	1928	1929	1930	1931	1932	1933	1934	1935
790	1936	1937	1938	1939	1940	1941	1942	1943	1944	1945	1946	1947	1948	1949	1950	1951
7A0	1952	1953	1954	1955	1956	1957	1958	1959	1960	1961	1962	1963	1964	1965	1966	1967
7B0	1968	1969	1970	1971	1972	1973	1974	1975	1976	1977	1978	1979	1980	1981	1982	1983
7C0	1984	1985	1986	1987	1988	1989	1990	1991	1992	1993	1994	1995	1996	1997	1998	1999
7D0	2000	2001	2002	2003	2004	2005	2006	2007	2008	2009	2010	2011	2012	2013	2014	2015
7E0	2016	2017	2018	2019	2020	2021	2022	2023	2024	2025	2026	2027	2028	2029	2030	2031
7F0	2032	2033	2034	2035	2036	2037	2038	2039	2040	2041	2042	2043	2044	2045	2046	2047
800	2048	2049	2050	2051	2052	2053	2054	2055	2056	2057	2058	2059	2060	2061	2062	2063
810	2064	2065	2066	2067	2068	2069	2070	2071	2072	2073	2074	2075	2076	2077	2078	2079
820	2080	2081	2082	2083	2084	2085	2086	2087	2088	2089	2090	2091	2092	2093	2094	2095
830	2096	2097	2098	2099	2100	2101	2102	2103	2104	2105	2106	2107	2108	2109	2110	2111
840	2112	2113	2114	2115	2116	2117	2118	2119	2120	2121	2122	2123	2124	2125	2126	2127
850	2128	2129	2130	2131	2132	2133	2134	2135	2136	2137	2138	2139	2140	2141	2142	2143
860	2144	2145	2146	2147	2148	2149	2150	2151	2152	2153	2154	2155	2156	2157	2158	2159
870	2160	2161	2162	2163	2164	2165	2166	2167	2168	2169	2170	2171	2172	2173	2174	2175
880	2176	2177	2178	2179	2180	2181	2182	2183	2184	2185	2186	2187	2188	2189	2190	2191
890	2192	2193	2194	2195	2196	2197	2198	2199	2200	2201	2202	2203	2204	2205	2206	2207
8A0	2208	2209	2210	2211	2212	2213	2214	2215	2216	2217	2218	2219	2220	2221	2222	2223
8B0	2224	2225	2226	2227	2228	2229	2230	2231	2232	2233	2234	2235	2236	2237	2238	2239

Table 17-10. Hexadecimal to Decimal Integer Conversions (cont'd).

	0	1	2	3	4	5	6	7	8	9	A	B	C	D	E	F
8C0	2240	2241	2242	2243	2244	2245	2246	2247	2248	2249	2250	2251	2252	2253	2254	2255
8D0	2256	2257	2258	2259	2260	2261	2262	2263	2264	2265	2266	2267	2268	2269	2270	2271
8E0	2272	2273	2274	2275	2276	2277	2278	2279	2280	2281	2282	2283	2284	2285	2286	2287
8F0	2288	2289	2290	2291	2292	2293	2294	2295	2296	2297	2298	2299	2300	2301	2302	2303
900	2304	2305	2306	2307	2308	2309	2310	2311	2312	2313	2314	2315	2316	2317	2318	2319
910	2320	2321	2322	2323	2324	2325	2326	2327	2328	2329	2330	2331	2332	2333	2334	2335
920	2336	2337	2338	2339	2340	2341	2342	2343	2344	2345	2346	2347	2348	2349	2350	2351
930	2352	2353	2354	2355	2356	2357	2358	2359	2360	2361	2362	2363	2364	2365	2366	2367
940	2368	2369	2370	2371	2372	2373	2374	2375	2376	2377	2378	2379	2380	2381	2382	2383
950	2384	2385	2386	2387	2388	2389	2390	2391	2392	2393	2394	2395	2396	2397	2398	2399
960	2400	2401	2402	2403	2404	2405	2406	2407	2408	2409	2410	2411	2412	2413	2414	2415
970	2416	2417	2418	2419	2420	2421	2422	2423	2424	2425	2426	2427	2428	2429	2430	2431
980	2432	2433	2434	2435	2436	2437	2438	2439	2440	2441	2442	2443	2444	2445	2446	2447
990	2448	2449	2450	2451	2452	2453	2454	2455	2456	2457	2458	2459	2460	2461	2462	2463
9A0	2464	2465	2466	2467	2468	2469	2470	2471	2472	2473	2474	2475	2476	2477	2478	2479
9B0	2480	2481	2482	2483	2484	2485	2486	2487	2488	2489	2490	2491	2492	2493	2494	2495
9C0	2496	2497	2498	2499	2500	2501	2502	2503	2504	2505	2506	2507	2508	2509	2510	2511
9D0	2512	2513	2514	2515	2516	2517	2518	2519	2520	2521	2522	2523	2524	2525	2526	2527
9E0	2528	2529	2530	2531	2532	2533	2534	2535	2536	2537	2538	2539	2540	2541	2542	2543
9F0	2544	2545	2546	2547	2548	2549	2550	2551	2552	2553	2554	2555	2556	2557	2558	2559
A00	2560	2561	2562	2563	2564	2565	2566	2567	2568	2569	2570	2571	2572	2573	2574	2575
A10	2576	2577	2578	2579	2580	2581	2582	2583	2584	2585	2586	2587	2588	2589	2590	2591
A20	2592	2593	2594	2595	2596	2597	2598	2599	2600	2601	2602	2603	2604	2605	2606	2607

	0	1	2	3	4	5	6	7	8	9	A	B	C	D	E	F
A30	2608	2609	2610	2611	2612	2613	2614	2615	2616	2617	2618	2619	2620	2621	2622	2623
A40	2624	2625	2626	2627	2628	2629	2630	2631	2632	2633	2634	2635	2636	2637	2638	2639
A50	2640	2641	2642	2643	2644	2645	2646	2647	2648	2649	2650	2651	2652	2653	2654	2655
A60	2656	2657	2658	2659	2660	2661	2662	2663	2664	2665	2666	2667	2668	2669	2670	2671
A70	2672	2673	2674	2675	2676	2677	2678	2679	2680	2681	2682	2683	2684	2685	2686	2687
A80	2688	2689	2690	2691	2692	2693	2694	2695	2696	2697	2698	2699	2700	2701	2702	2703
A90	2704	2705	2706	2707	2708	2709	2710	2711	2712	2713	2714	2715	2716	2717	2718	2719
AA0	2720	2721	2722	2723	2724	2725	2726	2727	2728	2729	2730	2731	2732	2733	2734	2735
AB0	2736	2737	2738	2739	2740	2741	2742	2743	2744	2745	2746	2747	2748	2749	2750	2751
AC0	2752	2753	2754	2755	2756	2757	2758	2759	2760	2761	2762	2763	2764	2765	2766	2767
AD0	2768	2769	2770	2771	2772	2773	2774	2775	2776	2777	2778	2779	2780	2781	2782	2783
AE0	2784	2785	2786	2787	2788	2789	2790	2791	2792	2793	2794	2795	2796	2797	2798	2799
AF0	2800	2801	2802	2803	2804	2805	2806	2807	2808	2809	2810	2811	2812	2813	2814	2815
B00	2816	2817	2818	2819	2820	2821	2822	2823	2824	2825	2826	2827	2828	2829	2830	2831
B10	2832	2833	2834	2835	2836	2837	2838	2839	2840	2841	2842	2843	2844	2845	2846	2847
B20	2848	2849	2850	2851	2852	2853	2854	2855	2856	2857	2858	2859	2860	2861	2862	2863
B30	2864	2865	2866	2867	2868	2869	2870	2871	2872	2873	2874	2875	2876	2877	2878	2879
B40	2880	2881	2882	2883	2884	2885	2886	2887	2888	2889	2890	2891	2892	2893	2894	2895
B50	2896	2897	2898	2899	2900	2901	2902	2903	2904	2905	2906	2907	2908	2909	2910	2911
B60	2912	2913	2914	2915	2916	2917	2918	2919	2920	2921	2922	2923	2924	2925	2926	2927
B70	2928	2929	2930	2931	2932	2933	2934	2935	2936	2937	2938	2939	2940	2941	2942	2943
B80	2944	2945	2946	2947	2948	2949	2950	2951	2952	2953	2954	2955	2956	2957	2958	2959
B90	2960	2961	2962	2963	2964	2965	2966	2967	2968	2969	2970	2971	2972	2973	2974	2975
BA0	2976	2977	2978	2979	2980	2981	2982	2983	2984	2985	2986	2987	2988	2989	2990	2991
BB0	2992	2993	2994	2995	2996	2997	2998	2999	3000	3001	3002	3003	3004	3005	3006	3007
BC0	3008	3009	3010	3011	3012	3013	3014	3015	3016	3017	3018	3019	3020	3021	3022	3023

Table 17-10. Hexadecimal to Decimal Integer Conversions (cont'd).

	0	1	2	3	4	5	6	7	8	9	A	B	C	D	E	F
BD0	3024	3025	3026	3027	3028	3029	3030	3031	3032	3033	3034	3035	3036	3037	3038	3039
BE0	3040	3041	3042	3043	3044	3045	3046	3047	3048	3049	3050	3051	3052	3053	3054	3055
BF0	3056	3057	3058	3059	3060	3061	3062	3063	3064	3065	3066	3067	3068	3069	3070	3071
C00	3072	3073	3074	3075	3076	3077	3078	3079	3080	3081	3082	3083	3084	3085	3086	3087
C10	3088	3089	3090	3091	3092	3093	3094	3095	3096	3097	3098	3099	3100	3101	3102	3103
C20	3104	3105	3106	3107	3108	3109	3110	3111	3112	3113	3114	3115	3116	3117	3118	3119
C30	3120	3121	3122	3123	3124	3125	3126	3127	3128	3129	3130	3131	3132	3133	3134	3135
C40	3136	3137	3138	3139	3140	3141	3142	3143	3144	3145	3146	3147	3148	3149	3150	3151
C50	3152	3153	3154	3155	3156	3157	3158	3159	3160	3161	3162	3163	3164	3165	3166	3167
C60	3168	3169	3170	3171	3172	3173	3174	3175	3176	3177	3178	3179	3180	3181	3182	3183
C70	3184	3185	3186	3187	3188	3189	3190	3191	3192	3193	3194	3195	3196	3197	3198	3199
C80	3200	3201	3202	3203	3204	3205	3206	3207	3208	3209	3210	3211	3212	3213	3214	3215
C90	3216	3217	3218	3219	3220	3221	3222	3223	3224	3225	3226	3227	3228	3229	3230	3231
CA0	3232	3233	3234	3235	3236	3237	3238	3239	3240	3241	3242	3243	3244	3245	3246	3247
CB0	3248	3249	3250	3251	3252	3253	3254	3255	3256	3257	3258	3259	3260	3261	3262	3263
CC0	3264	3265	3266	3267	3268	3269	3270	3271	3272	3273	3274	3275	3276	3277	3278	3279
CD0	3280	3281	3282	3283	3284	3285	3286	3287	3288	3289	3290	3291	3292	3293	3294	3295
CE0	3296	3297	3298	3299	3300	3301	3302	3303	3304	3305	3306	3307	3308	3309	3310	3311
CF0	3312	3313	3314	3315	3316	3317	3318	3319	3320	3321	3322	3323	3324	3325	3326	3327
D00	3328	3329	3330	3331	3332	3333	3334	3335	3336	3337	3338	3339	3340	3341	3342	3343
D10	3344	3345	3346	3347	3348	3349	3350	3351	3352	3353	3354	3355	3356	3357	3358	3359
D20	3360	3361	3362	3363	3364	3365	3366	3367	3368	3369	3370	3371	3372	3373	3374	3375
D30	3376	3377	3378	3379	3380	3381	3382	3383	3384	3385	3386	3387	3388	3389	3390	3391

D40	3392	3393	3394	3395	3396	3397	3398	3399	3400	3401	3402	3403	3404	3405	3406	3407
D50	3408	3409	3410	3411	3412	3413	3414	3415	3416	3417	3418	3419	3420	3421	3422	3423
D60	3424	3425	3426	3427	3428	3429	3430	3431	3432	3433	3434	3435	3436	3437	3438	3439
D70	3440	3441	3442	3443	3444	3445	3446	3447	3448	3449	3450	3451	3452	3453	3454	3455
D80	3456	3457	3458	3459	3460	3461	3462	3463	3464	3465	3466	3467	3468	3469	3470	3471
D90	3472	3473	3474	3475	3476	3477	3478	3479	3480	3481	3482	3483	3484	3485	3486	3487
DA0	3488	3489	3490	3491	3492	3493	3494	3495	3496	3497	3498	3499	3500	3501	3502	3503
DB0	3504	3505	3506	3507	3508	3509	3510	3511	3512	3513	3514	3515	3516	3517	3518	3519
DC0	3520	3521	3522	3523	3524	3525	3526	3527	3528	3529	3530	3531	3532	3533	3534	3535
DD0	3536	3537	3538	3539	3540	3541	3542	3543	3544	3545	3546	3547	3548	3549	3550	3551
DE0	3552	3553	3554	3555	3556	3557	3558	3559	3560	3561	3562	3563	3564	3565	3566	3567
DF0	3568	3569	3570	3571	3572	3573	3574	3575	3576	3577	3578	3579	3580	3581	3582	3583
E00	3584	3585	3586	3587	3588	3589	3590	3591	3592	3593	3594	3595	3596	3597	3598	3599
E10	3600	3601	3602	3603	3604	3605	3606	3607	3608	3609	3610	3611	3612	3613	3614	3615
E20	3616	3617	3618	3619	3620	3621	3622	3623	3624	3625	3626	3627	3628	3629	3630	3631
E30	3632	3633	3634	3635	3636	3637	3638	3639	3640	3641	3642	3643	3644	3645	3646	3647
E40	3648	3649	3650	3651	3652	3653	3654	3655	3656	3657	3658	3659	3660	3661	3662	3663
E50	3664	3665	3666	3667	3668	3669	3670	3671	3672	3673	3674	3675	3676	3677	3678	3679
E60	3680	3681	3682	3683	3684	3685	3686	3687	3688	3689	3690	3691	3692	3693	3694	3695
E70	3696	3697	3698	3699	3700	3701	3702	3703	3704	3705	3706	3707	3708	3709	3710	3711
E80	3712	3713	3714	3715	3716	3717	3718	3719	3720	3721	3722	3723	3724	3725	3726	3727
E90	3728	3729	3730	3731	3732	3733	3734	3735	3736	3737	3738	3739	3740	3741	3742	3743
EA0	3744	3745	3746	3747	3748	3749	3750	3751	3752	3753	3754	3755	3756	3757	3758	3759
EB0	3760	3761	3762	3763	3764	3765	3766	3767	3768	3769	3770	3771	3772	3773	3774	3775
EC0	3776	3777	3778	3779	3780	3781	3782	3783	3784	3785	3786	3787	3788	3789	3790	3791
ED0	3792	3793	3794	3795	3796	3797	3798	3799	3800	3801	3802	3803	3804	3805	3806	3807

Table 17-10. Hexadecimal to Decimal Integer Conversions (cont'd).

	0	1	2	3	4	5	6	7	8	9	A	B	C	D	E	F
EE0	3808	3809	3810	3811	3812	3813	3814	3815	3816	3817	3818	3819	3820	3821	3822	3823
EF0	3824	3825	3826	3827	3828	3829	3830	3831	3832	3833	3834	3835	3836	3837	3838	3839
F00	3840	3841	3842	3843	3844	3845	3846	3847	3848	3849	3850	3851	3852	3853	3854	3855
F10	3856	3857	3858	3859	3860	3861	3862	3863	3864	3865	3866	3867	3868	3869	3870	3871
F20	3872	3873	3874	3875	3876	3877	3878	3879	3880	3881	3882	3883	3884	3885	3886	3887
F30	3888	3889	3890	3891	3892	3893	3894	3895	3896	3897	3898	3899	3900	3901	3902	3903
F40	3904	3905	3906	3907	3908	3909	3910	3911	3912	3913	3914	3915	3916	3917	3918	3919
F50	3920	3921	3922	3923	3924	3925	3926	3927	3928	3929	3930	3931	3932	3933	3934	3935
F60	3936	3937	3938	3939	3940	3941	3942	3943	3944	3945	3946	3947	3948	3949	3950	3951
F70	3952	3953	3954	3955	3956	3957	3958	3959	3960	3961	3962	3963	3964	3965	3966	3967
F80	3968	3969	3970	3971	3972	3973	3974	3975	3976	3977	3978	3979	3980	3981	3982	3983
F90	3984	3985	3986	3987	3988	3989	3990	3991	3992	3993	3994	3995	3996	3997	3998	3999
FA0	4000	4001	4002	4003	4004	4005	4006	4007	4008	4009	4010	4011	4012	4013	4014	4015
FB0	4016	4017	4018	4019	4020	4021	4022	4023	4024	4025	4026	4027	4028	4029	4030	4031
FC0	4032	4033	4034	4035	4036	4037	4038	4039	4040	4041	4042	4043	4044	4045	4046	4047
FD0	4048	4049	4050	4051	4052	4053	4054	4055	4056	4057	4058	4059	4060	4061	4062	4063
FE0	4064	4065	4066	4067	4068	4069	4070	4071	4072	4073	4074	4075	4076	4077	4078	4079
FF0	4080	4081	4082	4083	4084	4085	4086	4087	4088	4089	4090	4091	4092	4093	4094	4095

A more detailed hexadecimal to decimal integer conversion arrangement is shown in Table 17-10. The numbers and letters shown horizontally in the row across the top are hexadecimal and so are the numbers in the column at the left. All the other numbers are decimal equivalents.

☐ **Example:**

What is the decimal equivalent of $IC9_{16}$?

In the left-hand column find IC0. This represents the first two digits of the hexadecimal number. Move across to the column having digit 9 as its heading. The number shown is 0457. This is the decimal equivalent. Hence, $IC9_{16} = 457_{10}$.

The subscript 16 indicates the number is hexadecimal; the subscript 10 means the number is decimal. These subscripts are for identification only and are not involved in the conversion process.

BINARY TO HEXADECIMAL

There are several ways of moving from binary to hexadecimal. One method would be to convert binary to decimal form and then go from decimal to hexadecimal, using tables for both conversions. Another method would be to use the listing shown in Table 17-11. This list extends from binary 0 to binary 01100100.

BINARY-CODED DECIMAL TO HEXADECIMAL

Because of its arrangement, binary coded decimal numbers are much easier to handle than straightforward binaries. Table 17-12 lists the conversion of BCD numbers to hexadecimal.

HEXADECIMAL TO DECIMAL FRACTION CONVERSIONS

While the preceding tables indicate whole numbers only, there are hexadecimal fractions just as there are decimal fractions. Table 17-13 lists decimal values and their hexadecimal equivalents.

☐ **Example:**

Find the decimal equivalent of 0.3D5. In column 1, locate .3 in the first column at the left. In the adjacent column to the right, you will see .1875. In the second column, locate the letter D, shown in the table as .0D. The number zero is used just to indicate position. Immediately to the right of .0D is its decimal equivalent, .0507 8125. Finally, determine the decimal value of 5, the rightmost number in hexadecimal 0.3D5. This is in the third column and is

Table 17-11. Binary to Hexadecimal.

Binary	Hexadecimal	Binary	Hexadecimal
00000000	0	00100110	26
00000001	1	00100111	27
00000010	2	00101000	28
00000011	3	00101001	29
00000100	4	00101010	2A
00000101	5	00101011	2B
00000110	6	00101100	2C
00000111	7	00101101	2D
00001000	8	00101110	2E
00001001	9	00101111	2F
00001010	A	00110000	30
00001011	B	00110001	31
00001100	C	00110010	32
00001101	D	00110011	33
00001110	E	00110100	34
00001111	F	00110101	35
00010000	10	00110110	36
00010001	11	00110111	37
00010010	12	00111000	38
00010011	13	00111001	39
00010100	14	00111010	3A
00010101	15	00111011	3B
00010110	16	00111100	3C
00010111	17	00111101	3D
00011000	18	00111110	3E
00011001	19	00111111	3F
00011010	1A	01000000	40
00011011	1B	01000001	41
00011100	1C	01000010	42
00011101	1D	01000011	43
00011110	1E	01000100	44
00011111	1F	01000101	45
00100000	20	01000110	46
00100001	21	01000111	47
00100010	22	01001000	48
00100011	23	01001001	49
00100100	24	01001010	4A
00100101	25	01001011	4B

Table 17-11. Binary to Hexadecimal (cont'd).

Binary	Hexadecimal	Binary	Hexadecimal
01001100	4C	01011001	59
01001101	4D	01011010	5A
01001110	4E	01011011	5B
01001111	4F	01011100	5C
01010000	50	01011101	5D
01010001	51	01011110	5E
01010010	52	01011111	5F
01010011	53	01100000	60
01010100	54	01100001	61
01010101	55	01100010	62
01010110	56	01100011	63
01010111	57	01100100	64
01011000	58		

Table 17-12. BCD to Hexadecimal.

BCD		Hexadecimal	BCD		Hexadecimal
0000	0000	0	0010	0001	15
0000	0001	1	0010	0010	16
0000	0010	2	0010	0011	17
0000	0011	3	0010	0100	18
0000	0100	4	0010	0101	19
0000	0101	5	0010	0110	1A
0000	0110	6	0010	0111	1B
0000	0111	7	0010	1000	1C
0000	1000	8	0010	1001	1D
0000	1001	9	0011	0000	1E
0001	0000	A	0011	0001	1F
0001	0001	B	0011	0010	20
0001	0100	C	0011	0011	21
0001	0011	D	0011	0100	22
0001	0100	E	0011	0101	23
0001	0101	F	0011	0110	24
0001	0110	10	0011	0111	25
0001	0111	11	0011	1000	26
0001	1000	12	0011	1001	27
0001	1001	13	0100	0000	28
0010	0000	14	0100	0001	29

Table 17-12. BCD to Hexadecimal (cont'd).

BCD		Hexadecimal	BCD		Hexadecimal
			0111	0001	47
0100	0010	2A	0111	0010	48
0100	0011	2B	0111	0011	49
0100	0100	2C	0111	0100	4A
0100	0101	2D	0111	0101	4B
0100	0110	2E	0111	0110	4C
0100	0111	2F	0111	0111	4D
0100	1000	30	0111	1000	4E
0100	1001	31	0111	1001	4F
0101	0000	32	1000	0000	50
0101	0001	33	1000	0001	51
0101	0010	34	1000	0010	52
0101	0011	35	1000	0011	53
0101	0100	36	1000	0100	54
0101	0101	37	1000	0101	55
0101	0110	38	1000	0110	56
0101	0111	39	1000	0111	57
0101	1000	3A	1000	1000	58
0101	1001	3B	1000	1001	59
0110	0000	3C	1001	0000	5A
0110	0001	3D	1001	0001	5B
0110	0010	3E	1001	0010	5C
0110	0011	3F	1001	0011	5D
0110	0100	40	1001	0100	5E
0110	0101	41	1001	0101	5F
0110	0110	42	1001	0110	60
0110	0111	43	1001	0111	61
0110	1000	44	1001	1000	62
0110	1001	45	1001	1001	63
0111	0000	46	0001 0000	0000	64

shown as .005. Again, the zeros are used just for positioning. The decimal equivalent is .0012 2070 3125. To get the final decimal equivalent value, we need to add the decimal numbers.

$$
\begin{array}{rl}
0.3 & = .1875 \\
.0D & = .0507\ 8125 \\
\underline{.005} & = \underline{.0012\ 2070\ 3125} \\
.3D5 & = .2395\ 0195\ 3125
\end{array}
$$

Table 17-13. Hexadecimal to Decimal Fraction Conversions.

HEX	0123 Dec
.0	.0000
.1	.0625
.2	.1250
.3	.1875
.4	.2500
.5	.3125
.6	.3750
.7	.4375
.8	.5000
.9	.5625
.A	.6250
.B	.6875
.C	.7500
.D	.8125
.E	.8750
.F	.9375

1

HEX	4567 Decimal
.000	.0000 0000
.001	.0039 0625
.002	.0078 1250
.003	.0117 1875
.004	.0156 2500
.005	.0195 3125
.006	.0234 3750
.007	.0273 4375
.008	.0312 5000
.009	.0351 5625
.00A	.0390 6250
.00B	.0429 6875
.00C	.0468 7500
.00D	.0507 8125
.00E	.0546 8750
.00F	.0585 9375

2

HEX	0123 Decimal
.0000	.0000 0000 0000
.0001	.0002 4414 0625
.0002	.0004 8828 1250
.0003	.0007 3242 1875
.0004	.0009 7656 2500
.0005	.0012 2070 3125
.0006	.0014 6484 3750
.0007	.0017 0898 4375
.0008	.0019 5312 5000
.0009	.0021 9726 5625
.000A	.0024 4140 6250
.000B	.0026 8554 6875
.000C	.0029 2968 7500
.000D	.0031 7382 8125
.000E	.0034 1796 8750
.000F	.0036 6210 9375

3

HEX	4567 Decimal Equivalent
.0000	.0000 0000 0000 0000
.0001	.0000 1525 8789 0625
.0002	.0000 3051 7578 1250
.0003	.0000 4577 6367 1875
.0004	.0000 6103 5156 2500
.0005	.0000 7629 3945 3125
.0006	.0000 9155 2734 3750
.0007	.0001 0681 1523 4375
.0008	.0001 2207 0312 5000
.0009	.0001 3732 9101 5625
.000A	.0001 5258 7890 6250
.000B	.0001 6784 6679 6875
.000C	.0001 8310 5468 7500
.000D	.0001 9836 4257 8125
.000E	.0002 1362 3046 8750
.000F	.0002 2888 1835 9375

4

The binary, decimal, and hexadecimal number systems are just a few of those that can be used. Each of these has a different base, that is, each uses a different amount of symbols. There are only two symbols in binary, 0 and 1, hence it is called a base 2 system. Decimal uses 10 symbols, and so is a base 10 system. Hexadecimal has 16 symbols, including the digits 0 through 9, and letters A through F.

BASE 3 SYSTEM

A base 3 system is one which uses three symbols, 0, 1, and 2. Table 17-14 shows numbers in the base 3 system and their decimal equivalents.

Table 17-14. Base 3 Numbering System.

Base 3	Decimal Equivalent	Base 3	Decimal Equivalent
001	1	100	9
002	2	101	10
010	3	102	11
011	4	110	12
012	5	111	13
020	6	112	14
021	7	120	15
022	8		

In the base 3 system, the digit at the extreme right has the same value as in the decimal system. The center number is multiplied by 3. The number at the extreme left is multiplied by 9. Thus, a trinary number such as 221 is the same as $(2 \times 9) + (2 \times 3) + 1$ or $18 + 6 + 1 = 25$. Hence, 221 in trinary form is equivalent to 25 in decimal form.

BASE 4 SYSTEM

In the quaternary, or base 4, system there are four symbols. These symbols are 0, 1, 2, and 3. Table 17-15 shows quaternary numbers and their decimal equivalents.

In the quaternary system, the rightmost digit has the same value as the same digit in the same position in the decimal system. However, that digit cannot have a value greater than 3. The center digit is equivalent to that digit multiplied by 4. The leftmost digit is equivalent to that digit multiplied by 16. For example, 322 in quaternary is equal to $(3 \times 16) + (2 \times 4) + (2 \times 1) = 48 + 8 + 2 = 58$. Hence, trinary 322 = decimal 58.

Table 17-15. Base 4 Numbering System.

Quaternary	Decimal	Quaternary	Decimal
000	00	031	13
001	01	032	14
002	02	033	15
003	03	100	16
010	04	101	17
011	05	102	18
012	06	103	19
013	07	110	20
020	08	111	21
021	09	112	22
022	10	113	23
023	11	120	24
030	12	121	25

BASE 5 SYSTEM

A base 5 system makes use of five symbols and these are 0, 1, 2, 3, and 4. Table 17-16 shows base five numbers (also called the quinary system) and their decimal equivalents. As in the preceding number systems, the rightmost digit is directly equivalent in decimal. The center digit is to be multiplied by 5 and the left-hand digit by 25. Thus, quinary 214 is equivalent to $(2 \times 25) + (5 \times 1) + (4 \times 1) = 59$. This number, 59, is in decimal form.

Table 17-16. Base 5 Numbering System.

Base 5	Decimal Equivalent	Base 5	Decimal Equivalent
000	000	013	008
001	001	014	009
002	002	020	010
003	003	021	011
004	004	022	012
010	005	023	013
011	006	024	014
012	007	030	015

The most commonly used numbering systems, though, are the binary, octonary (octal), decimal, and hexadecimal. The binary has just two number symbols, 0 and 1, octonary has eight number symbols, 0, 1, 2, 3, 4, 5, 6, and 7, the decimal has ten number symbols, 0, 1, 2, 3, 4, 5, 6, 7, 8, and 9, while the hexadecimal has 16.

Table 17-17. Powers of 8.

8^0	=	1		
8^1	=	8	8^{-1} = 0.125	
8^2	=	64	8^{-2} = 0.0156	
8^3	=	512	8^{-3} = 0.00195	
8^4	=	4096	8^{-4} = 0.000244	
8^5	=	32.768		
8^6	=	262.144		

OCTAL NUMBER SYSTEM

The octal number system uses 8 symbols, 0, 1, 2, 3, 4, 5, 6, and 7. Just as we can have powers of 2, or 10, or 16, so too can we have powers of 8. These are shown in Table 17-17.

□ **Example:**

Convert 254_8 to its decimal equivalent. The subscript, 8, in connection with 254 indicates that the number is in the octal system.

Table 17-18. Decimal to Octal Conversion.

Decimal	Octal	Decimal	Octal	Decimal	Octal	Decimal	Octal
00	00	24	30	49	61	75	113
01	01	25	31	50	62	76	114
02	02	26	32	51	63	77	115
03	03	27	33	52	64	78	116
04	04	28	34	53	65	79	117
04	04	29	35	54	66	80	120
05	05	30	36	55	67	81	121
06	06	31	37	56	70	82	122
07	07	32	40	57	71	83	123
08	10	33	41	58	72	84	124
09	11	34	42	59	73	85	125
10	12	35	43	60	74	86	126
11	13	36	44	61	75	87	127
12	14	37	45	62	76	88	130
13	15	38	46	63	77	89	131
14	16	39	47	64	100	90	132
15	17	40	50	65	101	91	133
16	20	41	51	66	102	92	134
17	21	42	52	67	103	93	135
18	22	43	53	68	104	94	136
19	23	44	54	69	105	95	137
20	24	45	55	70	106	96	140
21	25	46	56	71	107	97	141
22	26	47	57	72	110	98	142
23	27	48	60	73	111	99	143
				74	112	100	144

Table 17-19. Decimal Exponential Values in Octal.

Decimal	Octal			
10^0	000	000	000	000
10^1	000	000	000	012
10^2	000	000	000	144
10^3	000	000	0001	750
10^4	000	000	023	420
10^5	000	000	303	240
10^6	000	003	641	100
10^7	000	046	113	200
10^8	000	575	360	400
10^9	007	346	545	000
10^{10}	112	402	762	000

Table 17-20. Mathematical Constants in the Octal System.

$\pi =$	3.11037552411	$\sqrt{8} =$	2.650117146402
$2\pi =$	6.220773250413	$\sqrt{10} =$	3.123054072667
$1/\pi =$	0.242763015564	$e =$	2.557605213053
$1/2\pi =$	0.121371406672	$e^{-1} =$	0.274265306615
$\sqrt{\pi} =$	1.61337611067	$\sqrt{e} =$	1.51411230704
$\ln \pi =$	1.112064044344	$\log_{10} e =$	0.336267542512
$\log_{10} \pi =$	0.376424666307	$\log_2 e =$	1.34252166245
$\log \pi =$	1.51544163223	$\log_2 10 =$	3.24464741136
$\sqrt{2} =$	1.324047463201	$\ln 10 =$	2.232730673533
$\sqrt{3} =$	1.566636564132	$\ln 2 =$	0.542710277600
$\sqrt{5} =$	2.170673633460	$\gamma =$	0.44742147707
$\sqrt{6} =$	2.346107024023	$\ln \gamma =$	-0.43127233602
$\sqrt{7} =$	2.512477651650	$\log_2 \gamma =$	-0.62573030645

Table 17-21. Octal to Hexadecimal Conversion.

Octal	Hexadecimal	Octal	Hexadecimal
00	0	10	8
01	1	11	9
02	2	12	A
03	3	13	B
04	4	14	C
05	5	15	D
06	6	16	E
07	7	17	F

Table 17-22. Conversion of Octal Fractions to Decimal Fractions.

Octal	Decimal	Octal	Decimal
.000000	000000	.000025	000080
.000001	000003	.000026	000083
.000002	000007	.000027	000087
.000003	000011	.000030	.000091
.000004	.000015	.000031	000095
.000005	000019	.000032	000099
.000006	.000022	.000033	.000102
.000007	.000026	.000034	000106
.000010	.000030	.000035	000110
.000011	.000034	.000036	000114
.000012	.000038	.000037	000118
.000013	.000041	.000040	000122
.000014	.000045	.000041	000125
.000015	.000049	.000042	.000129
.000016	.000053	.000043	000133
.000017	.000057	.000044	000137
.000020	.000061	.000045	000141
.000021	000064	.000046	.000144
.000022	000068	.000047	.000148
.000023	.000072	.000050	000152
.000024	.000076	.000051	000156

Table 17-23. Place Values in Various Number Systems.

Decimal		Duodecimal		Octal		Binary				
tens	ones	twelves	ones	eights	ones	sixteens	eights	fours	twos	ones
	0		0		0					0
	1		1		1					1
	2		2		2				1	0
	3		3		3				1	1
	4		4		4			1	0	0
	5		5		5			1	0	1
	6		6		6			1	1	0
	7		7		7			1	1	1
	8		8	1	0		1	0	0	0
	9		9	1	1		1	0	0	1
1	0		A	1	2		1	0	1	0
1	1		B	1	3		1	0	1	1
1	2	1	0	1	4		1	1	0	0
1	3	1	1	1	5		1	1	0	1
1	4	1	2	1	6		1	1	1	0
1	5	1	3	1	7		1	1	1	1
1	6	1	4	2	0	1	0	0	0	0
1	7	1	5	2	1	1	0	0	0	1
1	8	1	6	2	2	1	0	0	1	0
1	9	1	7	2	3	1	0	0	1	1
.2	0	1	8	2	4	1	0	1	0	0
2	1	1	9	2	5	1	0	1	0	1
2	2	1	A	2	6	1	0	1	1	0
2	3	1	B	2	7	1	0	1	1	1
2	4	2	0	3	0	1	1	0	0	0

$$254_8 = (2 \times 8^2) + (5 \times 8^1) + (4 \times 8^0)$$
$$= (2 \times 64) + (5 \times 8) + (4 \times 1)$$
$$= \quad 128 \quad + \quad 40 \quad + \quad 4$$
$$= 172_{10}$$

Table 17-18 is a listing of decimal and octal equivalents from decimal 0 to decimal 100.

Table 17-19 is a listing of decimal exponential values in octal

Table 17-24. EBCDIC (Extended Binary-Coded Decimal Interchange Code) for Graphic Characters.

Graphic Character	EBCDIC 8-Bit Code Bit Positions 0123 4567	Hex Equiv-alent	Punched-Card Code	Graphic Character	EBCDIC 8-Bit Code Bit Positions 0123 4567	Hex Equiv-alent	Punched-Card Code
blank	0100 0000	40	no punches	u	1010 0100	A4	11-0-4
¢	0100 1010	4A	12-8-2	v	1010 0101	A5	11-0-5
.	0100 1011	4B	12-8-3	w	1010 0110	A6	11-0-6
(	0100 1101	4D	12-8-5	x	1010 0111	A7	11-0-7
+	0100 1110	4E	12-8-6	y	1010 1000	A8	11-0-8
&	0101 0000	50	12	z	1010 1001	A9	11-0-9
!	0101 1010	5A	11-8-2	A	1100 0001	C1	12-1
$	0101 1011	5B	11-8-3	B	1100 0010	C2	12-2
*	0101 1100	5C	11-8-4	C	1100 0011	C3	12-3
)	0101 1101	5D	11-8-5	D	1100 0100	C4	12-4
;	0101 1110	5E	11-8-6	E	1100 0101	C5	12-5
.	0110 0000	60	11	F	1100 0110	C6	12-6
,	0110 1011	6B	0-8-3	G	1100 0111	C7	12-7
%	0110 1100	6C	0-8-4	H	1100 1000	C8	12-8
?	0110 1111	6F	0-8-7	I	1100 1001	C9	12-9
:	0111 1010	7A	8-2	J	1101 0001	D1	11-1
#	0111 1011	7B	8-3	K	1101 0010	D2	11-2
@	0111 1100	7C	8-4	L	1101 0011	D3	11-3
'	0111 1101	7D	8-5	M	1101 0100	D4	11-4
=	0111 1110	7E	8-6	N	1101 0101	D5	11-5
"	0111 1111	7F	8-7	O	1101 0110	D6	11-6
a	1000 0001	81	12-0-1	P	1101 0111	D7	11-7
b	1000 0010	82	12-0-2	Q	1101 1000	D8	11-8
c	1000 0011	83	12-0-3	R	1101 1001	D9	11-9
d	1000 0100	84	12-0-4	S	1110 0010	E2	0-2
e	1000 0101	85	12-0-5	T	1110 0011	E3	0-3
f	1000 0110	86	12-0-6	U	1110 0100	E4	0-4
g	1000 0111	87	12-0-7	V	1110 0101	E5	0-5
h	1000 1000	88	12-0-8	W	1110 0110	E6	0-6
i	1000 1001	89	12-0-9	X	1110 0111	E7	0-7
j	1001 0001	91	12-11-1	Y	1110 1000	E8	0-8
k	1001 0010	92	12-11-2	Z	1110 1001	E9	0-9
l	1001 0011	93	12-11-3	0	1111 0000	F0	0
m	1001 0100	94	12-11-4	1	1111 0001	F1	1
n	1001 0101	95	12-11-5	2	1111 0010	F2	2
o	1001 0110	96	12-11-6	3	1111 0011	F3	3
p	1001 0111	97	12-11-7	4	1111 0100	F4	4
q	1001 1000	98	12-11-8	5	1111 0101	F5	5
r	1001 1001	99	12-11-9	6	1111 0110	F6	6
s	1010 0010	A2	11-0-2	7	1111 0111	F7	7
t	1010 0011	A3	11-0-3	8	1111 1000	F8	8
				9	1111 1001	F9	9

Table 17-25. ASCII Code.

Units

Tens	0	1	2	3	4	5	6	7	8	9
0		SOH	STX	ETX	EOT	ENQ	ACK	BEL	BS	HT
10	LF	VT	FF	CR	SO	SI	DLE	DC1	DC2	DC3
20	DC4	NAK	SYN	ETB	CAN	EM	SUB	ESC	FS	GS
30	RS	US	SP	!	"	#	$	%	&	'
40	(	)	*	+	,	-	.	/	0	1
50	2	3	4	5	6	7	8	9	:	;
60	<	=	>	?	@	A	B	C	D	E
70	F	G	H	I	J	K	L	M	N	O
80	P	Q	R	S	T	U	V	W	X	Y
90	Z	[	\	]	^	_	`	a	b	c
100	d	e	f	g	h	i	j	k	l	m
110	n	o	p	q	r	s	t	u	v	w
120	x	y	z	{	|	}	~			

ranging from 10^0 to 10^{10}. As in the decimal system, various mathematical constants are used in the octal system (Table 17-20). It isn't necessary (although it is sometimes helpful) to go through the decimal system in converting from one system to another. Figure 17-21 is a short listing of octal to hexadecimal conversions. Figure 17-22 consists of the conversion of octal fractions to decimal fractions.

One of the basic differences in the various numbering systems is in the place values. In the decimal system, the rightmost number is in the units column, the next number to its left is in the 10s column, and the one to the left of that in the 100s column. Table 17-23 lists place values for the decimal system, the duodecimal, octal, and binary systems.

AMERICAN STANDARD CODE FOR INFORMATION EXCHANGE

Abbreviated as ASCII, numbers from 0 to 120 are represented by special characters, as indicated in Table 17-25. In this code, the letter A is 65, B is 66, and so on. Also see Table 26-1, U.S. American Standard Code for Information Interchange.

SYSTEM FLOWCHART SYMBOLS

Figure 17-27 illustrates the symbols used in preparing a flowchart.

Table 17-26. Teletype Code.

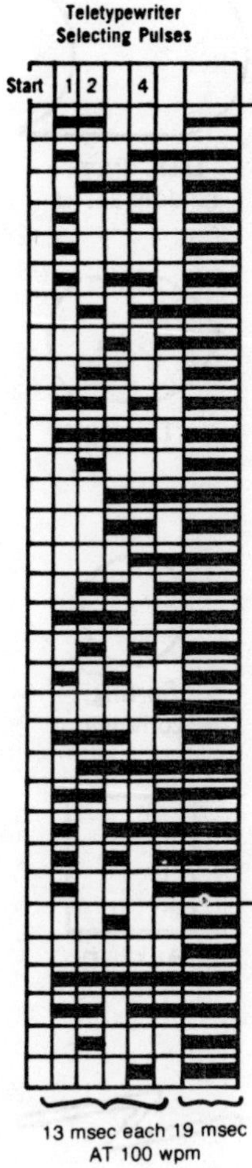
Teletypewriter
Selecting Pulses
Start
1
2
4
13 msec each 19 msec
AT 100 wpm

Letters	Figures	Octal	
A	—	30	
B	?	23	
C	:	16	
D	$	22	
E	3	20	
F	!	26	
G	&	13	
H		05	
I	8	14	
J	'	32	
K	(	36	
L	)	11	
M		07	
N	.	06	
O	9	03	
P	ϕ	15	
Q	1	35	
R	4	12	
S	Bell (🔔•)	24	
T	5	01	
U	7	34	
V	;	17	
W	2	31	
X	/	27	
Y	6	25	
Z	"	21	
Space	■•	04	
Blank	≀ •	00	
Letters	↓•	37	**Machine Functions**
Figures	↑•	33	
Line feed	■•	10†	
Carriage return	<•	02†	

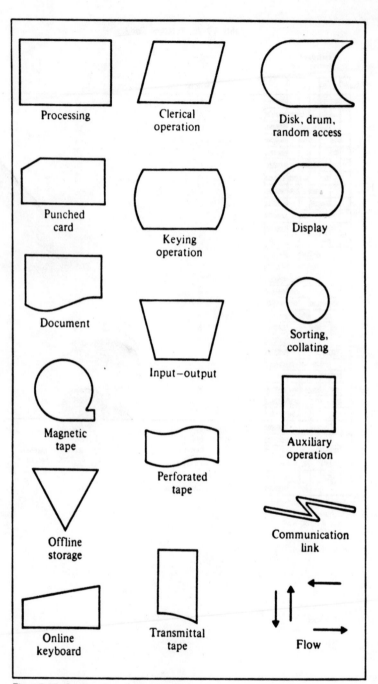

Fig. 17-27. System flowchart symbols.

18

Wire

GAUGE

The cross-sectional area of wire approximately doubles for every three gauge numbers (Table 18-1). At the same time the current carrying capacity is doubled. Every wire size becomes 12.3 percent greater in diameter as the wire gauge number is decreased by one. Wire gauges are based on this computation:

Wire gauge diameter (mils) × 1.123 = diameter of next larger size.

☐ **Example:**

Gauge 22 wire has a diameter of 25.347 mils

25.347 × 1.123 = 28.464681 mils = diameter of No. 21 wire

FUSING CURRENTS

Table 18-2 gives the fusing currents in amperes for five commonly used types of wires. The current I in amperes at which a wire will melt can be calculated from $I = Kd^{3/2}$ where d is the wire diameter in inches and K is a constant that depends on the metal concerned. A wide variety of factors influence the rate of heat loss and these figures must be considered as approximations.

CIRCULAR MIL AREA AND SQUARE MIL AREA

The cross-sectional area of a wire can be stated in square mils

Table 18-1. Diameter in Mils and Area in Circular Mils of Bare Copper Wire.

AWG	Diameter, mils, d	Area, circular mils, d²	Ohms per 1000 ft. at 20°C., or 68°F.
0000	460.00	211,600	0.04901
000	409.64	167,805	0.06180
00	364.80	133,079	0.07793
0	324.86	105,534	0.09827
1	289.30	83,694	0.1239
2	257.63	66,373	0.1563
3	229.42	52,634	0.1970
4	204.31	41,743	0.2485
5	181.94	33,102	0.3133
6	162.02	26,250	0.3951
7	144.28	20,817	0.4982
8	129.49	16,768	0.6282
9	114.43	13,094	0.7921
10	101.89	10,382	0.9989
11	90.742	8,234.1	1.260
12	80.808	6,529.9	1.588
13	71.961	5,178.4	2.003
14	64.084	4,106.8	2.525
15	57.068	3,256.8	3.184
16	50.820	2,582.7	4.016
17	45.257	2,048.2	5.064
18	40.303	1,624.3	6.385
19	35.890	1,288.1	8.051
20	31.961	1,021.5	10.15
21	28.465	810.10	12.80
22	25.347	642.47	16.14
23	22.571	509.45	20.36
24	20.100	404.01	25.67
25	17.900	320.41	32.37
26	15.940	254.08	40.81
27	14.195	201.50	51.47
28	12.641	159.79	64.90
29	11.257	126.72	81.83
30	10.025	100.50	103.2
31	8.928	79.71	130.1
32	7.950	63.20	164.1
33	7.080	50.13	206.9
34	6.305	39.75	260.9
35	5.615	31.53	329.0
36	5.000	25.00	414.8
37	4.453	19.83	523.1
38	3.965	15.72	059.6
39	3.531	12.47	831.8
40	3.145	9.89	1049

or circular mils. The cross-sectional area of a square wire given in square mils is equal to the square of any side, while the cross-sectional area of a circular wire given in circular mils is equal to the square of the diameter.

Table 18-2. Fusing Currents of Wires.

AWG B & S gauge	d in inches	Copper K = 10,244	Aluminum K = 7585	German silver K = 5320	Iron K = 3148	Tin K = 1642
40	0.0031	1.77	1.31	0.90	0.54	0.28
38	0.0039	2.50	1.85	1.27	0.77	0.40
36	0.0050	3.62	2.68	1.85	1.11	0.58
34	0.0063	5.12	3.79	2.61	1.57	0.82
32	0.0079	7.19	5.32	3.67	2.21	1.15
30	0.0100	10.2	7.58	5.23	3.15	1.64
28	0.0126	14.4	10.7	7.39	4.45	2.32
26	0.0159	20.5	15.2	10.5	6.31	3.29
24	0.0201	29.2	21.6	14.9	8.97	4.68
22	0.0253	41.2	30.5	21.0	12.7	6.61
20	0.0319	58.4	43.2	29.8	17.9	9.36
19	0.0359	69.7	51.6	35.5	21.4	11.2
18	0.0403	82.9	61.4	42.3	25.5	13.3
17	0.0452	98.4	72.9	50.2	30.2	15.8
16	0.0508	117	86.8	59.9	36.0	18.8
15	0.0571	140	103	71.4	43.0	22.4
14	0.0641	166	123	84.9	51.1	26.6
13	0.0719	197	146	101	60.7	31.7
12	0.0808	235	174	120	72.3	37.7
11	0.0907	280	207	143	86.0	44.9
10	0.1019	333	247	170	102	53.4
9	0.1144	396	293	202	122	63.5
8	0.1285	472	349	241	145	75.6
7	0.1443	561	416	287	173	90.0
6	0.1620	668	495	341	205	107

A circular wire having a diameter of 1 mil will have an area of 1 circular mil. A square wire having a side of 1 mil will have an area of 1 square mil. However, an area of 1 circular mil is somewhat smaller than an area of 1 square mil, as indicated in Fig. 18-1.

Table 18-3. Circular Mil Area vs. Square Mil Area (cont'd).

Cir. Mils	Sq. Mils	Cir. Mils	Sq. Mils	Cir. Mils	Sq. Mils
1	0.7854	11	8.6394	21	16.4934
2	1.5708	12	9.4248	22	17.2788
3	2.3562	13	10.2102	23	18.0642
4	3.1416	14	10.9956	24	18.8496
5	3.9270	15	11.7810	25	19.6350
6	4.7124	16	12.5664	26	20.4204
7	5.4978	17	13.3518	27	21.2058
8	6.2832	18	14.1372	28	21.9912
9	7.0686	19	14.9226	29	22.7766
10	7.8540	20	15.7080	30	23.5620
31	24.3474	56	43.9824	76	59.6904
32	25.1328	57	44.7678	77	60.4758
33	25.9182	58	45.5532	78	61.2612
34	26.7036	59	46.3386	79	62.0466
35	27.4890	60	47.1240	80	62.8320
36	28.2744	61	47.9094	81	63.6174
37	29.0598	62	48.6948	82	64.4028
38	29.8452	63	49.4802	83	65.1882
39	30.6306	64	50.2656	84	65.9736
40	31.4160	65	51.0510	85	66.7590
41	32.2014	66	51.8364	86	67.5444
42	32.9868	67	52.6218	87	68.3298
43	33.7722	68	53.4072	88	69.1152
44	34.5576	69	54.1926	89	69.9006
45	35.3430	70	54.9780	90	70.6860
46	36.1284	71	55.7634	91	71.4714
47	36.9138	72	56.5488	92	72.2568
48	37.6992	73	57.3342	93	73.0422
49	38.4846	74	58.1196	94	73.8276
50	39.2700	75	58.9050	95	74.6130
51	40.0554			96	75.3984
52	40.8408			97	76.1838
53	41.6262			98	76.9692
54	42.4116			99	77.7546
55	43.1970			100	78.5400

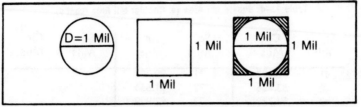

Fig. 18-1. Relationship of circular and square mil area. The shaded area in the drawing indicates the larger cross-sectional area of the square wire.

Table 18-4. Square Mil Area vs. Circular Mil Area.

Square Mils	Circular Mils	Sq. Mils	Cir. Mils	Sq. Mils	Cir. Mils
1	1.273	6	7.638	11	14.003
2	2.546	7	8.911	12	15.276
3	3.819	8	10.184	13	16.549
4	5.092	9	11.457	14	17.822
5	6.366	10	12.730	15	19.095
16	20.368	46	58.558	76	96.748
17	21.641	47	59.831	77	98.021
18	22.914	48	61.104	78	99.294
19	24.187	49	62.377	79	100.567
20	25.460	50	63.650	80	101.840
21	26.733	51	64.923	81	103.113
22	28.006	52	66.196	82	104.386
23	29.279	53	67.469	83	105.659
24	30.552	54	68.472	84	106.932
25	31.825	55	70.015	85	108.205
26	33.098	56	71.288	86	109.478
27	34.371	57	72.561	87	110.751
28	35.644	58	73.834	88	112.024
29	36.917	59	75.107	89	113.297
30	38.190	60	76.380	90	114.570
31	39.463	61	77.653	91	115.843
32	40.736	62	78.926	92	117.116
33	42.009	63	80.199	93	118.389
34	43.282	64	81.472	94	119.662
35	44.555	65	82.745	95	120.935
36	45.828	66	84.018	96	122.208
37	47.101	67	85.291	97	123.481
38	48.374	68	86.564	98	124.754
39	49.647	69	87.837	99	126.027
40	50.920	70	89.110	100	127.300

261

Table 18-4. Square Mil Area vs. Circular Mil Area (cont'd).

Sq. Mils	Cir. Mils	Sq. Mils	Cir. Mils	Sq. Mils	Cir. Mils
41	52.193	71	90.383		
42	53.466	72	91.656		
43	54.739	73	92.929		
44	56.012	74	94.202		
45	57.285	75	95.475		

Square mils = circular mils × 0.7854
Circular mils = square mils/0.7854

Tables 18-3 and 18-4 supply conversion data for circular and square wires.

19

Color Codes

EIA COLOR CODE

Resistors may be identified by using the color code recommended by the Electronics Industries Association (EIA). Each color indicates a number. See Table 19-1.

Table 19-1. The Basic Color Code.

Black	0	Green	5
Brown	1	Blue	6
Red	2	Violet	7
Orange	3	Gray	8
Yellow	4	White	9

These colors are used to indicate the values of resistors and they are also used to represent tolerance. Thus, brown is a tolerance of plus or minus 1 percent, red is plus or minus 2 percent, orange is plus or minus 3 percent, and yellow is plus or minus 4 percent. However, gold is used for plus or minus 5 percent and silver for plus or minus 10 percent. Absence of color means plus or minus 20 percent.

RESISTORS

The value of a resistor in ohms is coded by rings of color placed

Table 19-2. Color Code for Resistors.

Color	First Digit	Second Digit	Multiplier	Tolerance (±)
Black	—	0	1	—
Brown	1	1	10	1%
Red	2	2	100	2%
Orange	3	3	1,000	3%
Yellow	4	4	10,000	4%
Green	5	5	100,000	
Blue	6	6	1,000,000	
Violet	7	7	10,000,000	
Gray	8	8	100,000,000	
White	9	9	1,000,000,000	
Silver				10%
Gold				5%
No color				20%

Table 19-3. Color Code for Power Transformers.

Primary	Secondary Rectifier	Secondary Filament Windings	Color
Primary (no taps)	—	—	Black
Primary (tapped);			
Common	—	—	Black
Tap No. 1	—	—	Black-yellow
Tap No. 2	—	—	Black-orange
Finish	—	—	Black-red
—	Plate	—	Red
	center tap		Red-yellow
—	Filament	—	Yellow
	center tap		Yellow-blue
—	—	Filament No. 1	Green
		Center tap	Green-yellow
		Filament No. 2	Brown
		Center tap	Brown-yellow
—	—	Filament No. 3	Slate
		Center tap	Slate-yellow

at one end. The first color is the first significant figure or digit of the resistance value. The second color is the second significant figure or digit of the resistance value. The third color, known as the decimal multiplier, represents the number of zeros which follow the first two digits. Resistors having a tolerance of plus or minus 20 percent have only three bands of color. A fourth color, if used, is the tolerance. This fourth color is either silver or gold. Table 19-2 lists the resistor color code.

☐ **Example:**

What is the value of a resistor that is color coded yellow, violet, and orange?

The first color, yellow, is 4. The second color, violet, is 7. The final color, orange, is 3 and represents three zeros. The resistance is 47,000 ohms.

DIODES

Crystal diodes might be marked according to the EIA color code (Fig. 19-1) given in Table 19-1. The color bands start from the cathode end and represent the digits following the 1N prefix. For example, the diode in the drawing is a 1N627. (Other manufacturers use other number and letter combinations in place of 1N.) See also Fig. 19-2.

POWER TRANSFORMERS

Table 19-3 supplies the color code for power transformers. A bare wire (if used) represents the connection to the Faraday shield (electrostatic screen) between the primary and secondary windings. This wire should be grounded.

ELECTROLYTIC CAPACITORS

Electrolytic capacitors are marked showing the rated capacitance in microfarads, the rated dc working voltage, and terminal identification. The data are generally printed or stamped directly on the case of the capacitor.

BUTTON MICA CAPACITORS

The color coding for button mica capacitors in Fig. 19-3 follows the coding for resistors supplied in Table 19-4.

FIXED MICA CAPACITORS

These are available in two basic types: those having six color

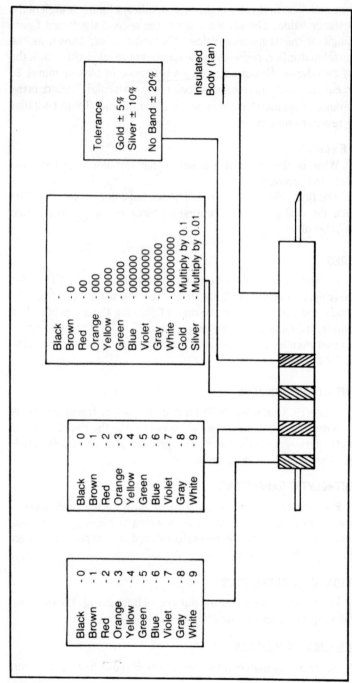

Fig. 19-1. Resistor markings.

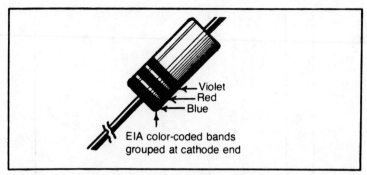

Violet
Red
Blue

EIA color-coded bands
grouped at cathode end

Fig. 19-2. Crystal diodes are marked according to the same basic color code used for resistors.

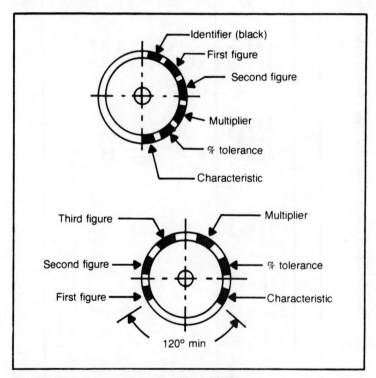

Identifier (black)
First figure
Second figure
Multiplier
% tolerance
Characteristic

Third figure
Second figure
First figure
Multiplier
% tolerance
Characteristic

120° min

Fig. 19-3. Coding system for button mica capacitors.

dots, as shown in Fig. 19-4A and those that have three dots (Fig. 19-4B). There is also a third type that combines the information supplied with types A and B. These are nine dot capacitors, with six dots on the front and three dots on the back. The color coding follows that supplied for resistors, shown in Table 19-4.

Table 19-4. Color Code for Class 1 and Class 2 Ceramic Dielectric Capacitors.

Color	Digit	Multiplier	Class 1						Class 2		
			Capacitance Tolerance		Temperature Coefficient ppm/°C (5-Dot System)	Temperature Coefficient Significant Figure (6-Dot System)	Temperature Coefficient Multiplier (6-Dot System)		Capacitance Tolerance (%)	Temperature Range (°C)	Maximum Capacitance Change Over Temperature Range (%)
			10 pF or less (pF)	Over 10 pF (%)							
Black	0	1	±2.0	+20	0	0.0	−1		+20	−	±2.2
Brown	1	10	±0.1	±1	−33	−	−10		−	+10 to +85	±3.3
Red	2	100	−	±2	−75	1.0	−100		−	−55 to +125	±4.7
Orange	3	1000	−	±3	−150	1.5	−1000		−	+10 to +65	±7.5
Yellow	4	10000	−	−	−220	2.2	−10 000		GMV	−	±10
Green	5	−	±0.5	±5	−330	3.3	+1		±5	−	±15
Blue	6	−	−	−	−470	4.7	+10		−	−	±22
Violet	7	−	−	−	−750	7.5	+100		−	−	+22, −33
Gray	8	0.01	±0.25	−	+150 to −1500	(−1000 to −5200 ppm/°C. With Black Multiplier)	+1 000		+80, −20	−	+22, −56
White	9	0.1	±1.0	±10	+100 to −750	−	+10 000		±10	−30 to +85	+22, −82
Silver	−	−	−	−	−	−	−		−	−55 to +85	±1.5
Gold	−	−	−	−	−	−	−		−	−	±1

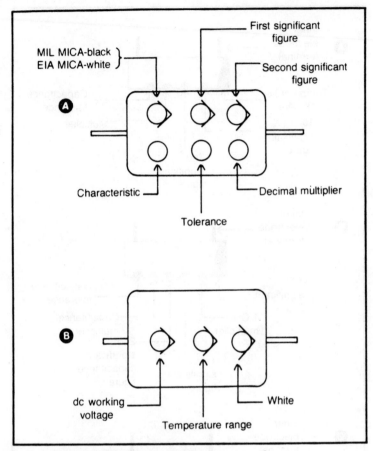

Fig. 19-4. Coding system for mica capacitors.

CERAMIC DIELECTRIC CAPACITORS

These are available as EIA Class 1 and EIA Class 2. Class 1 capacitors are available in five or six dot types. See Fig. 19-5. The color coding is described in Table 19-4.

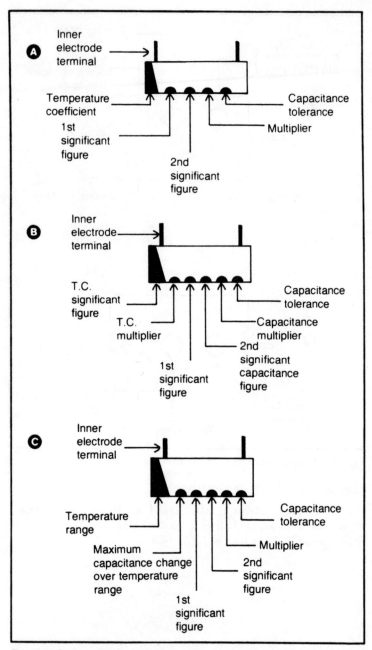

Fig. 19-5. Coding systems for ceramic dielectric capacitors: (A) five-dot system; (B) six-dot system. Both are EIA Class-1. Bottom drawing (C) is for EIA Class-2 ceramic dielectric.

Properties of Matter

HEAT AND ENERGY

Table 20-1 shows the relationship between heat and its equivalent in work or energy.

Table 20-1. Heat vs. Energy.

Amount of heat	Equivalent amount of work or energy
1 calorie	4.186 joules
1 BTU	778 foot-pounds
0.239 calorie	1 joule

METALS

Table 20-2 lists the metallic elements and their physical properties. Metals are good conductors of electricity and heat.

Table 20-2. Metallic Elements.

Element and Symbol		Color	Atomic Weight	Specific Gravity or Density	Specific Heat	Melting-point (°Celsius)	Coefficient of Linear Expansion
Aluminum	Al	Tin-white	27 1	2 67	0 2140	657	0 0000231
Antimony	Sb	Bluish-white	120 2	6 71-6 86	0 0508	630	0 0000105
Arsenic	As	Steel-grey	75 0	5 72	0 081	450	0 0000055
Barium	Ba	Pinkish-grey	137 4	3 8	0 068	850	—

Table 20-2. Metallic Elements (cont'd).

Element and Symbol	Color	Atomic Weight	Specific Gravity or Density	Specific Heat	Melting-point (°Celsius)	Coefficient of Linear Expansion
Beryllium .... Gl	Silver-white	9 1	1 9	0 5820	—	—
Bismuth.......... Bi	Pinkish-white	208 0	9 823	0 0305	268	0 000014
Bromine. .... Br	—	79 6	—	—	—	—
Cadmium .... Cd	Tin-white	112 4	8 546-8 667	0 0548	322	0 000027
Caesium .... Cs	Silver-white	132 8	1 9	0 048	27	—
Calcium .... Ca	Yellow	40 1	1 578	0 1700	800	0 0000269
Cerium .... Ce	Grey	140 2	7 64	0 0448	623	—
Chromium .... Cr	Grey	52 0	6 81-7 3	0 1200	1,700	—
Cobalt .... Co	Greyish-white	59.0	8 5-8.7	0 1070	1,490	0.0000123
Columbium .... Cb (see Niobium)						
Copper .... Cu	Red	63 6	8 92-8 95	0 0952	1,100	0 0000167
Erbium .... E	—	166 0	—	—	—	—
Gadolinium .... Gd	—	156 0	—	—	—	—
Gallium .... Ga	Bluish-white	69 9	5.9	0 079	30	—
Germanium .... Ge	Bluish-white	72 5	5.5	0 074	900	0 0000167
Gold........... Au	Yellow	197 2	19 265	0 0324	1,065	0 0000136
Indium............... In	White	114 8	7 42	0 0570	176	0 0000417
Indium............... Ir	Steel-white	193 1	22 38	0 0326	2,250	0 0000065
Iron .... Fe	Silver-white	55 9	7 84	0 1140	1,550	0 0000116
Lanthanum.................. La	Grey	139 0	6 163	0 0449	826	—
Lead .... Pb	Bluish-white	207 1	11 254-11 38	0 0314	328	0.000027
Lithium Li	Silver-white	7 02	0 589-0 598	0 9410	180	—
Magnesium .... Mg	silver-white	24.3	1 75	0.2500	632	0.0000269
Manganese .... Mn	Reddish-grey	55 0	8 0	0 1220	1,245	—
Mercury .... Hg	Bluish-white	200 0	13.594	0 0319	−40	0.0000610
Molybdenum .... Mo	Silver-white	96.0	8.6	0.0722	2,450	—
Neodymium.................. Nd	—	143.6	7 0	—	840	—
Nickel.................... Ni	—	58.7	8.9	0.1080	1,450	0.0000127
Niobium...................... Nb	Steel-grey	93.5	12.1	0.071	1,950	—
Osmium...................... Os	Bluish-white	190 9	22.5	0 0311	2,500	0.0000065
Palladium.................. Pd	Tin-white	106 7	11.4	0 0593	1,549	0.0000117
Platinum .... Pt	—	195.2	21 5	0 0324	1,780	0.0000089
Potassium.................. K	Silver-white	39 10	0.875	0.1660	60	0.0000841
Praseodymium .... Pr	—	140.5	6.5	—	940	—
Radium.................. Ra	—	225.0	—	—	—	—
Rhodium....................... Rh	Tin-white	102.9	12.1	0 0580	2,000	0.0000085
Rubidium.................. Rb	Silver-white	85.5	1 52	0.077	38 5	—
Ruthenium.................. Ru	—	101.7	12.261	0.0611	2,400	0.0000096
Samarium .... Sm	—	150.3	7.7	—	1,350	—
Scandium.................. Sc	—	44 1	—	—	—	—
Silver .... Ag	White	107 9	10.4-10.57	0 0560	962	0.0000192
Sodium.................. Na	Silver-white	23.0	0.98	0.293	96	0.000071
Strontium.................. Sr	Yellow	87.6	2.5	—	800	—
Tantalum.................. Ta	Black	181 6	16.8	0 0365	2,910	0.0000079
Tellurium.................. Te	—	127 5	6.25	0.049	452	0.0000167
Terbium .... Tb	—	160	—	—	—	—
Thallium .... Tl	Bluish-white	204.0	11 8	0 0335	303	0.0000302
Thorium.................. Th	Grey	232.4	11.2	0.0276	1,690	—
Thulium.................. Tm	—	171	—	—	—	—
Tin.................. Sn	White	119.0	7.293	0.0559	232	0.0000203
Titanium.................. Ti	Dark grey	48.1	3.6	0.13	1,800	—
Tungsten .... W	Light grey	184.0	19 129	0.0334	3,000	—
Uranium.................. U	Greyish-white	238.5	18.33	0.0277	1,500	—
Vanadium.................. V	Whitish-grey	51.1	5.9	0.125	1,680	—
Ytterbium.................. Yb	—	173.0	—	—	—	—
Yttrium.................. Yt	Grey	89.0	3.80	—	—	—
Zinc.................. Zn	Bluish-white	65.4	7 1	0.0935	419	0.0000274
Zirconium.................. Zr	Grey	90.6	4.15	0.0662	1,300	—

DENSITY

Density is a physical description of the mass of a substance per unit volume. Table 20-3 lists the densities of solids and liquids.

Table 20-3. Densities of Solids and Liquids.

Aluminum	2.58 g. per cub. cm.	1.61.1 lb. per cub. ft.
Copper	8.9 g. per cub. cm.	555.4 lb. per cub. ft.
Gold	19.3 g. per cub. cm.	1,205.0 lb. per cub. ft.
Ice	0.9167 g. per cub. cm.	57.2 lb. per cub. ft.
Iron	7.87 g. per cub. cm.	491.3 lb. per cub. ft.
Lead	11.0 g. per cub. cm.	686.7 lb. per cub. ft.
Mercury	13.596 g. per cub. cm.	848.7 lb. per cub. ft.
Nickel	8.80 g. per cub. cm.	549.4 lb. per cub. ft.
Platinum	21.50 g. per cub. cm.	1,342.2 lb. per cub. ft.
Sea Water	1.025 g. per cub. cm.	64.0 lb. per cub. ft.
Silver	10.5 g. per cub. cm.	655.5 lb. per cub. ft.
Tin	7.18 g. per cub. cm.	448. lb. per cub. ft.
Tungsten	18.6 g. per cub. cm.	1,161.2 lb. per cub. ft.
Uranium	18.7 g. per cub. cm.	1,167.4 lb. per cub. ft.
Water	1.000 g. per cub. cm.	62.4 lb. per cub. ft.
Zinc	7.19 g. per cub. cm.	448.6 lb. per cub. ft.

CRYSTALS

A crystal is a body formed by the solidification of a chemical element, a compound, or a mixture. It has a regularly repeating internal arrangement of its atoms. Table 20-4 provides information on crystals.

Table 20-4. Crystal Data.

Crystals and Their Symbols		Crystal Combinations
Bornite	$3Cu_2S_3Fe_2S_3$	Carborundum with Steel
Carborundum	SiC	Iron Pyrites with Silicon
Cassiterite (tinstone)	SnO_2	Galena with Tellurium
Copper pyrites	$Cu_2S_2FeS_2$	
Galena	PbS	Tellurium with Zincite
Graphite	C	Carborundum with Silicon
Hertzite	PbS	Copper Pyrites with
Iron Pyrites	FeS_2	Tellurium
Malachite	$CuCo_2CuH_2O$	
Molybdenite	MoS_2	
Silicon	Si	
Tellurium	Te	
Zincite	ZnO	

THERMOCOUPLES

A thermocouple consists of two dissimilar metals with one at a higher temperature than the other. A voltage is produced under

Table 20-5. Potentials of Thermocouples.

| Substance | Point | Degrees Celsius | Degrees Fahrenheit | EMF in millivolts | |
				Platinum-rhodium couple	Copper-constantan couple
Tin	melting	231.9	449.0	1.706	11.009
Cadmium	melting	320.9	609.6	2.503	16.083
Zinc	melting	419.4	786.9	3.430	
Sulphur	boiling	444.5	920.1	3.672	
Aluminum	melting	658.7	1217.6	5.827	
Silver	melting	960.2	1760.3	9.111	
Copper	melting	1082.8	1981.0	10.534	
Nickel	melting	1452.6	2646.6	14.973	
Platinum	melting	1755	3191	18.608	

these circumstances, with the amount of current generated proportional to the temperature difference. Table 20-5 shows voltages produced by thermocouple materials with the cold junction kept at 0° C. In Table 20-5 the column at the extreme left shows one element of the thermocouple; the other element is either platinum-rhodium or copper-constantan.

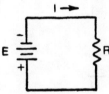

Formulas

DC CIRCUITS

$$I = E/R = P/E = \sqrt{P/R}$$
$$R = E/I = P/I^2 = E^2/P$$
$$E = IR = P/I = \sqrt{PR}$$
$$P = EI = E^2/R = I^2R$$

Ohms's law
R is in ohms; I in
amperes; E in volts and
P in watts

Resistors in Series

$$R_t = R1 + R2 + R3 \ldots$$

Resistance, R, is in
ohms, or identical
multiples

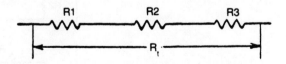

Two Resistors in Parallel

$$R_t = \frac{R1 \times R2}{R1 + R2}$$

$$\frac{1}{R_t} = \frac{1}{R1} + \frac{1}{R2}$$

$$R_t = \frac{R1}{2} \text{ or } \frac{R2}{2} \quad \text{(when both resistors are identical)}$$

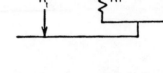

Three or More Resistors in Parallel

$$\frac{1}{R_t} = \frac{1}{R1} + \frac{1}{R2} + \frac{1}{R3} \cdots$$

$$R_t = \frac{1}{\dfrac{1}{R1} + \dfrac{1}{R2} + \dfrac{1}{R3}}$$

$$R_t = \frac{R1}{1 + R1/R2 + R1/R3 + R1/R4}$$

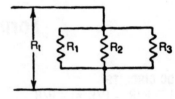

Resistors in Series-Parallel

$$R_t = R1 + \frac{1}{\dfrac{1}{R2} + \dfrac{1}{R3} + \dfrac{1}{R4}}$$

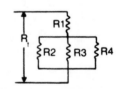

$$R_t = R1 + \frac{R2 \times R3}{R2 + R3} + R4 + R5 + \frac{R6 \times R7}{R6 + R7}$$

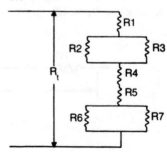

Parallel Voltage Drops

$$E = E_1 = E_2$$

$$I_1 = \frac{I_2 \times R2}{R1}$$

$$R1 = \frac{I_2 \times R2}{I_1}$$

$$I_2 = \frac{I_1 \times R1}{R2}$$

$$R2 = \frac{I_1 \times R1}{I_2}$$

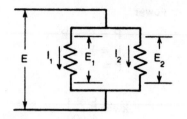

Conductances in Parallel

$G_t = G_1 + G_2 + G_3 \ldots$ Ohm's law for conductance
G in mhos or Siemens
$E = I/G$ E is in volts, I in amperes

$I = E/G$

$$G = \frac{1}{R}$$

G in mhos or Siemens
R in ohms

$$R = \frac{1}{G}$$

Resistance of a Conductor

$$R = K \frac{L}{D_2}$$

R is ohms; K is
resistance of a mil-foot
(specific resistance);
L is length in feet and
D is diameter in mils.

Power

$$P = I^2 \times R$$
$$P = E^2/R$$
$$P = E \times I$$

$$kw = \frac{E \times I}{1,000}$$

Kilowatts = Watts/1,000
Watts = Kilowatts × 1,000
Watt-hours = Watts × Hours
Kilowatt-hours = Kilowatts × Hours

Current

$$I = \sqrt{P/R}$$
$$I = P/E$$
$$I^2 = P/R$$

Voltage

$$E = \sqrt{P \times R}$$
$$E^2 = P \times R$$
$$E = P/I$$

Percentage Voltage Regulation

$$\text{Voltage regulation} = \frac{E_{nl} - E_{fl}}{E_{fl}} \times 100$$

E_{nl} is the no-load voltage; E_{fl} is the full-load voltage.

Resistance

$$R = P/I^2$$
$$R = E^2/P$$

Shunt Law

$$I_1 \times R1 = I_2 \times R2$$

$$I_1 = \frac{I_2 \times R2}{R1}$$

$$I_2 = \frac{I_1 \times R1}{R2}$$

$$R1 = \frac{I_2 \times R2}{I_1}$$

$$R2 = \frac{I_1 \times R1}{I_2}$$

I_1 and I_2 are currents in amperes.

R1 and R2 are resistance in ohms.

Kirchhoff's Current Law

$$I_1 + I_2 = I_3 + I_4 + I_5$$

$$I_1 + I_2 - I_3 - I_4 - I_5 = 0$$

J = junction

The sum of the currents flowing to and from a junction is zero.

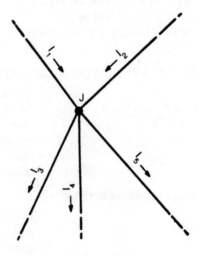

Kirchhoff's Voltage Law

$$E_1 + E_2 + E_3 = 0$$

The sum of the voltage drops in a closed loop is equal to zero.

Time Constant for Series RL

$t = L/R$

t is the time (in seconds) for current to reach 63.2% of peak; L is in henrys and R is in ohms.

Time Constant for Series RC

$t = R \times C$

t is the time (in seconds) for the voltage across the capacitor to reach 63.2% of peak; R is in ohms and C is in farads.

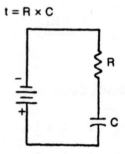

METERS AND MEASURING CIRCUITS

$I_s + R_s = I_m \times R_m$

I_s is shunt current: I_m is meter current: both must be in same current units. R_m is meter resistance. R_s is the shunt resistance. Both must be in same resistance unit.

Meter Multiplier Resistance

$R = R_m (n - 1)$

R is value of multiplier resistance: R_m is meter resistance and n is the multiplication factor.

Meter Sensitivity

$$M_s = \frac{R_m + R1}{E}$$

M_s is meter sensitivity in ohms per volt: R_m is meter resistance and R1 is multiplier resistance. E is full-scale reading in volts.

280

Wheatstone Bridge

$$R_x = \frac{R3 \times R2}{R1}$$

R_x is unknown resistance value:
R1, R2, and R3 are bridge resistance elements. G is a galvanometer.

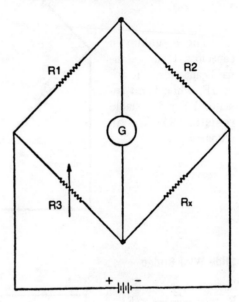

Inductance Bridge

$$L_x = R1 \times R2 \times C1$$

R1 and R2 are in ohms
C1 is in farads
L_x is in henrys

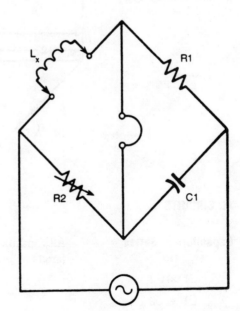

Capacitance Bridge

$$C_x = C1 \times \frac{R1}{R2}$$

The unknown capacitor, C_x and standard capacitor, C1, are in μF, while fixed resistor R1 and variable resistor R2 are in ohms.

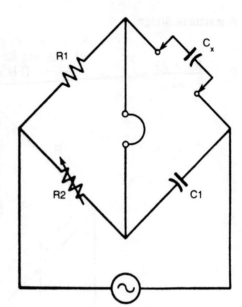

Slide-Wire Bridge

$$R_x = \frac{L2}{L1 - L2} \times R1$$

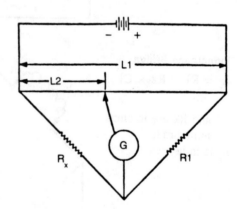

AC CIRCUITS

Capacitors in Series

All capacitance units must be in farads or in identical submultiples.

$$C_t = \frac{C1 \times C2}{C1 + C2}$$

$$\frac{1}{C_t} = \frac{1}{C1} + \frac{1}{C2} + \frac{1}{C3} \cdots$$

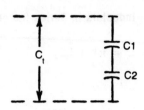

$$C_t = \frac{1}{\dfrac{1}{C1} + \dfrac{1}{C2} + \dfrac{1}{C3}}$$

Capacitors in Parallel

$$C_t = C1 + C2 + C3 \ldots$$

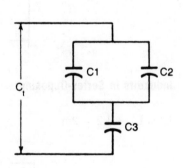

Capacitors in Series-Parallel

$$C_t = \frac{(C1 + C2)\,(C3)}{C1 + C2 + C3}$$

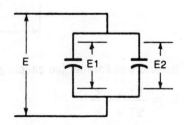

Parallel Capacitor Voltages

$$E = E1 = E2$$

283

Inductors in Series (no coupling)

$$L_t = L1 + L2$$

All inductance units must be in henrys or identical submultiples.

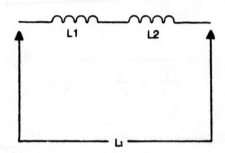

Inductors in Series-Aiding

$$L_t = L1 + L2 + 2m$$

m is mutual inductance, in henrys.

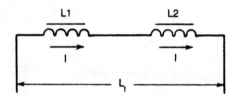

Inductors in Series-Opposing

$$L_t = L1 + L2 - 2m$$

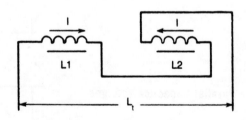

Inductors in Parallel (no coupling)

$$L_t = \frac{L1 \times L2}{L1 + L2}$$

284

Frequency-Wavelength of Sine Wave

$$\lambda = \frac{300,000}{f}$$

λ is wavelength in meters,
f is frequency in kHz.

$$\lambda = \frac{30}{f}$$

λ is wavelength in centimeters,
f is frequency in gigahertz.

$$\lambda = \frac{984,000}{f}$$

λ is wavelength in feet,
f is frequency in kilohertz.

$$\lambda = \frac{984}{f}$$

λ is wavelength in feet,
f is frequency in megahertz.

$$\lambda = \frac{11.8}{f}$$

λ is wavelength in inches,
f is frequency in gigahertz.

$\overset{\circ}{A} = 3.937 \times 10^{-9}$ inch
$\overset{\circ}{A} = 1 \times 10^{-10}$ meter
$\overset{\circ}{A} = 1 \times 10^{-4}$ micron

$\overset{\circ}{A}$ is 1 angstrom unit.

The micron was formerly referred to as a micrometer.

1 micron $= 3.937 \times 10^{-5}$ inch
1 micron $= 1 \times 10^{-6}$ meter
1 micron $= 1 \times 10^{4} \ \overset{\circ}{A}$

Period of a Sine Wave

$t = 1/f$

t is time in seconds: f is frequency in Hz.

Sine Wave Voltages or Currents

Average value of voltage or current sine wave.

$$E_{av} = 0.637 \times E_{peak}$$

Peak value of voltage or current sine wave.

$$E_{peak} = \frac{E_{av}}{0.637}$$

285

Peak value of voltage or current sine wave.

$$E_{peak} = 1.57 \times E_{av}$$

Peak-to-peak value of voltage or current sine wave.

$$E_{p-p} = 2 \times E_{peak}$$

Peak value of voltage or current sine wave.

$$E_{p} = E_{p-p}/2$$

Average value of voltage or current sine wave.

$$E_{av} = E_{p-p} \times 0.3185$$

Average value of voltage or current sine wave.

$$E_{av} = (E_{p-p}/2) \times 0.637$$

Average value of voltage or current sine wave.

$$E_{av} = E_{p-p} \times 0.5 \times 0.637$$

Average value of voltage or current sine wave.

$$E_{av} = \frac{E_{p-p} \times 1}{2} \times 0.637 = E_{p-p} \times \tfrac{1}{2} \times 0.637$$

Peak-to-peak value of voltage or current sine wave.

$$E_{p-p} = \frac{E_{av}}{0.3185}$$

Rms (effective) value of voltage or current sine wave.

$$I_{eff} = 0.707 \times I_{peak}$$

286

Rms value of voltage or current sine wave.

$$I_{eff} = \frac{I_{peak}}{1.414}$$

Peak value of voltage or current sine wave.

$$I_{peak} = \frac{I_{eff}}{0.707}$$

Peak-to-peak value of voltage or current sine wave.

$$E_{p-p} = 2.828 \times E_{eff}$$

Rms value of voltage or current sine wave.

$$E_{eff} = \frac{E_{p-p}}{2.828} = \frac{E_{peak}}{\sqrt{Z}} = \frac{E_{peak}}{1.414} = 0.707\, E_{peak}$$

Susceptance

$$B = \frac{X}{R^2 + X^2}$$

B = susceptance in mhos or siemans
R = resistive ohms
X = reactive ohms

$$B = 1/X$$

B = susceptance in mhos or siemans
X = reactive ohms

Apparent Power in an ac Circuit

$$P = E \times I$$

P = apparent power in watts
E = voltage in volts and I is current in amperes

True Power: $P = EI \cos\phi = EI \times pf$
$pf = P/EI = \cos\phi$ pf = power factor
$\cos\phi$ = true power/apparent power

$$E = I \times Z$$

$$E = \frac{P}{I \times \cos\phi}$$

$$E = \sqrt{\frac{P \times Z}{\cos \phi}}$$

$$Z = \frac{E}{I}$$

$$Z = \frac{E^2 \times \cos \phi}{P}$$

$$Z = \frac{P}{I^2 \times \cos \phi}$$

$$pf = \frac{R}{Z}$$

$$pf = \frac{watts}{volt\text{-}amperes}$$

True Power

$$P = \frac{E^2 \times \cos \phi}{Z}$$

$$P = I^2 \times Z \times \cos \phi$$

$$I = \sqrt{\frac{P}{E \times \cos \phi}}$$

$$I = \sqrt{\frac{P}{Z \times \cos \phi}}$$

$$I = \sqrt{\frac{P}{R}}$$

$$E = \frac{P}{I \times \cos \phi}$$

$$P = I^2 \times R$$

$$P = E \times I \times \cos \phi$$

$$P = \frac{E^2 \times \cos \phi}{Z}$$

$$P = I^2 \times Z \times \cos \phi$$

$$I = \frac{E}{Z}$$

$$I = \frac{P}{E \times \cos \phi}$$

$$I = \sqrt{\frac{P}{Z \times \cos \phi}}$$

$$I = \sqrt{\frac{P}{R}}$$

Average ac power in resistive circuit.

$$P_{av} = I_{av}^2 \times R$$

Peak ac power in resistive circuit.

$$P_{peak} = I_{peak}^2 \times R$$

Average ac power in resistive circuit.

$$P_{av} = \frac{E_{av}^2}{R}$$

Ac Voltage

Ac voltage across a capacitor or coil.

$$E = I \times X$$

Source voltage.

$$E_{source} = \sqrt{E_R^2 + E_x^2}$$

Applied voltage in R-C circuit.

$$E_{source} = I \times \sqrt{R^2 + X_c^2}$$

Applied voltage in R-L circuit.

$$E_{source} = I \times \sqrt{R^2 + X_L^2}$$

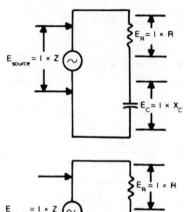

Voltages in series RC circuits.

$E = I \times Z$ = source voltage
$I = E/Z$
$Z = E/I$

Ac voltage across a coil.

$$E_L = I_L \times X_L$$

Current through a coil.

$$I_L = \frac{E_L}{X_L}$$

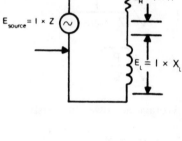

Reactance

Inductive reactance.

$$X_L = 2\pi fL = 6.28\, fL$$

Capacitive reactance.

$$X_c = \frac{1}{2\pi fC}$$

$$= \frac{1}{6.28\, fC}$$

$$X_L = \frac{E_L}{I_L}$$

$\pi = 3.14159$ approximately,
f is frequency in Hz; C is
capacitance in farads
L is inductance in henrys.

Voltages in series RL circuits.

$E = I \times Z$ = source voltage.
$I = E/Z$
$Z = E/I$

Phase angle of series RL
circuit.

$$\theta = arc\ tan\ \frac{X_L}{R}$$

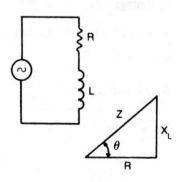

Phase angle of series RC circuit.

$$\theta = \text{arc tan } \frac{X_c}{R}$$

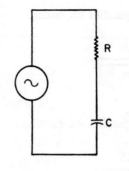
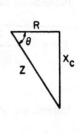

Impedance of R circuit.

$$Z = R$$

Impedance of series RC circuit.

$$Z = \sqrt{R^2 + X_c^2}$$

Impedance of series RL circuit.

$$Z = \sqrt{R^2 + X_L^2}$$

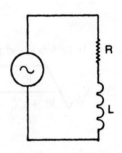

Impedance of series RLC circuit.

$$Z = \sqrt{R^2 + (X_L - X_c)^2} \quad X_L \text{ is larger than } X_c$$
$$Z = \sqrt{R^2 + (X_c - X_L)^2} \quad X_c \text{ is larger than } X_L$$
$$Z = \sqrt{R^2 + X^2}$$

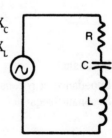

Vector addition of reactances.

$$X = X_L - X_c$$
$$X = X_c - X_L$$

Impedance of parallel RL circuit.

$$Z = \frac{RX_L}{\sqrt{R^2 + X_L^2}}$$

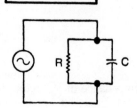

Impedance of parallel RC circuit.

$$Z = \frac{RX_c}{\sqrt{R^2 + X_c^2}}$$

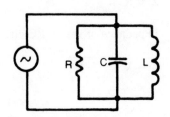

Impedance of parallel RLC circuit.

$$Z = \frac{RX_L X_c}{\sqrt{X_L^2 X_c^2 + R^2 (X_L - X_C)^2}}$$

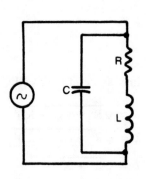

Impedance and phase angle of series RL shunted by C.

$$Z = X_c \sqrt{\frac{R^2 X_L^2}{R^2 + (X_L - X_c)^2}}$$

$$\phi = \text{arc tan} \; \frac{X_L (X_c - X_L) - R^2}{RX_c}$$

Resonance

Impedance at resonance: $X_L = X_c$.
Resonant frequency.

$$f_o = \frac{1}{2\pi \sqrt{LC}}$$

f_o is the frequency in hertz.
L is in henrys; C in farads.

Series resonance.

$$Z \text{ (at any frequency)} = R + J(X_L - X_c)$$
$$Z \text{ (at resonance)} = R$$

Parallel resonance.

$$Z_{max} \text{ (at resonance)} = \frac{X_L X_c}{R} = \frac{X_L^2}{R}$$

$$(X_L = X_c) = QX_L = \frac{L}{RC}$$

TRANSFORMERS

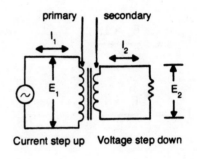

primary   secondary

Current step up   Voltage step down

Transformer primary turns.

$$N_1 = \frac{E_1 \times N_2}{E_2}$$

Transformer secondary turns.

$$N_2 = \frac{E_2 \times N_1}{E_1}$$

Transformer primary current.

$$I_1 = \frac{I_2 \times N_2}{N_1}$$

Transformer secondary current.

$$I_2 = \frac{I_1 \times N_1}{N_2}$$

Transformer primary turns.

$$N_1 = \frac{I_2 \times N_2}{I_1}$$

Transformer secondary turns.

$$N_2 = \frac{I_1 \times N_1}{I_2}$$

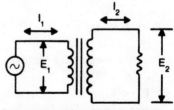

Current step down   Voltage step up

For all transformer turns ratios:
$P = E \times I$

$E_1 \times I_1 = E_2 \times I_2$
(disregarding transformer
losses)

293

Voltage.

$$E_1 = \frac{E_2 \times I_2}{I_1}$$

$$E_2 = \frac{E_1 \times I_1}{I_2}$$

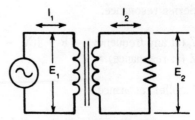

Primary current times primary
voltage = secondary current
times secondary voltage

Current.

$$I_1 = \frac{E_2 \times I_2}{E_1}$$

$$I_2 = \frac{E_1 \times I_1}{E_2}$$

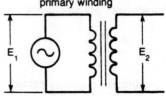

Small magnetizing
current flows in
primary winding

Secondary is open-circuited.
No current flows.
No power transferred from
primary to secondary.
Voltage exists across secondary.

Transformer primary voltage.

$$E_1 = \frac{E_2}{T_r}$$

T_r = turns ratio.

Transformer secondary voltage.

$$E_2 = E_1 \times T_r$$

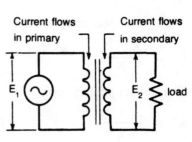

Current flows
in primary

Current flows
in secondary

load

Transformer primary voltage.

$$E_1 = \frac{E_2 \times N_1}{N_2}$$

Transformer secondary voltage.

$$E_2 = \frac{E_1 \times N_2}{N_1}$$

$$E_R = I_2 \times R = E_2$$

Hysteresis loss.

$$\rho = kvf\, B^x_{max}$$

> ρ = power loss, watts
> k = a constant for a given specimen
> v = volume of iron, meters^3
> f = frequency, hertz
> B = flux density, tesla
> x = an index between 1.5 and 2.3, often taken as 2

Eddy current loss.

$$\rho = kt^2 f^2 B^2$$

> ρ = power loss, watts
> k = a constant for a given specimen
> t = thickness of laminations, meters
> f = frequency, hertz
> B = flux density, tesla

Maximum ac magnetizing force.

$$H = \frac{NI\sqrt{2}}{l}$$

> H = magnetizing force, amperes/meter
> N = number of turns in magnetizing coil
> I = rms current, amperes
> l = magnetic path length, meters

Incremental Permeability

$$\mu\Delta = \left(\frac{10^9 1}{8\pi^2 N \Delta^2 fa}\right) \; Z$$

$$Z = \sqrt{R^2 + \omega^2 L^2}$$

$$\omega = 2\pi f = 6.28f$$

$$\Theta = \tan^{-1}\left(\frac{R}{\omega L}\right)$$

Θ = phase angle of complex permeability, degrees
R = resistance, ohms
ω = angular frequency
l = magnetic path length, meters
$N\Delta$ = number of turns
f = frequency, hertz
L = inductance, henrys
a = cross-sectional area, meter^2

Conversion Factors.

One ampere-turn
per meter = $4\pi \times 10^{-3}$ oersted
One weber per meter^2 = 10^4 gauss
One weber/meter^2 = 1 tesla
One watt per
kilogram = 0.454 watt per pound
One watt per pound = 2.2 watts per kilogram

Transformer and magnet formulas.

$$H = \frac{Nl}{L}$$

H = magnetizing force, amperes/meter
N = number of turns in magnetizing coil
I = current, amperes
L = magnetic path length, meters

Permeability.

$$B = \mu H$$

B = flux density, tesla
μ = permeability
H = magnetizing force, amperes/meter

Resistivity.

$$\rho = \frac{Ra}{L}$$

  ρ = resistivity, ohm meters
  R = resistance, ohms
  a = area, meters^2
  L = length, meters

Transformer equation.

$$E = 4.44\ Nf\Phi$$

  E = induced voltage
  N = number of turns
  f = frequency, hertz
  Φ = magnetic flux, webers

Area.
$A = L \times W$

L = length
W = width

Volume.
$V = L \times W \times D$

V = volume in cubic units
A = area in square units

$$\frac{E_P}{E_S} = \frac{N_P}{N_S}$$

$$N_S = N_P\ \frac{E_S}{E_P}$$

$$N_p = \frac{E_p N_s}{E_s}$$

E_P = primary voltage
E_S = secondary voltage

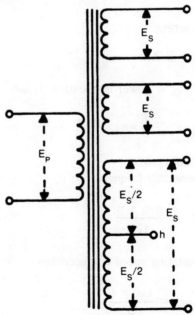

N_P = turns, primary winding
N_S = turns, secondary winding

$$\frac{I_P}{I_S} = \frac{N_S}{N_P}$$

I_P = primary current
I_S = secondary current
voltages and currents must be in basic units or similar multiples

$$T_r = \sqrt{Z_S/Z_P} = \frac{N_S}{N_P}$$

T_r = turns ratio: Z_S and Z_P are secondary and primary impedances in ohms: N_P = primary turns: N_S = secondary turns

Ohm's law for ac.

$$E_C = I \times X_c$$
$$E_L = I \times X_L$$
$$E_R = I \times R$$
$$E_{source} = I \times Z$$

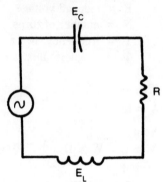

Impedance.
$$Z = E/I$$

Current.

$$I = E/Z$$

Peak ac power in resistive circuit.

$$P_{peak} = \frac{E^2_{peak}}{R}$$

Generators
Generator frequency.

f = frequency in hertz
p = number of poles
rpm = revolutions per minute

$$f = \frac{rpm \times p}{120}$$

Rotating speed of a generator.

$$rpm = \frac{120 \times f}{p}$$

298

Number of poles of generator.

$$p = \frac{120 \times f}{rpm}$$

Star-wound ac generator.

$$E_{line} = 1.732 \times E_{2\text{-phase}}$$

Power in single-phase reactive circuit.

$$P = E \times I \times pf \qquad \qquad pf = \text{power factor}$$

Power in three-phase reactive circuit.

$$P = E \times I \times pf \times 1.732$$

Current in three-phase reactive circuit.

$$I = \frac{P}{E \times pf \times 1.732}$$

Voltage in three-phase reactive circuit.

$$E = \frac{P}{I \times pf \times 1.732}$$

Current (single-phase).

$$I = \frac{746 \times hp}{E \times eff \times pf} \qquad \qquad eff = \text{efficiency}$$

Current (three-phase).

$$I = \frac{746 \times hp}{E \times eff \times pf \times 1.732} \qquad hp = \text{horsepower}$$

Single-phase kva.

$$kva = \frac{E \times I}{1,000} \qquad \qquad kva = \text{kilavolt-amperes}$$

Three-phase kva.

$$kva = \frac{E \times I \times 1.732}{1,000}$$

Two or three-wire distribution system.

$E_{load} = E_{source} \pm$ feeder drops

ILLUMINATION

lumens = candlepower × 12.57
candlepower = lumens/12.57

CROSS-SECTIONAL AREA

Square Mils

Sq. mils = any side^2

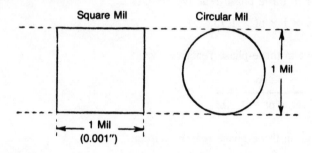

Square Mil   Circular Mil

1 Mil

1 Mil
(0.001")

Circular Mils

$CM = D \times D = D^2$
Circular mil area = Square mils/0.7854
Square Mil Area = Circular Mil Area × 0.7854

Specific Resistance

$R = \rho \dfrac{l}{a}$ at 20° C for drawn copper

Mil Foot

Wire standard consisting of wire whose length is 1 foot and whose diameter is 1 mil.

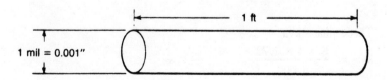

1 ft

1 mil = 0.001"

EFFICIENCY

Output = Input × Efficiency   (efficiency expressed as a decimal)

$$\text{Efficiency} = \frac{\text{Output}}{\text{Input}}$$ Output and Input in Watts

Percentage Efficiency.

$$\% \text{ Efficiency} = \frac{\text{Output}}{\text{Input}} \times 100$$

$$\text{Input} = \frac{\text{Output}}{\text{Efficiency}}$$

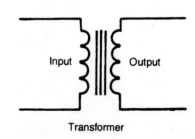

Transformer

Temperature Coefficiency of Resistance

$R_t = R_0(1 + aT)$ R_0 = resistance at 0°C
a = temperature coefficient of resistance of copper wire at 0°C, or 0.00427
T = temperature in degrees C

HEATING EFFECT OF A CURRENT

$H = 0.057168 \times R$ H is heat in calories/sec;
R is resistance in ohms.

Temperature

$F = (C \times 9/5) + 32$ F is temperature in degrees
$C = (F - 32) \times 5/9$ Fahrenheit.
C is temperature in degrees Celsius (formerly Centigrade).

$n = P_o/P_i$ n is efficiency,
P_o is output power,
P_i is input power,
multiply answer by 100 to obtain, efficiency in terms of percentage rather than a decimal.

Dc Voltage Drop

$$e = E \times \frac{R1}{R1 + R2}$$

E is the dc source voltage,
R1 and R2 are series resistors,
e is voltage across R1.

Voltage Divider

$$E_t = I(R1 + R2 + R3)$$
$$E_t = E_1 + E_2 + E_2 \ldots$$

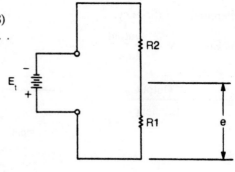

HORSEPOWER

1 hp = 746 watts = 550 ft lbs/sec
1 watt = 1/746 hp = 0.001341 hp   hp is horsepower

$$kW = \frac{hp \times 746}{10^3}$$

$$hp = \frac{No.\ of\ ft\text{-}lbs\ per\ minute}{33,000}$$

$$hp = \frac{E \times I}{746}$$
   E = voltage in volts
   I = current in amperes

Ampere-hours
Ampere-hours = amperes × hours

Current (dc).

$$I = \frac{746 \times hp}{E \times eff}$$ eff = efficiency

Current (ac).

$$I = \frac{746 \times hp}{E \times eff \times pf \times 1.732}$$

302

BANDWIDTH

$$\Delta = \frac{f_o}{Q} = \frac{R}{2\pi L}$$

TUBE CHARACTERISTICS

$$\mu = \frac{\Delta e_p}{\Delta e_\mu} \text{ (}i_p \text{ constant)}$$

μ is amplification factor

$$\mu = g_m \times r_p$$

g_m is grid-plate transconductance. r_p is ac plate resistance.

$$r_p = \frac{\Delta e_p}{\Delta i_p} \text{ (}e_g \text{ constant)}$$

$$gm = \frac{\Delta i_p}{\Delta e_g} \text{ (}e_p \text{ constant)}$$

TRANSISTOR CHARACTERISTICS

$$\alpha = \frac{\Delta i_c}{\Delta i_e}$$

α is current gain: i_c is collector current: i_e is emitter current.

Resistance Gain

$$r_g = \frac{r_o}{r_i}$$

r_g is resistance gain: r_o is output resistance: r_i is input resistance.

Voltage Gain

$$v_g = \frac{e_o}{e_i} = \frac{i_c \, r_o}{i_e \, r_i} = \alpha \, \frac{r_o}{r_i}$$

e_o is output voltage, e_i is input voltage.

Power Gain

$$P_g = \alpha^2 \, \frac{r_o}{r_i}$$

303

Base Current Amplification Factor

$$\beta = \frac{\Delta i_c}{\Delta i_b}$$

i_c is collector current,
β is base current amplification factor.

22

Time Constants

RC TIME CONSTANTS

RC networks are used in a variety of applications, including differentiating and integrating networks; emphasis and de-emphasis circuits; time delay elements in radio control receivers; oscillators; electronic switches and light flashers.

The time required to charge or discharge a capacitor depends on the amount of capacitance and the value of the associated resistor. The formula for the charge or discharge of a capacitor through a resistor is t equals R × C in which the resistance is in ohms and the capacitance is in farads. The time, t, is the time in seconds for the capacitor to charge to 63 percent of its maximum, or for the charge on the capacitor to drop to 37 percent of its maximum. Multiples are shown in Table 22-1.

Table 22-2 supplies the time constants for a capacitance range of 0.1 to 1 μf and from 0.1 to 10 seconds. Note that increasing the value of R or C will increase the time constant proportionately.

Although the formula for RC time constants requires basic units, multiples can be used if conversions are made in accordance with Table 22-1.

□ **Example:**
How long will it take to charge a capacitor having a value of 0.2 μF through a 500 K resistor?

Table 22-1. Multiples of Units for Time Constants.

Where R is in	C is in	L is in	T is in
ohms	farads	henrys	seconds
ohms	μf	μh	μsec
kilohms	μf	henrys	msec
kilohms	—	mh	μsec
megohms	μf	—	seconds
megohms	pf	henrys	μsec

Locate 0.2 in the capacitance column in Table 22-1. Immediately beneath this value locate 500 K. Move to the left and you will see that it will take 0.1 second.

☐ **Example**:

What combination of resistance and capacitance can be used to obtain a one-second time constant?

Locate one second in the *time in seconds* column. If you will now move to the right you will see that a number of possible combinations are available—10 megohms and 0.1 μF; 5 megohms and 0.2 μF; 3.3 megohms and 0.3 μF, etc.

Time constants can be had for values other than those shown in the left-hand column of Table 22-2. Thus if the value of capacitance is reduced by half or if the value of resistance is similarly reduced (but not both at the same time) the value of the time constant is also lowered by half.

☐ **Example**:

What is the time constant for a value of capacitance of .05 μF and a resistance of 1 megohm?

There is no column headed by .05 μF but there is a column marked 0.1. Divide this by 2 and you will have .05 μF. Move down from 0.1 (now considered as .05) until you reach 1.0 megohm, the given value of resistance. Move directly to the left and you will see a time constant of 0.1 second. Divide this by 2 and the time constant for the values of R and C that were given will be .05 second. To get a time constant of .05 second either reduce the resistance 1.0 megohm to 500 K ohms or the capacitance to .05 microfarads. If we reduce both R and C by 50 percent, the value of the time constant becomes one-fourth of the value given in Table 22-2. Thus, for a

Table 22-2. RC Time Constants.

Time (sec)	Capacitance (microfarads)				
	0.1	0.2	0.3	0.4	0.5
0.1	1.0M	500K	333K	250K	200K
0.15	1.5M	750K	500K	375K	300K
0.2	2.0M	1.00M	666K	500K	400K
0.25	2.5M	1.25M	833K	625K	500K
0.3	3.0M	1.50M	1.00M	750K	600K
0.35	3.5M	1.75M	1.17M	875K	700K
0.4	4.0M	2.00M	1.33M	1.00M	800K
0.45	4.5M	2.25M	1.50M	1.13M	900K
0.5	5.0M	2.50M	1.67M	1.25M	1.0M
0.55	5.5M	2.75M	1.83M	1.38M	1.1M
0.6	6.0M	3.00M	2.00M	1.50M	1.2M
0.65	6.5M	3.25M	2.17M	1.63M	1.3M
0.7	7.0M	3.50M	2.33M	1.75M	1.4M
0.75	7.5M	3.75M	2.50M	1.88M	1.5M
0.8	8.0M	4.00M	2.67M	2.00M	1.6M
0.85	8.5M	4.25M	2.83M	2.13M	1.7M
0.9	9.0M	4.50M	3.00M	2.25M	1.8M
0.95	9.5M	4.75M	3.17M	2.38M	1.9M
1.0	10.0M	5.00M	3.33M	2.50M	2.0M
1.5	15.0M	7.50M	5.00M	3.75M	3.0M
2.0	20.0M	10.00M	6.66M	5.00M	4.0M
2.5	25.0M	12.50M	8.33M	6.25M	5.0M
3.0	30.0M	15.00M	10.00M	7.50M	6.0M
3.5	35.0M	17.50M	11.66M	8.75M	7.0M
4.0	40.0M	20.00M	13.33M	10.00M	8.0M
4.5	45.0M	22.50M	15.00M	11.25M	9.0M
5.0	50.0M	25.00M	16.67M	12.50M	10.0M
5.5	55.0M	27.50M	18.33M	13.75M	11.0M
6.0	60.0M	30.00M	20.00M	15.00M	12.0M
6.5	65.0M	32.50M	21.67M	16.25M	13.0M
7.0	70.0M	35.00M	23.33M	17.50M	14.0M
7.5	75.0M	37.50M	25.00M	18.75M	15.0M
8.0	80.0M	40.00M	26.67M	20.00M	16.0M
9.0	90.0M	45.00M	30.00M	22.50M	18.0M
10.0	100.0M	50.00M	33.33M	25.00M	20.0M

K = kilohms   M = megohms

Table 22-2. RC Time Constants (cont'd).

Time (sec)	Capacitance (microfarads)				
	0.6	0.7	0.8	0.9	1.0
0.1	166K	143K	125K	111K	100K
0.15	250K	214K	188K	167K	150K
0.2	333K	286K	250K	222K	200K
0.25	417K	357K	313K	278K	250K
0.3	500K	429K	375K	333K	300K
0.35	583K	500K	438K	389K	350K
0.4	666K	571K	500K	444K	400K
0.45	750K	643K	563K	500K	450K
0.5	833K	714K	625K	555K	500K
0.55	917K	786K	688K	611K	550K
0.6	1.00M	857K	750K	666K	600K
0.65	1.08M	929K	813K	722K	650K
0.7	1.17M	1.00M	875K	778K	700K
0.75	1.25M	1.07M	938K	833K	750K
0.8	1.33M	1.14M	1.00M	889K	800K
0.85	1.42M	1.21M	1.06M	944K	850K
0.9	1.50M	1.29M	1.13M	1.00M	900K
0.95	1.58M	1.36M	1.19M	1.06M	950K
1.0	1.67M	1.43M	1.25M	1.11M	1.0M
1.5	2.50M	2.14M	1.88M	1.67M	1.5M
2.0	3.33M	2.86M	2.50M	2.22M	2.0M
2.5	4.17M	3.57M	3.13M	2.78M	2.5M
3.0	5.00M	4.29M	3.75M	3.33M	3.0M
3.5	5.83M	5.00M	4.38M	3.89M	3.5M
4.0	6.66M	5.71M	5.00M	4.44M	4.0M
4.5	7.50M	6.43M	5.63M	5.00M	4.5M
5.0	8.33M	7.14M	6.25M	5.55M	5.0M
5.5	9.17M	7.86M	6.88M	6.11M	5.5M
6.0	10.00M	8.57M	7.50M	6.66M	6.0M
6.5	10.83M	9.29M	8.13M	7.22M	6.5M
7.0	11.67M	10.00M	8.75M	7.78M	7.0M
7.5	12.50M	10.71M	9.38M	8.33M	7.5M
8.0	13.33M	11.43M	10.00M	8.89M	8.0M
9.0	15.00M	12.86M	11.25M	10.00M	9.0M
10.0	16.66M	14.28M	12.50M	11.11M	10.0M

K = kilohms   M = megohms

value of 500K ohms and a capacitance of .05 μF the time constant is .025 second.

RL TIME CONSTANTS

RL circuits—circuits consisting of a resistor in series with an inductor—are used in timing circuits or in relays where the relay must make or break at predetermined times. The time constant of an RL circuit is based on the formula t equals L/R, where t is the time in seconds, L is the inductance in henrys, and R is the resistance in ohms. The time in seconds is that required for the current to reach 63 percent of its maximum value, or to fall to 37 percent of its maximum. Time constants of RL circuits are in Table 22-3.

The resistance of the coil wire wound must be considered when calculating the time constant. This resistance is regarded as acting in series with the coil. Thus, if the resistance of a coil is 10 ohms and a 100-ohm series resistor is required for an RL circuit, a value close to 90 ohms should be used. As a rule of thumb, if the resistance of the coil is 10 percent or more of the required resistance, it should be taken into consideration. In the case just mentioned, subtracting the 10 ohms of the coil from 100 ohms shows that a 90-ohm resistor would be required. If the value of the external resistor is in the order of kilohms, a coil whose resistance is just a few ohms would not seriously affect the time constant.

☐ **Example:**

A relay coil whose internal resistance is negligible, and whose inductance is 0.05 henry, will make (contacts will close) when the current through the coil reaches 63 percent of its peak. What value of series resistor is needed to have the relay close 0.008 second after the circuit is on?

Note that Table 22-3 does not include a time of 0.008 second nor an inductance of 0.05 henry. However, it is easy to extend the table by moving the decimal point. Change 0.008 to 8 by moving its decimal point three places to the right. The inductance, 0.05 henry, will then become 50. Locate the number 8 in the time column. Move across to the right to reach the 50 column. The answer is 6.3 ohms. The decimal point does not need to be moved in the answer. The reason for this is based on the time-constant formula, T = L/R. Equal increases or decreases in L and R will have no effect on the time constant.

Table 22-3. RL Time Constants.

Time (sec)	Inductance (henrys)				
	10	20	30	40	50
0.1	100.0	200.0	300.0	400.0	500.0
0.15	66.7	133.3	200.0	266.7	333.3
0.2	50.0	100.0	150.0	200.0	250.0
0.25	40.0	80.0	120.0	160.0	200.0
0.3	33.3	66.7	100.0	133.3	166.7
0.35	28.6	57.1	86.6	114.3	142.9
0.4	25.0	50.0	75.0	100.0	125.0
0.45	22.2	44.4	66.7	88.9	111.1
0.5	20.0	40.0	60.0	80.0	100.0
0.55	18.2	36.4	54.5	72.7	90.9
0.6	16.7	33.3	50.0	66.7	83.3
0.65	15.4	30.8	46.2	61.5	76.9
0.7	14.3	28.6	42.9	57.1	71.4
0.75	13.3	26.7	40.0	53.3	66.7
0.8	12.5	25.0	37.5	50.0	62.5
0.85	11.8	23.5	35.3	47.1	58.8
0.9	11.1	22.2	33.3	44.4	55.5
0.95	10.5	21.1	31.6	42.1	52.6
1.0	10.0	20.0	30.0	40.0	50.0
1.5	6.7	13.3	20.0	26.7	33.3
2.0	5.0	10.0	15.0	20.0	25.0
2.5	4.0	8.0	12.0	16.0	20.0
3.0	3.3	6.7	10.0	13.3	16.7
3.5	2.9	5.7	8.7	11.4	14.3
4.0	2.5	5.0	7.5	10.0	12.5
4.5	2.2	4.4	6.7	8.9	11.1
5.0	2.0	4.0	6.0	8.0	10.0
5.5	1.8	3.6	5.5	7.3	9.1
6.0	1.7	3.3	5.0	6.7	8.3
6.5	1.5	3.1	4.6	6.2	7.7
7.0	1.4	2.9	4.3	5.7	7.1
7.5	1.3	2.7	4.0	5.3	6.7
8.0	1.2	2.5	3.8	5.0	6.3
9.0	1.1	2.2	3.3	4.4	5.5
10.0	1.0	2.0	3.0	4.0	5.0

All resistance values in ohms

Table 22-3. RL Time Constants (cont'd).

Time (sec)	Inductance (henrys)				
	60	70	80	90	100
0.1	600.0	700.0	800.0	900.0	1000.0
0.15	400.0	466.7	533.3	600.0	666.7
0.2	300.0	350.0	400.0	450.0	500.0
0.25	240.0	280.0	320.0	360.0	400.0
0.3	200.0	233.3	266.6	300.0	333.3
0.35	171.4	200.0	228.6	257.1	285.7
0.4	150.0	175.0	200.0	225.0	250.0
0.45	133.3	155.6	177.8	200.0	222.2
0.5	120.0	140.0	160.0	180.0	200.0
0.55	109.1	127.3	145.5	163.6	181.8
0.6	100.0	116.7	133.3	150.0	166.7
0.65	92.3	107.7	123.1	138.5	153.8
0.7	85.7	100.0	114.3	128.7	142.9
0.75	80.0	93.3	106.7	120.0	133.3
0.8	75.0	87.5	100.0	112.5	125.0
0.85	70.6	82.3	94.1	105.9	117.6
0.9	66.6	77.8	88.9	100.0	111.1
0.95	63.2	73.7	84.2	94.7	105.3
1.0	60.0	70.0	80.0	90.0	100.0
1.5	40.0	46.7	53.3	60.0	66.7
2.0	30.0	35.0	40.0	45.0	50.0
2.5	24.0	28.0	32.0	36.0	40.0
3.0	20.0	23.3	26.7	30.0	33.3
3.5	17.1	20.0	22.9	25.7	28.6
4.0	15.0	17.5	20.0	22.5	25.0
4.5	13.3	15.6	17.8	20.0	22.2
5.0	12.0	14.0	16.0	18.0	20.0
5.5	10.9	12.7	14.6	16.4	18.2
6.0	10.0	11.7	13.3	15.0	16.7
6.5	9.2	10.8	12.3	13.9	15.4
7.0	8.6	10.0	11.4	12.9	14.3
7.5	8.0	9.3	10.7	12.0	13.3
8.0	7.5	8.8	10.0	11.3	12.5
9.0	6.7	7.8	8.9	10.0	11.1
10.0	6.0	7.0	8.0	9.0	10.0

All resistance values in ohms

□ **Example:**

An RL circuit has an inductance of 30 henrys and a resistance of 42.9 ohms. What is its time constant?

Locate the column marked with the number 30 at the top. Move down in this column until you reach the number 42.9. Then move to the left to the time (sec) column and you will find the time constant of this circuit is 0.7 second.

If you were to do this problem by using the formula, $t = L/R$, you would have $t = 30/42.9 = 0.6993$ second. The difference between the two answers is $0.7 - 0.6993 = 0.0007$ second, an extremely small percentage of error.

Conversions

FREQUENCY-WAVELENGTH CONVERSION (KHZ TO METERS)

The relationship between the frequency of a wave (f) in hertz or cycles per second and its wavelength, in meters, is supplied by the formula λ equals $300,000,000/f$. The number $300,000,000$ in the numerator of the formula is the velocity of light (and of radio waves) in space, and is a constant. λ is the wavelength in meters.

Table 23-1 supplies the abbreviations and descriptions of waves whose frequency extends from 30 Hz to 3,000 GHz.

Table 23-1. Frequency Bands.

Frequency	Designation	Abbreviation
30 Hz to 300 Hz	extremely low frequencies (Audio)	ELF
20 Hz to 20 kHz	sound/frequencies	AF
20 kHz to 30 kHz	very-low frequency	VLF
30 to 300 kHz	low frequency	LF
300 to 3,000 kHz	medium frequency	MF
3,000 to 30,000 kHz	high frequency	HF
30 to 300 MHz	very-high frequency	VHF
300 to 3,000 MHz	ultra-high frequency	UHF
3,000 to 30,000 MHz	super-high frequency	SHF
30,000 to 300,000 MHz (30 GHz to 300 GHz)	extremely-high frequency	EHF
300 GHz to 3000 GHz	no designation	

The velocity of light is in meters per second. Although it is frequently rounded off to 300,000,000, its more probable value is 299,820,000 meters per second, the value used in Table 23-2. When the frequency is in hertz, wavelength in meters equals 299,820,000/f. When the frequency is in kilohertz, wavelength in meters equals 299,820/f. When the frequency is in megahertz, wavelength in meters equals 299.82/f.

Wavelength and frequency have an inverse relationship. As frequency increases, wavelength decreases. Conversely, a decrease in frequency means an increase in wavelength. In terms of formulas: λ equals 299,820,000 divided by f, or f equals 299,820,000 divided by λ. Consequently, the columns indicated in Table 23-2 are interchangeable. Thus, the first line in the first column indicates that a wavelength of 10 meters is equivalent to a frequency of 29,982 kilohertz, or 10 kilohertz corresponds to a wavelength of 29,982 meters.

□ **Example:**

A radio wave has a frequency of 500 kilohertz. What is its wavelength in meters?

Find the number 500 in Table 23-2. Move horizontally and you will see the corresponding wavelength of 599.6 meters.

Table 23-2. Kilohertz (kHz) to Meters (m) or Meters to Kilohertz.

kHz	m	kHz	m	kHz	m	kHz	m
1	299,820	80	3,748	240	1,249	400	749.6
2	149,910	90	3,331	250	1,199	410	731.7
3	99,940	100	2,998	260	1,153	420	714.3
4	74,955	110	2,726	270	1,110	430	697.7
5	59,964	120	2,499	280	1,071	440	681.4
6	49,970	130	2,306	290	1,034	450	666.7
7	42,831	140	2,142	300	999.4	460	652.2
8	37,478	150	1,999	310	967.7	470	638.3
9	33,313	160	1,874	320	937.5	480	624.6
10	29,982	170	1,764	330	908.1	490	612.2
20	14,991	180	1,666	340	882.4	500	599.6
30	9,994	190	1,578	350	859.1	510	588.2
40	7,495	200	1,499	360	833.3	520	576.9
50	5,996	210	1,428	370	810.8	530	565.7
60	4,997	220	1,363	380	789.5	540	555.6
70	4,283	230	1,304	390	769.2	550	545.4

Table 23-2. Kilohertz (kHz) to Meters (m) or Meters to Kilohertz (cont'd).

kHz	m	kHz	m	kHz	m
560	535.7	910	329.5	1,260	238.0
570	526.3	920	325.9	1,270	236.1
580	517.2	930	322.4	1,280	234.2
590	508.5	940	319.0	1,290	232.4
600	499.7	950	315.6	1,300	230.6
610	491.8	960	312.3	1,310	228.9
620	483.7	970	309.1	1,320	227.1
630	476.2	980	305.9	1,330	225.4
640	468.7	990	302.8	1,340	223.7
650	461.5	1,000	299.8	1,350	222.1
660	454.5	1,010	296.9	1,360	220.5
670	447.8	1,020	293.9	1,370	218.8
680	441.2	1,030	291.1	1,380	217.3
690	434.8	1,040	288.3	1,390	215.7
700	428.6	1,050	285.5	1,400	214.2
710	422.5	1,060	282.8	1,410	212.6
720	416.7	1,070	280.2	1,420	211.1
730	410.7	1,080	277.6	1,430	209.7
740	405.4	1,090	275.1	1,440	208.2
750	399.8	1,100	272.6	1,450	206.8
760	394.7	1,110	270.1	1,460	205.4
770	389.6	1,120	267.7	1,470	204.0
780	384.6	1,130	265.3	1,480	202.6
790	379.8	1,140	263.0	1,490	201.2
800	374.8	1,150	260.7	1,500	199.9
810	370.4	1,160	258.5	1,510	198.6
820	365.9	1,170	256.3	1,520	197.2
830	361.4	1,180	254.1	1,530	196.0
840	357.1	1,190	251.9	1,540	194.7
850	352.9	1,200	249.9	1,550	193.4
860	348.8	1,210	247.8	1,560	192.2
870	344.8	1,220	245.8	1,570	191.0
880	340.9	1,230	243.8	1,580	189.8
890	337.1	1,240	241.8	1,590	188.6
900	333.3	1,250	239.9	1,600	187.4

Table 23-2. Kilohertz (kHz) to Meters (m) or Meters to Kilohertz (cont'd).

kHz	m	kHz	m	kHz	m
1,610	186.2	1,910	157.0	2,210	135.7
1,620	185.1	1,920	156.2	2,220	135.1
1,630	183.9	1,930	155.3	2,230	134.4
1,640	182.8	1,940	154.5	2,240	133.8
1,650	181.7	1,950	153.8	2,250	133.3
1,660	180.6	1,960	153.0	2,260	132.7
1,670	179.5	1,970	152.2	2,270	132.1
1,680	178.5	1,980	151.4	2,280	131.5
1,690	177.4	1,990	150.7	2,290	130.9
1,700	176.4	2,000	149.9	2,300	130.4
1,710	175.3	2,010	149.2	2,310	129.8
1,720	174.3	2,020	148.4	2,320	129.2
1,730	173.3	2,030	147.7	2,330	128.7
1,740	172.3	2,040	147.0	2,340	128.1
1,750	171.3	2,050	146.3	2,350	127.6
1,760	170.4	2,060	145.5	2,360	127.0
1,770	169.4	2,070	144.8	2,370	126.5
1,780	168.4	2,080	144.1	2,380	126.0
1,790	167.5	2,090	143.5	2,390	125.4
1,800	166.6	2,100	142.8	2,400	124.9
1,810	165.6	2,110	142.1	2,410	124.4
1,820	164.7	2,120	141.4	2,420	123.9
1,830	163.8	2,130	140.8	2,430	123.4
1,840	162.9	2,140	140.1	2,440	122.9
1,850	162.1	2,150	139.5	2,450	122.4
1,860	161.2	2,160	138.8	2,460	121.9
1,870	160.3	2,170	138.2	2,470	121.4
1,880	159.5	2,180	137.5	2,480	120.9
1,890	158.6	2,190	136.9	2,490	120.4
1,900	157.8	2,200	136.3	2,500	119.9

Table 23-2. Kilohertz (kHz) to Meters (m) or Meters to Kilohertz (cont'd).

kHz	m	kHz	m	kHz	m
2,510	119.5	2,810	106.7	3,110	96.41
2,520	119.0	2,820	106.3	3,120	96.10
2,530	118.5	2,830	105.9	3,130	95.79
2,540	118.0	2,840	105.6	3,140	95.48
2,550	117.6	2,850	105.2	3,150	95.18
2,560	117.1	2,860	104.8	3,160	94.88
2,570	116.7	2,870	104.5	3,170	94.58
2,580	116.2	2,880	104.1	3,180	94.28
2,590	115.8	2,890	103.7	3,190	93.99
2,600	115.3	2,900	103.4	3,200	93.69
2,610	114.9	2,910	103.0	3,210	93.40
2,620	114.4	2,920	102.7	3,220	93.11
2,630	114.0	2,930	102.3	3,230	92.82
2,640	113.6	2,940	102.0	3,240	92.54
2,650	113.1	2,950	101.6	3,250	92.25
2,660	112.7	2,960	101.3	3,260	91.97
2,670	112.3	2,970	100.9	3,270	91.69
2,680	111.9	2,980	100.6	3,280	91.41
2,690	111.5	2,990	100.3	3,290	91.13
2,700	111.0	3,000	99.94	3,300	90.86
2,710	110.6	3,010	99.61	3,310	90.58
2,720	110.2	3,020	99.28	3,320	90.31
2,730	109.8	3,030	98.95	3,330	90.04
2,740	109.4	3,040	98.63	3,340	89.77
2,750	109.0	3,050	98.30	3,350	89.50
2,760	108.6	3,060	97.98	3,360	89.23
2,770	108.2	3,070	97.66	3,370	88.97
2,780	107.8	3,080	97.34	3,380	88.70
2,790	107.5	3,090	97.03	3,390	88.44
2.800	107.1	3,100	96.72	3,400	88.18

Table 23-2. Kilohertz (kHz) to Meters (m) or Meters to Kilohertz (cont'd).

kHz	m	kHz	m	kHz	m
3,410	87.92	3,710	80.81	4,010	74.77
3,420	87.67	3,720	80.60	4,020	74.58
3,430	87.41	3,730	80.38	4,030	74.40
3,440	87.16	3,740	80.17	4,040	74.21
3,450	86.90	3,750	79.95	4,050	74.03
3,460	86.65	3,760	79.74	4,060	73.85
3,470	86.40	3,770	79.53	4,070	73.67
3,480	86.16	3,780	79.32	4,080	73.49
3,490	85.91	3,790	79.11	4,090	73.31
3,500	85.66	3,800	78.90	4,100	73.13
3,510	85.42	3,810	78.69	4,110	72.95
3,520	85.18	3,820	78.49	4,120	72.77
3,530	84.93	3,830	78.28	4,130	72.60
3,540	84.69	3,840	78.08	4,140	72.42
3,550	84.46	3,850	77.88	4,150	72.25
3,560	84.22	3,860	77.67	4,160	72.07
3,570	83.98	3,870	77.47	4,170	71.90
3,580	83.75	3,880	77.27	4,180	71.73
3,590	83.52	3,890	77.07	4,190	71.56
3,600	83.28	3,900	76.88	4,200	71.39
3,610	83.05	3,910	76.68	4,210	71.22
3,620	82.82	3,920	76.48	4,220	71.05
3,630	82.60	3,930	76.29	4,230	70.88
3,640	82.37	3,940	76.10	4,240	70.71
3,650	82.14	3,950	75.90	4,250	70.55
3,660	81.92	3,960	75.51	4,260	70.38
3,670	81.69	3,970	75.52	4,270	70.22
3,680	81.47	3,980	75.33	4,280	70.05
3,690	81.25	3,990	75.14	4,290	69.89
3,700	81.03	4,000	74.96	4,300	69.73

Table 23-2. Kilohertz (kHz) to Meters (m) or Meters to Kilohertz (cont'd).

kHz	m	kHz	m	kHz	m
4,310	69.56	4,610	65.04	4,910	61.06
4,320	69.40	4,620	64.90	4,920	60.94
4,330	69.24	4,630	64.76	4,930	60.82
4,340	69.08	4,640	64.62	4,940	60.69
4,350	68.92	4,650	64.48	4,950	60.57
4,360	68.77	4,660	64.34	4,960	60.45
4,370	68.61	4,670	64.20	4,970	60.33
4,380	68.45	4,680	64.06	4,980	60.20
4,390	68.30	4,690	63.93	4,990	60.08
4,400	68.14	4,700	63.79	5,000	59.96
4,410	67.99	4,710	63.66	5,010	59.84
4,420	67.83	4,720	63.52	5,020	59.73
4,430	67.68	4,730	63.39	5,030	59.61
4,440	67.53	4,740	63.25	5,040	59.49
4,450	67.38	4,750	63.12	5,050	59.37
4,460	67.22	4,760	62.99	5,060	59.25
4,470	67.07	4,770	62.86	5,070	59.14
4,480	66.92	4,780	62.72	5,080	59.02
4,490	66.78	4,790	62.59	5,090	58.90
4,500	66.63	4,800	62.46	5,100	58.79
4,510	66.48	4,810	62.33	5,110	58.67
4,520	66.33	4,820	62.20	5,120	58.56
4,530	66.19	4,830	62.07	5,130	58.44
4,540	66.04	4,840	61.95	5,140	58.33
4,550	65.89	4,850	61.82	5,150	58.22
4,560	65.75	4,860	61.69	5,160	58.10
4,570	65.61	4,870	61.56	5,170	57.99
4,580	65.46	4,880	61.44	5,180	57.88
4,590	65.32	4,890	61.31	5,190	57.77
4,600	65.18	4,900	61.19	5,200	57.66

Table 23-2. Kilohertz (kHz) to Meters (m) or Meters to Kilohertz (cont'd).

kHz	m	kHz	m	kHz	m
5,210	57.55	5,510	54.41	5,810	51.60
5,220	57.44	5,520	54.32	5,820	52.52
5,230	57.33	5,530	54.22	5,830	51.43
5,240	57.22	5,540	54.12	5,840	51.34
5,250	57.11	5,550	54.02	5,850	51.25
5,260	57.00	5,560	53.92	5,860	51.16
5,270	56.89	5,570	53.83	5,870	51.08
5,280	56.78	5,580	53.73	5,880	50.99
5,290	56.68	5,590	53.64	5,890	50.90
5,300	56.57	5,600	53.54	5,900	50.82
5,310	56.46	5,610	53.44	5,910	50.73
5,320	56.36	5,620	53.35	5,920	50.65
5,330	56.25	5,630	53.25	5,930	50.56
5,340	56.15	5,640	53.16	5,940	50.47
5,350	56.04	5,650	53.07	5,950	50.39
5,360	55.94	5,660	52.97	5,960	50.31
5,370	55.83	5,670	52.88	5,970	50.22
5,380	55.73	5,680	52.79	5,980	50.14
5,390	55.63	5,690	52.69	5,990	50.05
5,400	55.52	5,700	52.60	6,000	49.97
5,410	55.42	5,710	52.51	6,010	49.89
5,420	55.32	5,720	52.42	6,020	49.80
5,430	55.22	5,730	52.32	6,030	49.72
5,440	55.11	5,740	52.23	6,040	49.64
5,450	55.01	5,750	52.14	6,050	49.56
5,460	54.91	5,760	52.05	6,060	49.48
5,470	54.81	5,770	51.96	6,070	49.39
5,480	54.71	5,780	51.87	6,080	49.31
5,490	54.61	5,790	51.78	6,090	49.23
5,500	54.51	5,800	51.69	6,100	49.15

Table 23-2. Kilohertz (kHz) to Meters (m) or Meters to Kilohertz (cont'd).

kHz	m	kHz	m	kHz	m
6,110	49.07	6,410	46.77	6,710	44.68
6,120	48.99	6,420	46.70	6,720	44.62
6,130	48.91	6,430	46.63	6,730	44.55
6,140	48.83	6,440	46.56	6,740	44.48
6,150	48.75	6,450	46.48	6,750	44.42
6,160	48.67	6,460	46.41	6,760	44.35
6,170	48.59	6,470	46.34	6,770	44.29
6,180	48.51	6,480	46.27	6,780	44.22
6,190	48.44	6,490	46.20	6,790	44.16
6,200	48.36	6,500	46.13	6,800	44.09
6,210	48.28	6,510	46.06	6,810	44.03
6,220	48.20	6,520	45.98	6,820	43.96
6,230	48.13	6,530	45.91	6,830	43.90
6,240	48.05	6,540	45.84	6,840	43.83
6,250	47.97	6,550	45.77	6,850	43.77
6,260	47.89	6,560	45.70	6,860	43.71
6,270	47.82	6,570	45.63	6,870	43.64
6,280	47.74	6,580	45.57	6,880	43.58
6,290	47.67	6,590	45.50	6,890	43.52
6,300	47.59	6,600	45.43	6,900	43.45
6,310	47.52	6,610	45.36	6,910	43.39
6,320	47.44	6,620	45.29	6,920	43.33
6,330	47.36	6,630	45.22	6,930	43.26
6,340	47.29	6,640	45.15	6,940	43.20
6,350	47.22	6,650	45.09	6,950	43.14
6,360	47.14	6,660	45.02	6,960	43.08
6,370	47.07	6,670	44.95	6,970	43.02
6,380	46.99	6,680	44.88	6,980	42.95
6,390	46.92	6,690	44.82	6,990	42.89
6,400	46.85	6,700	44.75	7,000	42.83

Table 23-2. Kilohertz (kHz) to Meters (m) or Meters to Kilohertz (cont'd).

kHz	m	kHz	m	kHz	m
7,010	42.77	7,310	41.02	7,610	39.40
7,020	42.71	7,320	40.96	7,620	39.35
7,030	42.65	7,330	40.90	7,630	39.29
7,040	42.59	7,340	40.85	7,640	39.24
7,050	42.53	7,350	40.79	7,650	39.19
7,060	42.47	7,360	40.74	7,660	39.14
7,070	42.41	7,370	40.68	7,670	39.09
7,080	42.35	7,380	40.63	7,680	39.04
7,090	42.29	7,390	40.57	7,690	38.99
7,100	42.23	7,400	40.52	7,700	38.94
7,110	42.17	7,410	40.46	7,710	38.89
7,120	42.11	7,420	40.41	7,720	38.84
7,130	42.05	7,430	40.35	7,730	38.79
7,140	41.99	7,440	40.30	7,740	38.74
7,150	41.93	7,450	40.24	7,750	38.69
7,160	41.87	7,460	40.10	7,760	38.64
7,170	41.82	7,470	40.14	7,770	38.59
7,180	41.76	7,480	40.08	7,780	38.54
7,190	41.70	7,490	40.03	7,790	38.49
7,200	41.64	7,500	39.98	7,800	38.44
7,210	41.58	7,510	39.92	7,810	38.39
7,220	41.53	7,520	39.87	7,820	38.34
7,230	41.47	7,530	39.82	7,830	38.29
7,240	41.41	7,540	39.76	7,840	38.24
7,250	41.35	7,550	39.71	7,850	38.19
7,260	41.30	7,560	39.66	7,860	38.15
7,270	41.24	7,570	39.61	7,870	38.10
7,280	41.18	7,580	39.55	7,880	38.05
7,290	41.13	7,590	39.50	7,890	38.00
7,300	41.07	7,600	39.45	7,900	37.95

Table 23-2. Kilohertz (kHz) to Meters (m) or Meters to Kilohertz (cont'd).

kHz	m	kHz	m	kHz	m
7,910	37.90	8,210	36.52	8,510	35.23
7,920	37.86	8,220	36.47	8,520	35.19
7,930	37.81	8,230	36.43	8,530	35.15
7,940	37.76	8,240	36.39	8,540	35.11
7,950	37.71	8,250	36.34	8,550	35.07
7,960	37.67	8,260	36.30	8,560	35.03
7,970	37.62	8,270	36.25	8,570	34.98
7,980	37.57	8,280	36.21	8,580	34.94
7,990	37.52	8,290	36.17	8,590	34.90
8,000	37.48	8,300	36.12	8,600	34.86
8,010	37.43	8,310	36.08	8,610	34.82
8,020	37.38	8,320	36.04	8,620	34.78
8,030	37.34	8,330	35.99	8,630	34.74
8,040	37.29	8,340	35.95	8,640	34.70
8,050	37.24	8,350	35.91	8,650	34.66
8,060	37.20	8,360	35.86	8,660	34.62
8,070	37.15	8,370	35.82	8,670	34.58
8,080	37.11	8,380	35.78	8,680	34.54
8,090	37.06	8,390	35.74	8,690	34.50
8,100	37.01	8,400	35.69	8,700	34.46
8,110	36.97	8,410	35.65	8,710	34.42
8,120	36.92	8,420	35.61	8,720	34.38
8,130	36.88	8,430	35.57	8,730	34.34
8,140	36.83	8,440	35.52	8,740	34.30
8,150	36.79	8,450	35.48	8,750	34.27
8,160	36.74	8,460	35.44	8,760	34.23
8,170	36.70	8,470	35.40	8,770	34.19
8,180	36.65	8,480	35.36	8,780	34.15
8,190	36.61	8,490	35.31	8,790	34.11
8,200	36.56	8,500	35.27	8,800	34.07

Table 23-2. Kilohertz (kHz) to Meters (m) or Meters to Kilohertz (cont'd).

kHz	m	kHz	m	kHz	m
8,810	34.03	9,110	32.91	9,410	31.86
8,820	33.99	9,120	32.88	9,420	31.83
8,830	33.95	9,130	32.84	9,430	31.79
8,840	33.92	9,140	32.80	9,440	31.76
8,850	33.88	9,150	32.77	9,450	31.73
8,860	33.84	9,160	32.73	9,460	31.69
8,870	33.80	9,170	32.70	9,470	31.66
8,880	33.76	9,180	32.66	9,480	31.63
8,890	33.73	9,190	32.62	9,490	31.59
8,900	33.69	9,200	32.59	9,500	31.56
8,910	33.65	9,210	32.55	9,510	31.53
8,920	33.61	9,220	32.52	9,520	31.49
8,930	33.57	9,230	32.48	9,530	31.46
8,940	33.54	9,240	32.45	9,540	31.43
8,950	33.50	9,250	32.41	9,550	31.39
8,960	33.46	9,260	32.38	9,560	31.36
8,970	33.42	9,270	32.34	9,570	31.33
8,980	33.39	9,280	32.31	9,580	31.30
8,990	33.35	9,290	32.27	9,590	31.26
9,000	33.31	9,300	32.24	9,600	31.23
9,010	33.28	9,310	32.20	9,610	31.20
9,020	33.24	9,320	32.17	9,620	31.17
9,030	33.20	9,330	32.14	9,630	31.13
9,040	33.17	9,340	32.10	9,640	31.10
9,050	33.13	9,350	32.07	9,650	31.07
9,060	33.09	9,360	32.03	9,660	31.04
9,070	33.06	9,370	32.00	9,670	31.01
9,080	33.02	9,380	31.96	9,680	30.97
9,090	32.98	9,390	31.93	9,690	30.94
9,100	32.95	9,400	31.90	9,700	30.91

Table 23-2. Kilohertz (kHz) to Meters (m) or Meters to Kilohertz (cont'd).

kHz	m	kHz	m	kHz	m
9,710	30.88	9,810	30.56	9,910	30.25
9,720	30.85	9,820	30.53	9,920	30.22
9,730	30.81	9,830	30.50	9,930	30.19
9,740	30.78	9,840	30.47	9,940	30.16
9,750	30.75	9,850	30.44	9,950	30.13
9,760	30.72	9,860	30.41	9,960	30.10
9,770	30.69	9,870	30.38	9,970	30.07
9,780	30.66	9,880	30.35	9,980	30.04
9,790	30.63	9,890	30.32	9,990	30.01
9,800	30.59	9,900	30.28	10,000	29.98

☐ **Example:**

One of the bands of a short-wave receiver covers the range from 3,000 kHz to 5,000 kHz. What wavelength range does this include?

Table 23-2 shows that 3,000 kHz corresponds to 99.94 meters and that 5,000 kHz corresponds to 59.96 meters. Thus, this particular band is from approximately 60 to 100 meters.

☐ **Example:**

What is the length, in feet, of a wave having a frequency of 4,280 kilohertz?

Locate this frequency (4280 kHz) in the kHz column. You will note it corresponds to a wavelength of 70.05 meters. However, the problem calls for the answer in feet. Consult Table 23-3 and you will see that 70 meters is 229.66 feet.

CONVERSION—METERS TO FEET

The distance from the start to the finish of a single cycle of a wave, called the wavelength, is usually specified in meters. Table 23-3 supplies data on the conversion of meters to feet. The Table can easily be extended by moving the decimal point. Move the decimal point an equal number of places in the same direction in both columns. Thus, a wavelength of 18 meters corresponds to 59.055 feet. And 180 meters corresponds to 590.5 feet, etc. Use Table 23-4 for converting feet to meters.

☐ **Example:**

What is the length, in feet, of a wave that is 36 meters long? Table 23-3 shows that 36 m equals 118.08 ft. (36 × 3.28).

Table 23-3. Meters to Feet.

Meters	Feet	Meters	Feet	Meters	Feet	Meters	Feet
1	3.2808	26	85.302	51	167.32	76	249.34
2	6.5617	27	88.583	52	170.60	75	256.62
3	9.8425	28	91.863	53	173.88	78	255.90
4	13.123	29	95.144	54	177.16	79	259.19
5	16.404	30	98.425	55	180.45	80	262.47
6	19.685	31	101.71	56	183.73	81	265.75
7	22.966	32	104.99	57	187.01	82	269.03
8	26.247	33	108.27	58	190.29	83	272.31
9	29.527	34	111.55	59	193.57	84	275.59
10	32.808	35	114.83	60	196.85	85	278.87
11	36.089	36	118.08	61	200.13	86	282.15
12	39.370	37	121.39	62	203.41	87	285.43
13	42.651	38	124.67	63	206.69	88	288.71
14	45.932	39	127.95	64	209.97	89	291.99
15	49.212	40	131.23	65	213.25	90	295.27
16	52.493	41	134.51	66	216.53	91	298.56
17	55.774	42	137.80	67	219.82	92	301.84
18	59.055	43	141.08	68	223.10	93	305.12
19	62.336	44	144.36	69	226.38	94	308.40
20	65.617	45	147.64	70	229.66	95	311.68
21	68.897	46	150.92	71	232.94	96	314.96
22	72.178	47	154.20	72	236.22	97	318.23
23	75.459	48	157.48	73	239.50	98	321.52
24	78.740	49	160.76	74	242.78	99	324.80
25	82.021	50	164.04	75	246.06	100	328.08

Table 23-4. Feet to Meters.

Feet	Meters	Feet	Meters	Feet	Meter	Feet	Meters
1	0.3048	26	7.9248	51	15.545	76	23.165
2	0.6096	27	8.2296	52	15.850	77	23.470
3	0.91440	28	8.5344	53	16.154	78	23.774
4	1.2192	29	8.8392	54	16.459	79	24.079
5	1.5240	30	9.1440	55	16.764	80	24.384
6	1.8288	31	9.4488	56	17.069	81	24.689
7	2.1336	32	9.7536	57	17.374	82	24.994
8	2.4384	33	10.058	58	17.678	83	25.298
9	2.7432	34	10.363	59	17.983	84	25.603
10	3.0480	35	10.668	60	18.288	85	25.908
11	3.3528	36	10.973	61	18.593	86	26.213
12	3.6576	37	11.278	62	18.898	87	26.518
13	3.9624	38	11.582	63	19.202	88	26.822
14	4.2672	39	11.887	64	19.507	89	27.127
15	4.5720	40	12.192	65	19.812	90	27.432
16	4.8768	41	12.497	66	20.117	91	27.737
17	5.1816	42	12.802	67	20.422	92	28.042
18	5.4864	43	13.106	68	20.726	93	28.346
19	5.7912	44	13.411	69	21.031	94	28.651
20	6.0960	45	13.716	70	21.336	95	28.956
21	6.4008	46	14.021	71	21.641	96	29.261
22	6.7056	47	14.326	72	21.946	97	29.566
23	7.0104	48	14.630	73	22.250	98	29.870
24	7.3152	49	14.935	74	22.555	99	30.175
25	7.6200	50	15.240	75	22.860	100	30.480

☐ **Example:**

What is the frequency of a wave whose wavelength is 20 feet?

Table 23-3 shows 19.68 feet equals 6 meters. For greater accuracy, using Table 23-4 (1 foot equals 0.3048006 meter), 20 ft. × 0.3048 equals 6.096 meters.

Table 23-2 does not have a value for 6 meters but it does have an amount that is very close. This is 60.08 meters, corresponding to a frequency of 4,990 kHz. Frequency varies inversely with wavelength and so we must multiply this answer by 10. 4,990 kHz becomes 49,900 kHz. For greater accuracy use the following formula:

$$f = 299,820,000/\lambda$$

$$= \frac{299,820,000}{6.09}$$

$$= 49,231,527 \text{ Hz}$$

$$= 49,231 \text{ kHz}$$

The conversion factor in Table 23-3 is based on 1 meter = 3.2808 feet.

The conversion factor in Table 23-4 is based on 1 foot = 0.3048 meter.

FREQUENCY TO WAVELENGTH CONVERSION FOR VERY HIGH FREQUENCIES

At very high frequencies certain components, such as the elements of receiving or transmitting antennas, become small enough to be measured easily. Knowing the frequency at which such elements work and using data such as that contained in Table 23-5 makes it easy to convert frequency into lengths in meters or centimeters.

In Table 23-5, λ represents the wavelength in centimeters or meters; MHz is the frequency in megahertz.

☐ **Example:**

What is the frequency of a wave whose length is 10 cm?

Using Table 23-5 find the number 10 under the heading of cm. As you move to the right you will note this is also listed as 0.1 meter or one-tenth of a meter. Continue moving to the right and locate the answer in the MHz column—3,000 MHz.

Table 23-5. Wavelength to Frequency (VHF and UHF).

m	MHz	m	MHz	m	MHz
0.1	3,000	4.1	73.2	8.1	37.0
0.2	1,500	4.2	71.4	8.2	36.6
0.3	1,000	4.3	69.8	8.3	36.1
0.4	750	4.4	68.2	8.4	35.7
0.5	600	4.5	66.7	8.5	35.3
0.6	500	4.6	65.2	8.6	34.9
0.7	429	4.7	63.8	8.7	34.5
0.8	375	4.8	62.5	8.8	34.1
0.9	333	4.9	61.2	8.9	33.7
1.0	300	5.0	60.0	9.0	33.3
1.1	273	5.1	58.8	9.1	33.0
1.2	250	5.2	57.7	9.2	32.6
1.3	231	5.3	56.6	9.3	32.3
1.4	214	5.4	55.6	9.4	31.9
1.5	200	5.5	54.5	9.5	31.6
1.6	188	5.6	53.6	9.6	31.3
1.7	176	5.7	52.6	9.7	30.9
1.8	167	5.8	51.7	9.8	30.6
1.9	158	5.9	50.8	9.9	30.3
2.0	150	6.0	50.0	10.0	30.0
2.1	143	6.1	49.2	10.1	29.7
2.2	136	6.2	48.4	10.2	29.4
2.3	130	6.3	47.6	10.3	29.1
2.4	125	6.4	46.9	10.4	28.8
2.5	120	6.5	46.2	10.5	28.6
2.6	115	6.6	45.5	10.6	28.3
2.7	111	6.7	44.8	10.7	28.0
2.8	107	6.8	44.1	10.8	27.8
2.9	103	6.9	43.5	10.9	27.5
3.0	100	7.0	42.9	11.0	27.3
3.1	96.8	7.1	42.3	11.1	27.0
3.2	93.8	7.2	41.7	11.2	26.8
3.3	90.9	7.3	41.1	11.3	26.5
3.4	88.2	7.4	40.5	11.4	26.3
3.5	85.7	7.5	40.0	11.5	26.1
3.6	83.3	7.6	39.5	11.6	25.9
3.7	81.1	7.7	39.0	11.7	25.6
3.8	78.9	7.8	38.5	11.8	25.4
3.9	76.9	7.9	38.0	11.9	25.2
4.0	75.0	8.0	37.5	12.0	25.0

Table 23-6. Decimal Inches to Millimeters.

1 inch = 25.40 millimeters

Inches	Milli-meters	Inches	Milli-meters	Inches	Milli-meters
0.001	0.0254	0.290	7.37	0.660	16.76
0.002	0.0508	0.300	7.62	0.670	17.02
0.003	0.0762	0.310	7.87	0.680	17.27
0.004	0.1016	0.320	8.13	0.690	17.53
0.005	0.1270	0.330	8.38	0.700	17.78
0.006	0.1524	0.340	8.64	0.710	18.03
0.007	0.1778	0.350	8.89	0.720	18.29
0.008	0.2032	0.360	9.14	0.730	18.54
0.009	0.2286	0.370	9.40	0.740	18.80
0.010	0.2540	0.380	9.65	0.750	19.05
0.020	0.5080	0.390	9.91	0.760	19.30
0.030	0.7620	0.400	10.16	0.770	19.56
0.040	1.016	0.410	10.41	0.780	19.81
0.050	1.270	0.420	10.67	0.790	20.07
0.060	1.524	0.430	10.92	0.800	20.32
0.070	1.778	0.440	11.18	0.810	20.57
0.080	2.032	0.450	11.43	0.820	20.83
0.090	2.286	0.460	11.68	0.830	21.08
0.100	2.540	0.470	11.94	0.840	21.34
0.110	2.794	0.480	12.19	0.850	21.59
0.120	3.048	0.490	12.45	0.860	21.84
0.130	3.302	0.500	12.70	0.870	22.10
0.140	3.556	0.510	12.95	0.880	22.35
0.150	3.810	0.520	13.21	0.890	22.61
0.160	4.064	0.530	13.46	0.900	22.86
0.170	4.318	0.540	13.72	0.910	23.11
0.180	4.572	0.550	13.97	0.920	23.37
0.190	4.826	0.560	14.22	0.930	23.62
0.200	5.080	0.570	14.48	0.940	23.88
0.210	5.334	0.580	14.73	0.950	24.13
0.220	5.588	0.590	14.99	0.960	24.38
0.230	5.842	0.600	15.24	0.970	24.64
0.240	6.096	0.610	15.49	0.980	24.89
0.250	6.350	0.620	15.75	0.990	25.15
0.260	6.604	0.630	16.00	1.000	25.40
0.270	6.858	0.640	16.26	....	....
0.280	7.112	0.650	16.51	....	....

☐ **Example:**
What is the wavelength of a wave whose frequency is 25 MHz?
In the MHz column you will see 25.0. Move to the left and you

will see that the corresponding wavelength is 12 meters.

CONVERSION—INCHES TO
MILLIMETERS AND MILLIMETERS TO INCHES

Table 23-6 supplies a convenient way of converting decimal inches to millimeters. Use Table 23-7 for converting millimeters to decimal inches. The range of both tables can easily be extended by moving the decimal point an equal number of places in the same direction, for both columns.

☐ **Example:**

What is the length in millimeters of a wave whose length is 0.280 inch?

Locate 0.280 in the inches column in Table 23-6. The answer, 7.112 millimeters, is shown in the column immediately to the right.

☐ **Example:**

A wave is approximately 33 millimeters long. What is its length in inches?

Locate 3.302 in the millimeters column in Table 23-6. By moving the decimal point one place to the right, you will have 33.02 millimeters. The corresponding distance in inches is 0.130, but we must move the decimal point one place to the right. The answer is 1.30 inches.

☐ **Example:**

A wave has a length of 0.92 millimeter. What is the corresponding length in inches?

Using Table 23-7 locate 0.92 in the millimeters column. The length in inches is 0.036156.

Table 23-7. Millimeters to Decimal Inches.

1 millimeter = 0.0393700 inch

millimeters	inches	millimeters	inches
0.01	0.0003937	0.36	0.014148
0.02	0.000786	0.37	0.014541
0.03	0.001179	0.38	0.014934
0.04	0.001572	0.39	0.015327
0.05	0.001965	0.40	0.015720
0.06	0.002358	0.41	0.016113
0.07	0.002751	0.42	0.016506
0.18	0.003144	0.43	0.016899

Table 23-7. Millimeters to Decimal Inches (cont'd).

millimeters	inches	millimeters	inches
0.09	0.003537	0.44	0.017292
0.10	0.003937	0.45	0.017685
0.11	0.004323	0.46	0.018078
0.12	0.004716	0.47	0.018471
0.13	0.005109	0.48	0.018864
0.14	0.005502	0.49	0.019257
0.15	0.005895	0.50	0.019650
0.16	0.006288	0.51	0.020043
0.17	0.006681	0.52	0.020436
0.18	0.007074	0.53	0.020829
0.19	0.007467	0.54	0.021222
0.20	0.007860	0.55	0.021615
0.21	0.008253	0.56	0.022008
0.22	0.008646	0.57	0.022401
0.23	0.009039	0.58	0.022794
0.24	0.009432	0.59	0.023187
0.25	0.009825	0.60	0.023580
0.26	0.010218	0.61	0.023973
0.27	0.010611	0.62	0.024366
0.28	0.011004	0.63	0.024759
0.29	0.011397	0.64	0.025152
0.30	0.011790	0.65	0.055450
0.31	0.012183	0.66	0.025938
0.32	0.012576	0.67	0.026331
0.33	0.012969	0.68	0.026724
0.34	0.013362	0.69	0.027117
0.35	0.013755	0.70	0.027510
0.71	0.027903	0.86	0.033798
0.72	0.028296	0.87	0.034191
0.73	0.028689	0.88	0.034584
0.74	0.029082	0.89	0.034977
0.75	0.029475	0.90	0.035370
0.76	0.029868	0.91	0.035763
0.77	0.030261	0.92	0.036156
0.78	0.030654	0.93	0.036549
0.79	0.031047	0.94	0.036942
0.80	0.031440	0.95	0.037335
0.81	0.031833	0.96	0.039370
0.82	0.032226	0.97	0.037728
0.83	0.032619	0.98	0.038121
0.84	0.033012	0.99	0.038514
0.85	0.033405	1.00	0.038907

FREQUENCY CONVERSIONS

Three common measures of frequency are used in electronics. To convert from one to the other, see Table 23-8.

Table 23-8. Frequency Conversion.

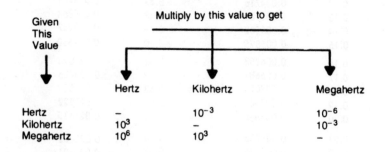

Given This Value	Multiply by this value to get		
	Hertz	Kilohertz	Megahertz
Hertz	–	10^{-3}	10^{-6}
Kilohertz	10^3	–	10^{-3}
Megahertz	10^6	10^3	–

FREQUENCY ALLOCATIONS

Table 23-9 shows frequency allocations from 15 MHz to 806 MHz.

DIELECTRIC CONSTANTS

The material placed between the plates of a capacitor is its dielectric. All dielectrics are compared to a reference, air, which has a dielectric constant (k) of 1. The capacitance of a capacitor is directly proportional to the dielectric constant. Values of these constants are only approximate since there can be considerable variations in the qualities of the materials used as dielectrics. See Table 23-10.

CAPACITANCE CONVERSIONS

Electronic formulas involving capacitance sometimes require changing capacitance units to some multiple or submultiple. Table 23-11 supplies the multiplication factors for making these conversions.

MEASURING SYSTEMS

Two measuring systems are in common use—the English and the metric. The advantage of the metric system is that it is decimal, and so it is easy to move from basic metric units to multiples or submultiples. The meter and the gram are bases for the metric system. Commonly used abbreviations are m for meters, mm for millimeters, cm for centimeters, dm for decimeters, dkm for decameters, hm for hectometers, km for kilometers, and mym for myriameters. The micron is represented by μ (Greek letter mu). Table 23-12 shows the relationships between English measure and metric equivalents.

Table 23-13 supplies data on length, volume, and mass in metric form.

PROPER (COMMON) FRACTIONS AND MILLIMETRIC EQUIVALENTS

Fractions are the usual form when measurements are made with a foot rule or yardstick. Table 23-14 shows the metric equivalents in millimeters and also decimal equivalents for fractions of an inch, from 1/64 inch to 1 inch, in 1/64-inch steps.

When the conversion involves a whole number plus a fraction, use a factor of 25.40005 millimeters for each inch.

☐ **Example:**

Convert 2-17/64 inches to millimeters.

Two inches equals 2 × 25.40005 millimeters, or 50.8001 millimeters. Table 23-14 shows that 17/64 inch equals 6.746875 millimeters. 50.8001 millimeters + 6.746875 millimeters equals 57.546975 mm.

TEMPERATURE CONVERSIONS

Components used in electronics, including wire, can have a positive temperature coefficient, with resistance increasing as temperature rises, or a negative temperature coefficient, with resistance varying inversely with temperature. Some components are specifically designed to have a zero temperature coefficient, with resistance not affected by temperature changes.

While resistance is often specified in degrees Fahrenheit, the use of the Celsius temperature scale (formerly known as Centigrade) is becoming more acceptable in electrical and electronic applications (see Fig. 23-1). The following tables aren't complete for all possible values of F degrees and C degrees but are intended for the more common values encountered in electrical and electronic applications.

Tables 23-15 and 23-16 are based on the formulas: $C = (F - 32) \times 5/9$ and $F = (C \times 9/5) + 32$.

All Celsius and Fahrenheit temperatures in Table 23-15 and Table 23-16 are plus unless otherwise indicated.

Table 23-9. FCC Frequency Allocations—15 MHz to 806 MHz.

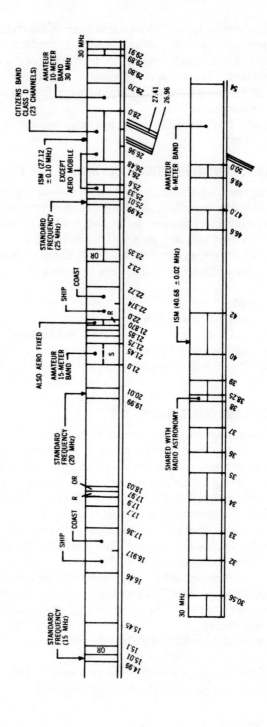

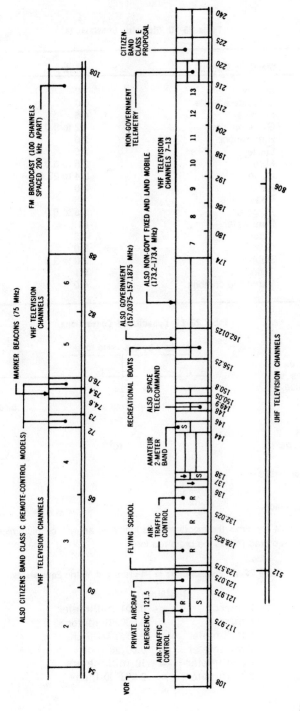

335

Table 23-10. Dielectric Constants (k).

Dielectric Material	k
Air, dry	1.0
Cellulose acetate	6.0
Formica	4.6
Glass, window	4.2 to 8.0
Hard rubber	2.0
Lucite	2.5
Mica	2.5 to 6.0
Nylon	3.4 to 22.4
Paper	2.0
Polystyrene	2.5
Porcelain	5.5 to 6.0
Pure water	81.0
Pyrex	4.5
Quartz	5.0
Rubber	2.0 to 3.0
Teflon	2.1
Varnished cambric	4.0

Table 23-11. Capacitance Conversions.

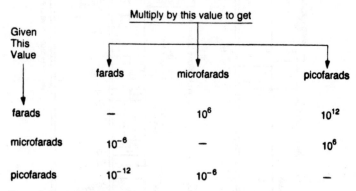

Given This Value	Multiply by this value to get		
	farads	microfarads	picofarads
farads	—	10^6	10^{12}
microfarads	10^{-6}	—	10^6
picofarads	10^{-12}	10^{-6}	—

Note: At one time, the picofarad was indicated as $\mu\mu$F or micromicrofarad. The picofarad (pF) is the preferred form.

Table 23-12. English Units of Measure and Metric Equivalents.

Distance

1 micron	=	0.001 millimeter
1 micron	=	0.000001 meter
1 millimeter	=	0.0393700 inch
1 millimeter	=	0.00328 foot
1 centimeter	=	10 millimeters
1 centimeter	=	0.393700 inch

Table 23-12. English Units of Measure and Metric Equivalents (cont'd).

1 centimeter	=	0.032808 foot
1 centimeter	=	0.01093611 yard
1 meter	=	39.3700 inches
1 meter	=	3.280833333 feet
1 meter	=	1.09361 yards
1 decimeter	=	10 centimeters
1 decimeter	=	3.937 inches
1 meter	=	10 decimeters
1 meter	=	100 centimeters
1 meter	=	1,000 millimeters
1 decameter	=	10 meters
1 decameter	=	393.7 inches
1 hectometer	=	10 decameters
1 hectometer	=	328 feet, 1 inch
1 kilometer	=	10 hectometers
1 kilometer	=	0.62137 mile
1 myriameter	=	10 kilometers
1 myriameter	=	6.2137 miles
1 inch	=	25.40005 millimeters
1 inch	=	2.540005 centimeters
1 inch	=	0.02540005 meter
1 foot	=	304.8006 millimeters
1 foot	=	30.48006 centimeters
1 foot	=	0.3048006 meter
1 yard	=	0.9144 meter
1 mile	=	1.609 kilometers

Area

1 sq. inch	=	6.452 sq. cm.
1 sq. foot	=	0.0929 sq. meter
1 sq. yard	=	0.84 sq. meter
1 sq. mile	=	2.589 sq. kilometers
1 sq. millimeter	=	0.00155 sq. in.
1 sq. centimeter	=	0.1550 sq. in.
1 sq. meter	=	1.196 sq. yd.
1 sq. kilometer	=	0.3861 sq. mi.

Volume

1 cu. in.	=	16.3872 cu. centimeters
1 cu. ft.	=	0.02832 cu. meter
1 cu. yd.	=	0.7646 cu. meter

Table 23-12. English Units of Measure and Metric Equivalents (cont'd).

1 liquid qt.	=	0.9464 liter
1 liquid gal.	=	3.785 liters
1 dry qt.	=	1.101 liters
1 cu. centimeter	=	0.06102 cu. in.
1 cubic meter	=	1.308 cu. yds.
1 milliliter	=	0.03381 liq. oz.
1 liter	=	1.057 liq. qts.
1 liter	=	0.9081 dry qt.

Weight

1 grain	=	0.0648 gram
1 oz. (avoir.)	=	28.35 grams
1 oz. (troy)	=	31.10 grains
1 lb. (avoir.)	=	0.4536 kilogram
1 lb. (troy)	=	0.3732 kilogram
1 ton (short)	=	0.9072 metric ton
1 gram	=	15.43 grains (troy)
1 gram	=	0.03215 oz. (troy)
1 gram	=	0.03527 oz. (avoir.)
1 kilog.	=	2.205 lbs. (avoir.)
1 kilog.	=	2.679 lbs. (troy)
1 metric ton	=	1.102 tons (short)

ELECTRONIC UNITS

Formulas in electronics use basic units—units such as the volt, ohm, ampere, farad, and henry. However, problems invariably supply information in terms of multiples or submultiples of these units. Thus, before any solution can be attempted, it is often necessary to convert multiples or submultiples to basic units.

Quite frequently the numbers used in formulas will be very large whole numbers or large decimals. In either case, it is highly advantageous to be able to use powers of ten and to be familiar with the rules for dividing and multiplying numbers using exponents. Table 23-17 shows the method of expressing small and large numbers using powers of ten.

SYMBOLS AND PREFIXES FOR POWERS OF 10

Numbers have names. The prefixes given in Table 23-18 are

Table 23-13. Length, Volume and Mass in Metric Form.

Prefix	Exponential Form	Symbol	Length	Volume	Mass
tera	10^{12}	T	terameter	teraliter	teragram
giga	10^9	G	gigameter	gigaliter	gigagram
mega	10^6	M	megameter	megaliter	megagram
kilo	10^3	k	kilometer	kiloliter	kilogram
hecto	10^2	h	hectometer	hectoliter	hectogram
deca	10^1	da	decameter	decaliter	decagram
	10^0		meter	liter	gram
deci	10^{-1}	d	decimeter	deciliter	decigram
centi	10^{-2}	c	centimeter	centiliter	centigram
milli	10^{-3}	m	millimeter	milliliter	milligram
micro	10^{-6}	μ	micrometer	microliter	microgram
nano	10^{-9}	n	nanometer	nanoliter	nanogram
pico	10^{-12}	p	picometer	picoliter	picogram
femto	10^{-15}	f	femtometer	femtoliter	femtogram
atto	10^{-18}	a	attometer	attoliter	attogram

Note: There are certain standards to observe in using these abbreviations. Thus, a microsecond could be abbreviated as μs. Conceivably, a thousandth of a microsecond could then be written as mμs or a millimicrosecond. The preferred form is ns or nanosecond.

339

Table 23-14. Fractional and Decimal Inches and Millimetric Equivalents.

(1 inch = 25.40005 millimeters)

Inches		
Fraction	Decimal	Millimeters
1/64	0.015625	0.396875
1/32	0.031250	0.793750
3/64	0.046875	1.190625
1/16	0.062500	1.587500
5/64	0.078125	1.984375
3/32	0.093750	2.381250
7/64	0.109375	2.778125
1/8	0.125000	3.175000
9/64	0.140625	3.571875
5/32	0.156250	3.968750
11/64	0.171875	4.365625
3/16	0.187500	4.762500
13/64	0.203125	5.159375
7/32	0.218750	5.556250
15/64	0.234375	5.953125
1/4	0.250000	6.350000
17/64	0.265625	6.746875
9/32	0.281250	7.143750
19/64	0.296875	7.540625
5/16	0.312500	7.937500
21/64	0.328125	8.334375
11/32	0.343750	8.731250
23/64	0.359375	9.128125
3/8	0.375000	9.525000
25/64	0.390625	9.921875
13/32	0.406250	10.318750
27/64	0.421875	10.715625
7/16	0.437500	11.112500
29/64	0.453125	11.509375
15/32	0.468750	11.906250
31/64	0.484375	12.303125
1/2	0.500000	12.700000
33/64	0.515625	13.096875
17/32	0.531250	13.493750
35/64	0.546875	13.890625
9/16	0.562500	14.287500
37/64	0.578125	14.684375
19/32	0.593750	15.081250
39/64	0.609375	15.478125
5/8	0.625000	15.875000
41/64	0.640625	16.271875
21/32	0.656250	16.668750

(1 inch = 25.40005 millimeters)

Inches

Fraction	Decimal	Millimeters
43/64	0.671875	17.065625
11/16	0.687500	17.462500
45/64	0.703125	17.859375
23/32	0.718750	18.256250
47/64	0.734375	18.653125
3/4	0.750000	19.050000
49/64	0.765625	19.446875
25/32	0.781250	19.843750
51/64	0.796875	20.240625
13/16	0.812500	20.637500
53/64	0.828125	21.034375
27/32	0.843750	21.431250
55/64	0.859375	21.828125
7/8	0.875000	22.225000
57/64	0.890625	22.621875
29/32	0.906250	23.018750
59/64	0.921875	23.415625
15/16	0.937500	23.812500
61/64	0.953125	24.209375
31/32	0.968750	24.606250
63/64	0.984375	25.003125
1	1.000000	25.400050

Table 23-15. Degrees Celsius to Degrees Fahrenheit.

°C	°F	°C	°F	°C	°F
−100	−148	5	41	105	221
−95	−139	10	50	110	230
−90	−130	15	59	115	239
−85	−121	20	68	120	248
−80	−112	25	77	125	257
−75	−103	30	86	130	266
−70	−94	35	93	135	275
−65	−85	40	104	140	284
−60	−76	45	113	145	293
−55	−67	50	122	150	302
−50	−58				
−45	−49	55	131	155	311
−40	−40	60	140	160	320
−35	−31	65	149	165	329

Table 23-15. Degrees Celsius to Degrees Fahrenheit (cont'd).

°C	°F	°C	°F	°C	°F
−30	−22	70	158	170	338
−25	−13	75	167	175	347
−20	−4	80	176	180	356
−15	5	85	185	185	365
−10	14	90	194	190	374
−5	23	95	203	195	383
0	32	100	212	200	392

helpful in identifying particular values of powers of 10. Thus, a gigahertz (prefix giga) corresponds to 10^9.

Table 23-19 supplies data on the conversion of electronic units from one form to another. Most of the conversion factors are supplied as powers of 10.

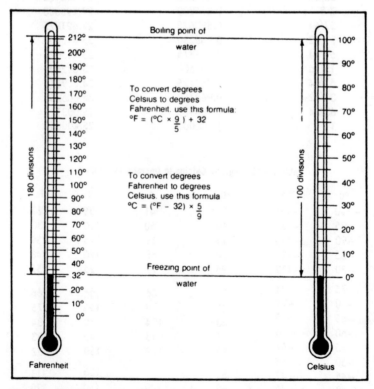

Fig. 23-1. Graphic representation of degrees Fahrenheit compared to degrees Celsius.

Table 23-16. Degrees Fahrenheit to Degrees Celsius.

°F	°C	°F	°C
-100	-73	70	21
-90	-68	75	24
-80	-62	80	27
-70	-57	85	29
-60	-51	90	32
-50	-46	95	35
-40	-40	100	38
-30	-34	105	41
-20	-29	110	43
-10	-23	115	46
-5	-21	120	49
0	-18	125	52
5	-15	130	54
10	-12	135	57
15	-9	140	60
20	-7	145	63
25	-4	150	66
30	-1	155	68
35	1.7	160	71
40	4	165	74
45	7	170	77
50	10	175	79
55	13	180	82
60	16	185	85
65	18	190	88
		195	91
		200	93

Table 23-17. Powers of 10.

10^0 = 1		10^0 = 1	
10^1 = 10		10^{-1} = .1	
10^2 = 100		10^{-2} = .01	
10^3 = 1,000		10^{-3} = .001	

Table 23-17. Powers of 10 (cont'd).

$10^4 = 10,000$	$10^{-4} = .0001$
$10^5 = 100,000$	$10^{-5} = .00001$
$10^6 = 1,000,000$	$10^{-6} = .000001$
$10^7 = 10,000,000$	$10^{-7} = .0000001$
$10^8 = 100,000,000$	$10^{-8} = .00000001$
$10^9 = 1,000,000,000$	$10^{-9} = .000000001$
$10^{10} = 10,000,000,000$	$10^{-10} = .0000000001$
$10^{11} = 100,000,000,000$	$10^{-11} = .00000000001$
$10^{12} = 1,000,000,000,000$	$10^{-12} = .000000000001$

Table 23-18. Powers of 10, Symbols, and Prefixes.

Power of ten	Prefix	Symbol
10^{12}	tera	T
10^9	giga	G
10^6	mega	M
10^3	kilo	k
10^2	hecto or hekto	h
10	deka	dk
10^{-1}	deci	d
10^{-2}	centi	c
10^{-3}	milli	m
10^{-6}	micro	μ
10^{-9}	nano	n
10^{-12}	pico (micromicro)	p
10^{-15}	femto	f
10^{-18}	atto	a

Table 23-19. Electronic Units.

Multiply	By	To Convert To
ampere turns	1.257	gilberts
amperes	10^{12}	micromicroamperes
amperes	10^6	microamperes
amperes	10^3	milliamperes
centimeters	10	millimeters
degrees	60	minutes
farads	10^{12}	picofarads (micromicrofarads)
farads	10^6	microfarads
farads	10^3	millifarads

Table 23-19. Electronic Units (cont'd).

Multiply	By	To Convert To
gauss	1	lines per square centimeter
gauss	6.452	lines per square inch
henrys	10^6	microhenrys
henrys	10^3	millihenrys
Hertz	10^{-6}	megahertz
Hertz	10^{-3}	kilohertz
Hertz	10^9	gigahertz (kilomegahertz)
horsepower	.7457	kilowatts
horsepower	745.7	watts
joules	1	watt-seconds
joules	10	
kilohertz	10^3	hertz
kilohertz	10^{-3}	megahertz
kilohertz	10^6	gigahertz
kilovolt-amperes	10^3	volt-amperes
kilovolts	10^3	volts
kilowatt-hours	3.6×10^6	joules
kilowatts	10^3	watts
kilowatts	1.341	horsepower
lines per square centimeter	1	gausses
lines per square centimeter	6.452	lines per square inch
megahertz	10^6	hertz
megahertz	10^3	kilohertz
megawatts	10^{-6}	watts
megohms	10^6	ohms
meters	10^2	centimeters
meters	3.281	feet
meters	39.37	inches
meters	10^3	millimeters
mhos	10^6	micromhos
mhos	10^3	millimhos
microamperes	10^{-6}	amperes
microamperes	10^{-3}	milliamperes
microfarads	10^{-6}	farads
microfarads	10^6	picofarads (micromicrofarads)
microhenrys	10^{-6}	henrys
microhenrys	10^{-3}	millihenrys
micromhos	10^{-6}	mhos

Table 23-19. Electronic Units (cont'd).

Multiply	By	To Convert To
micromicroamperes	10^6	microamperes
micromicrofarads (picofarads)	10^{-12}	farads
micromicro-ohms	10^{-12}	ohms
microvolts	10^{-3}	millivolts
microvolts	10^{-6}	volts
microwatts	10^{-6}	watts
microwatts	10^{-3}	milliwatts
milliamperes	10^{-3}	amperes
milliamperes	10^3	microamperes
millihenrys	10^{-3}	henrys
millimeters	10^{-3}	meters
millimeters	10^{-1}	centimeters
millimhos	10^{-3}	mhos
millivolts	10^{-3}	volts
millivolts	10^3	microvolts
millivolts	10^{-6}	kilovolts
milliwatts	10^{-3}	watts
milliwatts	10^3	microwatts
milliwatts	10^{-9}	megawatts
mils	10^{-3}	inches
minutes	60	seconds
minutes	1/60	degrees
ohms	10^{-3}	kilohms
ohms	10^{-6}	megohms
radians	57.3	degrees
seconds	1/3600	degrees
seconds	10^3	milliseconds
seconds	10^6	microseconds
seconds	1/60	minutes
square centimeters	$1,973 \times 10^2$	circular mils
square inches	$1,273 \times 10^3$	circular mils
square mils	1.273	circular mils
volt-amperes	1/1,000	kilovolt-amperes
volts	10^6	microvolts
volts	10^3	millivolts
volts	10^{-3}	kilovolts
watt-hours	36×10^2	joules
watt-seconds	1	joules
watts	10^6	microwatts

Table 23-19. Electronic Units (cont'd).

Multiply	By	To Convert To
watts	10^3	milliwatts
watts	10^{-3}	kilowatts
watts	10^{-6}	megawatts

Table 23-20. Electronic Unit Multiples and Submultiples.

Current

Microamperes   milliamperes x 1,000
Microamperes   amperes x 1,000,000
Milliamperes   amperes x 1,000
Milliamperes   microamperes/1,000
Amperes   microamperes/1,000,000
Amperes   milliamperes/1,000

Resistance

Ohms   kilohms x 1,000
Ohms   megohms x 1,000,000
Kilohms   ohms / 1,000
Kilohms   megohms x 1,000
Megohms   ohms / 1,000,000
Megohms   kilohms / 1,000

Voltage

Volts   kilovolts x 1,000
Volts   megavolts x 1,000,000
Kilovolts   volts/1,000
Kilovolts   megavolts x 1,000
Megavolts   volts/1,000,000
Megavolts   kilovolts/1,000
Millivolts   volts x 1,000
Millivolts   microvolts/1,000
Microvolts   volts x 1,000,000
Microvolts   millivolts x 1,000

Power

Watts = milliwatts/1,000
Watts = microwatts/1,000,000
Watts = kilowatts x 1,000
Watts = megawatts x 1,000,000
Kilowatts = watts/1,000
Kilowatts = megawatts x 1,000
Megawatts = kilowatts/1,000
Megawatts = watts/1,000,000
Milliwatts = watts x 1,000
Milliwatts = microwatts/1,000
Microwatts = milliwatts x 1,000
Microwatts = watts x 1,000,000

Capacitance

Microfarads = farads x 1,000,000

Table 23-20. Electronic Unit Multiples and Submultiples (cont'd).

Picofarads = farads x 1,000,000,000,000
Microfarads = picofarads/1,000,000
Picofarads = microfarads x 1,000,000
Farads = microfarads/1,000,000
Farads = picofarads/1,000,000,000,000

Inductance
Microhenrys = henrys x 1,000,000
Microhenrys = millihenrys x 1,000
Millihenrys = henrys x 1,000
Millihenrys = microhenrys/1,000
Henrys = millihenrys/1,000
Henrys = microhenrys/1,000,000

Frequency
Hertz (Hz) = cycles per second (cps)
Hertz = millihertz x 1,000
Hertz = megahertz x 1,000,000
Millihertz = hertz/1,000
Millihertz = megahertz x 1,000
Megahertz = hertz/1,000,000
Megahertz = millihertz/1,000

Table 23-21. Conversions Using Powers of 10.

Current
Microamperes = milliamperes $\times 10^3$ = milliamperes/10^{-3}
Microamperes = amperes $\times 10^6$ = amperes/10^{-6}
Milliamperes = amperes $\times 10^3$ = amperes/10^{-3}
Milliamperes = microamperes $\times 10^{-3}$ = microamperes/10^3
Amperes = microamperes $\times 10^{-6}$ = microamperes/10^6
Amperes = milliamperes $\times 10^{-3}$ = milliamperes/10^3

Resistance
Ohms = kilohms $\times 10^3$ = kilohms/10^{-3}
Ohms = megohms $\times 10^6$ = megohms/10^{-6}
Kilohms = ohms/10^3 = ohms $\times 10^{-3}$
Kilohms = megohms $\times 10^3$ = megohms/10^{-3}
Megohms = ohms/10^6 = ohms $\times 10^{-6}$
Megohms = kilohms/10^3 = kilohms $\times 10^{-3}$

Voltage
Volts = kilovolts $\times 10^3$ = kilovolts/10^{-3}
Volts = megavolts $\times 10^6$ = megavolts/10^{-6}
Kilovolts = volts/10^3 = volts $\times 10^{-3}$
Kilovolts = megavolts $\times 10^3$ = megavolts/10^{-3}
Megavolts = volts/10^6 = volts $\times 10^{-6}$
Megavolts = kilovolts/10^3 = kilovolts $\times 10^{-3}$
Millivolts = volts $\times 10^3$ = volts/10^{-3}
Millivolts = microvolts/10^3 = microvolts $\times 10^{-3}$
Microvolts = volts $\times 10^6$ = volts/10^{-6}
Microvolts = millivolts $\times 10^3$ = millivolts/10^{-3}

Table 23-21. Conversions Using Powers of 10 (cont'd).

Power
Watts = milliwatts/10^3 = milliwatts × 10^{-3}
Watts = microwatts/10^6 = microwatts × 10^{-6}
Watts = kilowatts × 10^3 = kilowatts/10^{-3}
Watts = megawatts × 10^6 = megawatts/10^{-6}
Kilowatts = watts/10^3 = watts × 10^{-3}
Kilowatts = megawatts × 10^3 = megawatts/10^{-3}
Megawatts = kilowatts/10^3 = kilowatts × 10^{-3}
Megawatts = watts/10^6 = watts × 10^{-6}
Milliwatts = watts × 10^3 = watts/10^{-3}
Milliwatts = microwatts × 10^3 microwatts × 10^{-3}
Microwatts = milliwatts × 10^3 = milliwatts/10^{-3}
Microwatts = microwatts × 10^6 = watts/10^{-6}

Capacitance
Microfarads = farads × 10^6 = farads/10^{-6}
Picofarads = farads × 10^{12} = farads/10^{-12}
Microfarads = picofarads/10^6 = picofarads × 10^{-6}
Picofarads = microfarads × 10^6 = microfarads/10^{-6}
Farads = microfarads/10^6 = microfarads × 10^{-6}
Farads = picofarads/10^{12} = picofarads × 10^{-12}

Inductance
Microhenrys = henrys × 10^6 = henrys/10^{-6}
Microhenrys = millihenrys × 10^3 = millihenrys/10^{-3}
Millihenrys = henrys × 10^3 = henrys/10^{-3}
Millihenrys = microhenrys/10^3 = microhenrys × 10^{-3}
Henrys = millihenrys/10^3 = millihenrys × 10^{-3}
Henrys = microhenrys/10^6 = microhenrys × 10^{-6}

Inches and Mils
Mils = inches/10^3 = inches × 10^{-3}
Inches = mils × 10^3 = mils/10^{-3}

While it is more convenient to handle electronic conversions using powers of 10, such conversions can also be made using whole numbers. Table 23-20 shows the conversion relationships for current, resistance, voltage, power, capacitance, inductance, and frequency.

Table 23-22. Conversion Factors for Electronic Multiples and Submultiples.

TO CONVERT THESE TO → / THESE, MULTIPLY BY THE FIGURES BELOW ↓

	Pico—	Nano—	Micro—	Milli—	Centi—	Deci—	Units	Deka—	Hekto—	Kilo—	Myria—	Mega—	Giga—	Tera—
Pico—		0.001	10^{-6}	10^{-9}	10^{-10}	10^{-11}	10^{-12}	10^{-13}	10^{-14}	10^{-15}	10^{-16}	10^{-18}	10^{-21}	10^{-24}
Nano—	1000		0.001	10^{-6}	10^{-7}	10^{-8}	10^{-9}	10^{-10}	10^{-11}	10^{-12}	10^{-13}	10^{-15}	10^{-18}	10^{-21}
Micro—	10^{6}	1000		0.001	0.0001	10^{-5}	10^{-6}	10^{-7}	10^{-8}	10^{-9}	10^{-10}	10^{-12}	10^{-15}	10^{-18}
Milli—	10^{9}	10^{6}	1000		0.1	0.01	0.001	0.0001	10^{-5}	10^{-6}	10^{-7}	10^{-9}	10^{-12}	10^{-15}
Centi—	10^{10}	10^{7}	10,000	10		0.1	0.01	0.001	0.0001	10^{-5}	10^{-6}	10^{-8}	10^{-11}	10^{-14}
Deci—	10^{11}	10^{8}	10^{5}	100	10		0.1	0.01	0.001	0.0001	10^{-5}	10^{-7}	10^{-10}	10^{-13}
Units	10^{12}	10^{9}	10^{6}	1000	100	10		0.1	0.01	0.001	0.0001	10^{-6}	10^{-9}	10^{-12}
Deka—	10^{13}	10^{10}	10^{7}	10,000	1000	100	10		0.1	0.01	0.001	10^{-5}	10^{-8}	10^{-11}
Hekto—	10^{14}	10^{11}	10^{8}	10^{5}	10,000	1000	100	10		0.1	0.01	0.0001	10^{-7}	10^{-10}
Kilo—	10^{15}	10^{12}	10^{9}	10^{6}	10^{5}	10,000	1000	100	10		0.1	0.001	10^{-6}	10^{-9}
Myria—	10^{16}	10^{13}	10^{10}	10^{7}	10^{6}	10^{5}	10,000	1000	100	10		0.01	10^{-5}	10^{-8}
Mega—	10^{18}	10^{15}	10^{12}	10^{9}	10^{8}	10^{7}	10^{6}	10^{5}	10,000	1000	100		0.001	10^{-6}
Giga—	10^{21}	10^{18}	10^{15}	10^{12}	10^{11}	10^{10}	10^{9}	10^{8}	10^{7}	10^{6}	10^{5}	1000		0.001
Tera—	10^{24}	10^{21}	10^{18}	10^{15}	10^{14}	10^{13}	10^{12}	10^{11}	10^{10}	10^{9}	10^{8}	10^{6}	1000	

24

Numbers

EXPONENTS

Formulas, charts, and tables are all involved with numbers. Often, to be able to get maximum advantage from formulas, charts, and tables, numbers must be manipulated, or changed from one form to another.

One way of expressing a number is through the use of exponents. An exponent indicates the number of times a number is to be multiplied by itself. Thus, 9^2 is an abbreviated way of writing 9×9. It is much easier and quicker to write 9^{18} than to write the digit 9 a total of 18 times including a long series of multiplication signs. Table 24-1 supplies numbers from 1 to 100 with exponents from 3 to 8.

☐ **Example:**
What is the value of 9^6?
Locate 9 in the left-hand column in Table 24-1. Move to the right to the column headed by n^6. The value of 9^6 is given as 531,441.

☐ **Example:**
What is the value of 23^7?
Locate 23 in the left-hand column. Move to the right to the column headed by n^7. The number shown here is 3.404825. However, this number must be multiplied by 10^9 as indicated in the

Table 24-1. Powers of Numbers.

n	n^3	n^4	n^5	n^6	n^7	n^8
1	1	1	1	1	1	1
2	8	16	32	64	128	256
3	27	81	243	729	2187	6561
4	64	256	1024	4096	16384	65536
5	125	625	3125	15625	78125	390625
6	216	1296	7776	46656	279936	1679616
7	343	2401	16807	117649	823543	5764801
8	512	4096	32768	262144	2097152	16777216
9	729	6561	59049	531441	4782969	43046721
						$\times 10^8$
10	1000	10000	100000	1000000	10000000	1.000000
11	1331	14641	161051	1771561	19487171	2.143589
12	1728	20736	248832	2985984	35831808	4.299817
13	2197	28561	371293	4826809	62748517	8.157307
14	2794	38416	537824	7529536	105413504	14.757891
15	3375	50625	759375	11390625	170859375	25.628906
16	4096	65536	1048576	16777216	268435456	42.949673
17	4913	83521	1419857	24137569	410338673	69.757574
18	5832	104976	1889568	34012224	612220032	110.199606
19	6859	130321	2476099	47045881	893871739	169.835630
					$\times 10^9$	$\times 10^{10}$
20	8000	160000	3200000	64000000	1.280000	2.560000
21	9261	194481	4084101	85766121	1.801089	3.782286
22	10648	234256	5153632	113379904	2.494358	5.487587
23	12167	279841	6436343	148035889	3.404825	7.831099
24	13824	331776	7962624	191102976	4.586471	11.007531
25	15625	390625	9765625	244140625	6.103516	15.258789
26	17576	456976	11881376	308915776	8.031810	20.882706
27	19683	531441	14348907	387420489	10.460353	28.242954
28	21952	614656	17210368	481890304	13.492929	37.780200
29	24389	707281	20511149	594823321	17.249876	50.024641
				$\times 10^8$	$\times 10^{10}$	$\times 10^{11}$
30	27000	810000	24300000	7.290000	2.187000	6.561000
31	29791	923521	28629151	8.875037	2.751261	8.528910
32	32768	1048576	33554432	10.737418	3.435974	10.995116
33	35937	1185921	39135393	12.914680	4.261844	14.064086
34	39304	1336336	45435424	15.448044	5.252335	17.857939

Table 24-1. Powers of Numbers (cont'd).

n	n^3	n^4	n^5	n^6	n^7	n^8
35	42875	1500625	52521875	18.382656	6.433930	22.518754
36	46656	1679616	60466176	21.767823	7.836416	28.211099
37	50653	1874161	69343957	25.657264	9.493188	35.124795
38	54872	2085136	79235168	30.109364	11.441558	43.477921
39	59319	2313441	90224199	35.187438	13.723101	53.520093
				$\times 10^9$	$\times 10^{10}$	$\times 10^{12}$
40	64000	2560000	102400000	4.096000	16.384000	6.553600
41	68921	2825761	115856201	4.750104	19.475427	7.984925
42	74088	3111696	130691232	5.489032	23.053933	9.682652
43	79507	3418801	147008443	6.321363	27.181861	11.688200
44	85184	3748096	164916224	7.256314	31.927781	14.048224
45	91125	4100625	184528125	8.303766	37.366945	16.815125
46	97336	4477456	205962976	9.474297	43.581766	20.047612
47	103823	4879681	229345007	10.779215	50.662312	23.811287
48	110592	5308416	254803968	12.230590	58.706834	28.179280
49	117649	5764801	282475249	13.841287	67.822307	33.232931
				$\times 10^9$	$\times 10^{11}$	$\times 10^{13}$
50	125000	6250000	312500000	15.625000	7.812500	3.906250
51	132651	6765201	345025251	17.596288	8.974107	4.576794
52	140608	7311616	380204032	19.770610	10.280717	5.345973
53	148877	7890481	418195493	22.164361	11.747111	6.225969
54	157464	8503056	459165024	24.794911	13.389252	7.230196
55	166375	9150625	503284375	27.680641	15.224352	8.373394
56	175616	9834496	550731776	30.840979	17.270948	9.671731
57	185193	10556001	601692057	34.296447	19.548975	11.142916
58	195112	11316496	656356768	38.068693	22.079842	12.806308
59	205379	12117361	714924299	42.180534	24.886515	14.683044
				$\times 10^8$	$\times 10^{10}$	$\times 10^{11}$	$\times 10^{13}$
60	216000	12960000	7.776000	4.665600	27.993600	16.796160
61	226981	13845841	8.445963	5.152037	31.427428	19.170731
62	238328	14776336	9.161328	5.680024	35.216146	21.834011
63	250047	15752961	9.924365	6.252350	39.389806	24.815578
64	262114	16777216	10.737418	6.871948	43.980465	28.147498
65	274625	17850625	11.602906	7.541889	49.022279	31.864481
66	287496	18974736	12.523326	8.265395	54.551607	36.004061
67	300763	20151121	13.501251	9.045838	60.607116	40.606768

Table 24-1. Powers of Numbers (cont'd).

n	n^3	n^4	n^5	n^6	n^7	n^8
68	314432	21381376	14.539336	9.886748	67.229888	45.716324
69	328509	22667121	15.640313	10.791816	74.463533	51.379837

			$\times 10^8$	$\times 10^{10}$	$\times 10^{12}$	$\times 10^{14}$
70	343000	24010000	16.807000	11.764900	8.235430	5.764801
71	357911	25411681	18.042294	12.810028	9.095120	6.457535
72	373248	26873856	19.349176	13.931407	10.030613	7.222041
73	389017	28398241	20.730716	15.133423	11.047399	8.064601
74	405224	29986576	22.190066	16.420649	12.151280	8.991947
75	421875	31640625	23.730469	17.797852	13.348389	10.011292
76	438976	33362176	25.355254	19.269993	14.645195	11.130348
77	456533	35153041	27.067842	20.842238	16.048523	12.357363
78	474552	37015056	28.871744	22.519960	17.565569	13.701144
79	493039	38950081	30.770564	24.308746	19.203909	15.171088

			$\times 10^8$	$\times 10^{10}$	$\times 10^{12}$	$\times 10^{14}$
80	512000	40960000	32.768000	26.214400	20.971520	16.777216
81	531441	43046721	34.867844	28.242954	22.876792	18.530202
82	551368	45212176	37.073984	30.400667	24.928547	20.441409
83	571787	47458321	39.390406	32.694037	27.136051	22.522922
84	592704	49787136	41.821194	35.129803	29.509035	24.787589
85	614125	52200625	44.370531	37.714952	32.057709	27.249053
86	636056	54700816	47.042702	40.456724	34.792782	29.921793
87	658503	57289761	49.842092	43.362620	37.725479	32.821167
88	681472	59969536	52.773192	46.440409	40.867560	35.963452
89	704969	62742241	55.840594	49.698129	44.231335	39.365888

			$\times 10^9$	$\times 10^{11}$	$\times 10^{13}$	$\times 10^{15}$
90	729000	65610000	5.904900	5.314410	4.782969	4.304672
91	753571	68574961	6.240321	5.678693	5.167610	4.702525
92	778688	71639296	6.590815	6.063550	5.578466	5.132189
93	804357	74805201	6.956884	6.469902	6.017009	5.595818
94	830584	78074896	7.339040	6.898698	6.484776	6.095689
95	857375	81450625	7.737809	7.350919	6.983373	6.634204
96	884736	84934656	8.153727	7.827578	7.514475	7.213896
97	912673	88529281	8.587340	8.329720	8.079828	7.837434
98	941192	92236816	9.039208	8.858424	8.681255	8.507630
99	970299	96059601	9.509900	9.414801	9.320653	9.227447
100	1000000	100000000	10.000000	10.000000	10.000000	10.000000

column heading. Hence, 23^7 equals 3.404825×10^9. To get the final answer, move the decimal point 9 places to the right. 3.404825×10^9 equals, $3,404,825,000$.

SQUARE ROOTS OF NUMBERS

Table 24-2 lists the square roots of numbers from 1 to 100.

☐ **Example:**

What is the square root of 19?
Use Table 24-2 to find the square roots of numbers.

Table 24-2. Square Roots of Numbers.

n	$\sqrt{n}$	n	$\sqrt{n}$	n	$\sqrt{n}$	n	$\sqrt{n}$
1	1.0000	26	5.0990	51	7.1414	76	8.7178
2	1.4142	27	5.1962	52	7.2111	77	8.7750
3	1.7321	28	5.2915	53	7.2801	78	8.8318
4	2.0000	29	5.3852	54	7.3485	79	8.8882
5	2.2361	30	5.4772	55	7.4162	80	8.9443
6	2.4495	31	5.5678	56	7.4833	81	9.0000
7	2.6458	32	5.6569	57	7.5498	82	9.0554
8	2.8284	33	5.7446	58	7.6158	83	9.1104
9	3.0000	34	5.8310	59	7.6811	84	9.1652
10	3.1623	35	5.9161	60	7.7460	85	9.2195
11	3.3166	36	6.0000	61	7.8102	86	9.2736
12	3.4641	37	6.0828	62	7.8740	87	9.3274
13	3.6056	38	6.1644	63	7.9373	88	9.3808
14	3.7417	39	6.2450	64	8.0000	89	9.4340
15	3.8730	40	6.3246	65	8.0623	90	9.4868
16	4.0000	41	6.4031	66	8.1240	91	9.5394
17	4.1231	42	6.4807	67	8.1854	92	9.5917
18	4.2426	43	6.5574	68	8.2462	93	9.6437
19	4.3589	44	6.6332	69	8.3066	94	9.6954
20	4.4721	45	6.7082	70	8.3666	95	9.7468
21	4.5826	46	6.7823	71	8.4261	96	9.7980
22	4.6904	47	6.8557	72	8.4853	97	9.8490
23	4.7958	48	6.9282	73	8.5440	98	9.8995
24	4.8990	49	7.0000	74	8.6023	99	9.9499
25	5.0000	50	7.0711	75	8.6603	100	10.0000

□ **Example:**

Find 19 in the column headed by the letter n (abbreviation for number). The square root of 19 appears immediately to the right in the column identified by $\sqrt{n}$. The square root of 19 is 4.3589.

CUBE ROOTS OF NUMBERS

Table 24-3 can be used to find the cube roots of numbers ranging from 1 to 10,000.

□ **Example:**

What is the cube root of 16?

Locate 16 in the column headed by the letter n. Move to the right to the column identified by $\sqrt[3]{n}$. The cube root of 16 is 2.519842.

Two additional columns are marked $\sqrt[3]{10n}$ and $\sqrt[3]{100n}$. These are used for extending the values in the n column. Thus, if n is 18, then 10n is 10 × 18 equals 180, and 100n is 100 × 18 equals 1800. Consequently, 18 in the n column can represent 18, 180, or 1,800.

Table 24-3. Cube Roots of Numbers.

n	$\sqrt[3]{n}$	$\sqrt[3]{10n}$	$\sqrt[3]{100n}$
1	1.000000	2.154435	4.641589
2	1.259921	2.714418	5.848035
3	1.442250	3.107233	6.694330
4	1.587401	3.419952	7.368063
5	1.709976	3.684031	7.937005
6	1.817121	3.914868	8.434327
7	1.912931	4.121285	8.879040
8	2.000000	4.308869	9.283178
9	2.080084	4.481405	9.654894
10	2.154435	4.641589	10.00000
11	2.223980	4.791420	10.32280
12	2.289428	4.932424	10.62659
13	2.351335	5.065797	10.91393
14	2.410142	5.192494	11.18689
15	2.466212	5.313293	11.44714
16	2.519842	5.428835	11.69607
17	2.571282	5.539658	11.93483
18	2.620741	5.646216	12.16440
19	2.668402	5.748897	12.38562
20	2.714418	5.848035	12.59921

Table 24-3. Cube Roots of Numbers (cont'd).

n	$\sqrt[3]{n}$	$\sqrt[3]{10n}$	$\sqrt[3]{100n}$
21	2.758924	5.943922	12.80579
22	2.802039	6.036811	13.00591
23	2.843867	6.126926	13.20006
24	2.884499	6.214465	13.38866
25	2.924018	6.299605	13.57209
26	2.962496	6.382504	13.75069
27	3.000000	6.463304	13.92477
28	3.036589	6.542133	14.09460
29	3.072317	6.619106	14.26043
30	3.107233	6.694330	14.42250
31	3.141381	6.767899	14.58100
32	3.174802	6.839904	14.73613
33	3.207534	6.910423	14.88806
34	3.239612	6.979532	15.03695
35	3.271066	7.047299	15.18294
36	3.301927	7.113787	15.32619
37	3.332222	7.179054	15.46680
38	3.361975	7.243156	15.60491
39	3.391211	7.306144	15.74061
40	3.419952	7.368063	15.87401
41	3.448217	7.428959	16.00521
42	3.476027	7.488872	16.13429
43	3.503398	7.547842	16.26133
44	3.530348	7.605905	16.38643
45	3.556893	7.663094	16.50964
46	3.583048	7.719443	16.63103
47	3.608826	7.774980	16.75069
48	3.634241	7.829735	16.86865
49	3.659306	7.883735	16.98499
50	3.684031	7.937005	17.09976
51	3.708430	7.989570	17.21301
52	3.732511	8.041452	17.32478
53	3.756286	8.092672	17.43513
54	3.779763	8.143253	17.54411
55	3.802952	8.193213	17.65174

Table 24-3. Cube Roots of Numbers (cont'd).

n	$\sqrt[3]{n}$	$\sqrt[3]{10n}$	$\sqrt[3]{100n}$
56	3.825862	8.242571	17.75808
57	3.848501	8.291344	17.86316
58	3.870877	8.339551	17.96702
59	3.892996	8.387207	18.06969
60	3.914868	8.434327	18.17121
61	3.936497	8.480926	18.27160
62	3.957892	8.527019	18.37091
63	3.979057	8.572619	18.46915
64	4.000000	8.617739	18.56636
65	4.020726	8.662391	18.66256
66	4.041240	8.706588	18.75777
67	4.061548	8.750340	18.85204
68	4.081655	8.793659	18.94536
69	4.101566	8.836556	19.03778
70	4.121285	8.879040	19.12931
71	4.140818	8.921121	19.21997
72	4.160168	8.962809	19.30979
73	4.179339	9.004113	19.39877
74	4.198336	9.045042	19.48695
75	4.217163	9.085603	19.57434
76	4.235824	9.125805	19.66095
77	4.254321	9.165656	19.74681
78	4.272659	9.205164	19.83192
79	4.290840	9.244335	19.91632
80	4.308869	9.283178	20.00000
81	4.326749	9.321698	20.08299
82	4.344481	9.359902	20.16530
83	4.362071	9.397796	20.24694
84	4.379519	9.435388	20.32793
85	4.396830	9.472682	20.40828
86	4.414005	9.509685	20.48800
87	4.431048	9.546403	20.56710
88	4.447960	9.582840	20.64560
89	4.464745	9.619002	20.72351
90	4.481405	9.654894	20.80084
91	4.497941	9.690521	20.87759
92	4.514357	9.725888	20.95379

Table 24-3. Cube Roots of Numbers (cont'd).

n	$\sqrt[3]{n}$	$\sqrt[3]{10n}$	$\sqrt[3]{100n}$
93	4.530655	9.761000	21.02944
94	4.546836	9.795861	21.10454
95	4.562903	9.830476	21.17912
96	4.578857	9.864848	21.25317
97	4.594701	9.898983	21.32671
98	4.610436	9.932884	21.39975
99	4.626065	9.966555	21.47229
100	4.641589	10.00000	21.54435

□ **Example:**

What is the cube root of 3,100?

Locate 31 in the n column. 3100, however, is 31 × 100, and so the answer will be found in the $\sqrt[3]{100n}$ column. Move horizontally to this column and the answer is given as 14.58100. The last two zeros of this number do not contribute to its value and so can be omitted. $\sqrt[3]{3100}$ equals 14.581.

NUMBERS AND RECIPROCALS

The reciprocal of a number is the inverse of that number, or the number divided into 1. The recriprocal of 5 is 1/5; the reciprocal of 87 is 1/87. Table 24-4 supplies reciprocals of numbers ranging from 0.1 to 100.

□ **Example:**

What is the reciprocal of 27?

Locate 27 in the n column. Move to the right and in the column headed by 1/n the answer is given as 0.0370.

Table 24-4 can be extended by moving the decimal point as required.

□ **Example:**

What is the reciprocal of 160?

160 does not appear directly in the table. Instead, locate 16 in the n column and change it to 160. In the 1/n column, the corresponding reciprocal value for 16 is 0.0625. Add another zero directly after the decimal point (equivalent to dividing the reciprocal by 10 and the answer becomes 0.00625.

To find the reciprocal of 1,600, divide the answer in the 1/n column (corresponding to 16) by 100, or insert two zeros after the decimal point. Hence the reciprocal of 1,600 is 0.000625.

Table 24-4. Numbers and Reciprocals.

n	1/n	n	1/n	n	1/n
0.1	10.0000	28	.0357	65	.0154
0.2	5.0000	29	.0345	66	.0152
0.3	3.3333	30	.0333	67	.0149
0.4	2.5000	31	.0323	68	.0147
0.5	2.0000	32	.0313	69	.0145
0.6	1.6666	33	.0303	70	.0143
0.7	1.4286	34	.0294	71	.0141
0.8	1.2500	35	.0286	72	.0139
0.9	1.1111	36	.0278	73	.0137
		37	.0270	74	.0135
1	1.0000	38	.0263	75	.0133
2	.5000	39	.0256	76	.0132
3	.3333	40	.0250	77	.0130
4	.2500	41	.0244	78	.0128
5	.2000	42	.0238	79	.0127
6	.1667	43	.0233	80	.0125
7	.1429	44	.0227	81	.0123
8	.1250	45	.0222	82	.0122
9	.1111	46	.0217	83	.0120
10	.1000	47	.0213	84	.0119
11	.0909	48	.0208	85	.0118
12	.0833	49	.0204	86	.0116
13	.0769	50	.0200	87	.0115
14	.0714	51	.0196	88	.0114
15	.0667	52	.0192	89	.0112
16	.0625	53	.0189	90	.0111
17	.0588	54	.0185	91	.0110
18	.0555	55	.0182	92	.0109
19	.0526	56	.0179	93	.0108
20	.0500	57	.0175	94	.0106
21	.0476	58	.0172	95	.0105
22	.0455	59	.0169	96	.0104
23	.0435	60	.0167	97	.0103
24	.0417	61	.0164	98	.0102
25	.0400	62	.0161	99	.0101
26	.0385	63	.0159	100	.0100
27	.0370	64	.0156		

Large numbers can be more simply written in exponential form as shown in Table 24-5. For whole numbers the exponent indicates the number of zeros following digit 1. Thus, 10^3 indicates the digit 1 is to be followed by three zeros (1,000). For decimal values the

Table 24-5. Powers of 10.

$$10^{-6} = 0.000001 = \frac{1}{1,000,000} = \frac{1}{10^6}$$

$$10^{-5} = 0.00001 = \frac{1}{100,000} = \frac{1}{10^5}$$

$$10^{-4} = 0.0001 = \frac{1}{10,000} = \frac{1}{10^4}$$

$$10^{-3} = 0.001 = \frac{1}{1,000} = \frac{1}{10^3}$$

$$10^{-2} = 0.01 = \frac{1}{100} = \frac{1}{10^2}$$

$$10^{-1} = 0.1 = \frac{1}{10} = \frac{1}{10^1}$$

$$10^0 = 1$$

$$10^1 = 10 = \frac{1}{10^{-1}}$$

$$10^2 = 100 = \frac{1}{10^{-2}}$$

$$10^3 = 1,000 = \frac{1}{10^{-3}}$$

$$10^4 = 10,000 = \frac{1}{10^{-4}}$$

$$10^5 = 100,000 = \frac{1}{10^5}$$

$$10^6 = 1,000,000 = \frac{1}{10^{-6}}$$

exponent indicates the number of digits following the decimal point. Thus, 10^{-2} indicates .01.

SIGNIFICANT FIGURES IN THE DECIMAL SYSTEM

Some of the tables that have been presented consist of whole numbers only, while others are comprised of whole numbers followed by decimals, with the two, whole numbers and decimals, separated by a decimal point.

The numbers used in these tables are known as significant figures. A number, such as 35, in one of the tables, indicates an accuracy of two significant figures. The number 35 could have been written as 35. but the decimal point is generally omitted, or is said to be "understood." This means that a decimal point, technically, should follow the two numbers to indicate the extent of accuracy.

If the number 35 is written as 35.0, from a numbers viewpoint it is the same as 35, but there is a difference. Thirty-five represents a two digit accuracy, or the number has an accuracy of two significant figures. If it is written 35.0, the accuracy has improved for the number now has three significant figures. 35.0 implies an order of accuracy to a tenth.

If we now take the same number and write it as 35.00, it has four significant figures. Again, from a numerical value viewpoint, 35.00 is the same as 35.0 and that is the same as 35. But 35.00 tells us that this number is accurate to the hundredths. If 35 represents a voltage, 35 means 35 volts. 35.0 means a more precise measurement while 35.00 is still more exact.

When using a formula or a table it isn't always desirable to have answers with a high degree of accuracy. It all depends if the component being used must have a value that is as precise as possible. For most applications this isn't necessary.

Formulas and tables do not take the need for a high order of accuracy into consideration. If, in using a formula such as Ohm's law, you are asked to determine the value of resistance and are told that the voltage is 65 volts, and the current is 165 milliamperes, you would calculate the resistance in the following manner:

$$R = E/I = 65/0.165 = 393.9393 \text{ ohms}$$

This number has seven significant figures. For most applications a resistor having a value of 400 ohms would be satisfactory. If this resistor has a tolerance of plus or minus 5 percent, then its uppermost limit would be 400 + 20, or 420 ohms. Its lowermost limit would be 400 − 20 or 380 ohms.

Table 24-6. Numerical Prefixes.

Number	Greek prefix	Latin prefix
½	hemi-	semi-
1	mono- or mon-	uni-
1½	———	sesqui-
2	di-	bi-; duo-
3	tri-	tri- or ter-
4	tetra- or tetr-	quadri- or quadr-
5	penta- or pent-	quinque- or quinqu-
6	hexa- or hex-	sexi- or sex-
7	hepta- or hept-	septi- or sept-
8	octa- or oct- or octo-	octo-
9	ennea- or enne-	nona-; novem-
10	deca- or dec-	decem-
11	hendeca- or hendec-	undeca- or undec-
12	dodeca- or dodec-	duodec-
13	trideca- or tridec-	tredec-
14	tetradeca- or tetradec-	quatuordec-
15	pentadeca- or pentadec-	quindec-
16	hexadeca- or hexadec-	sextodec-
17	heptadeca- or heptadec-	septendec-
18	octadeca- or octadec-	octodec-
19	nonadeca- or nonadec-	novemdec-
20	eicosa- or eicos-	viginti-
21	heneicosa- or heneicos-	
22	docosa- or docos-	
23	tricosa- or tricos-	
24	tetracosa- or tetracos-	
25	pentacosa- or pentacos-	
26	hexacosa- or hexacos-	
27	heptacosa- or heptacos-	
28	octacosa- or octacos-	
29	nonacosa- or nonacos-	
30	triaconta- or triacont-	triginti-
31	hentriaconta- or hentriacont-	
32	dotriaconta- or dotriacont-	
40	tetraconta- or tetracont-	quadragin-
50	pentaconta- or pentacont-	quinquagin-
60	hexaconta- or hexacont-	sexagin-

Table 24-7. Decimal Equivalents.

.0156— 1/64	.2656— 17/64	.5156— 33/64	.7656— 49/64
1/32— .0312	9/32— .2812	17/32— .5312	25/32— .7812
.0469— 3/64	.2969— 19/64	.5469— 35/64	.7969— 51/64
1/16— .0625	5/16— .3125	9/16— .5625	13/16— .8125
.0781— 5/64	.3281— 21/64	.5781— 37/64	.8281— 53/64
3/32— .0937	11/32— .3437	19/32— .5937	27/32— .8437
.1094— 7/64	.3594— 23/64	.6094— 39/64	.8594— 55/64
1/8— .1250	3/8— .3750	5/8— .6250	7/8— .8750
.1406— 9/64	.3906— 25/64	.6406— 41/64	.8906— 57/64
5/32— .1562	13/32— .4062	21/32— .6562	29/32— .9062
.1719— 11/64	.4219— 27/64	.6719— 43/64	.9219— 59/64
3/16— .1875	7/16— .4375	11/16— .6875	15/16— .9375
.2031— 13/64	.4531— 29/64	.7031— 45/64	.9531— 61/64
7/32— .2187	15/32— .4687	23/32— .7187	31/32— .9687
.2344— 15/16	.4844— 31/64	.7344— 47/64	.9844— 63/64
1/4— .2500	1/2— .5000	3/4— .7500	1—1.000

Table 24-8. Commonly Used Values in Electronic Formulas and Tables.

(1)	$\epsilon = 2.71828183$	(17)	$\log 8 = 0.903090$
(2)	$\dfrac{1}{\epsilon} = 0.36787944$	(18)	$\log 9 = 0.954243$
		(19)	$\log 10 = 1.000000$
(3)	$\log \epsilon = 0.43429448$	(20)	1 radian = $180°/\pi = 57°17'44.8''$
(4)	$\pi = 3.14159265$	(21)	$1° = \pi/180° = 0.01745329$ radian
(5)	$2\pi = 6.28318530$	(22)	$\sqrt{1} = 1.0000$
(6)	$\dfrac{1}{\pi} = 0.31830989$	(23)	$\sqrt{2} = 1.4142$
		(24)	$\sqrt{3} = 1.7321$
(7)	$\pi^2 = 9.8690440$	(25)	$\sqrt{4} = 2.0000$
(8)	$\sqrt{\pi} = 1.77245385$	(26)	$\sqrt{5} = 2.2361$
(9)	$\log \pi = 0.49714987$	(27)	$\sqrt{6} = 2.4495$
(10)	$\log 1 = 0.000000$	(28)	$\sqrt{7} = 2.6458$
(11)	$\log 2 = 0.301030$	(29)	$\sqrt{8} = 2.8284$
(12)	$\log 3 = 0.477121$	(30)	$\sqrt{9} = 3.0000$
(13)	$\log 4 = 0.602060$	(31)	$\sqrt{10} = 3.1623$
(14)	$\log 5 = 0.698970$	(32)	$j = \sqrt{-1}$
(15)	$\log 6 = 0.778151$		
(16)	$\log 7 = 0.845098$		

[logs are to base 10]

ABSOLUTE ERROR

The difference between a calculated value using a formula and the actual value of a component is known as absolute error. In the example given above, the absolute error would be $400 - 393 = 7$. Even here, the amount of absolute error is simply an approximation since the resistor could have an actual value, as indicated, of 380 to 420 ohms. Absolute error, then, is based on calculations, rather than measurements.

RELATIVE ERROR

If a table or the use of a formula supplies one answer, but the component you select has some other value, the ratio of the two is known as relative error.

Assume a table or your use of a formula calls for a resistor having a value of 87 ohms, but the actual value of the resistor as measured, is 90 ohms. $90 - 87 = 3$. If we now divide 3 by the exact value, 90 in this case, we will have $3/90 = 0.0333$. This can be converted to a percent by moving the decimal point two places to the right. The relative error in this example is 3.33 percent.

ROUNDING OFF NUMBERS

Quite often values of components used in electronics are

reasonable approximations. The greater the accuracy demanded of a component the more expensive it is, and so from a point of view of practical economics, there is no benefit in using a component having 1 percent precision when a tolerance of 20 percent will do.

Hence, while some of the tables that have been presented may indicate results having four or five figure significance, that much precision may not be required.

As an example, the number 3.1416 is often used in electronics problems. This number, having five significant figures, can be rounded off by reducing it to four significant figures, or even three.

The rule for rounding off is simple. If the last digit, the digit at the extreme right, has a value of less than 5, just discard it. However, if it has a value of 5 or more, discard it also, but make the digit to its immediate left larger by 1.

□ **Example:**

Round off the following numbers

712.1

This number has four significant figures. It can be rounded off by dropping the decimal. The rounded-off number then becomes 712.

921.19

This number has five significant figures. It can be rounded off by dropping the rightmost digit and then increasing the number to its left by 1.

921.2

Round off the following number to three significant figures.

612.3759

The rightmost digit has a value of 9. We discard it and increase the digit preceding it by 1.

612.376

The rightmost digit has a value greater than 5. We drop it and increase the digit preceding it by 1.

612.38

The number has five significant digits. Again, we can drop the last number and increase the digit preceding it by 1.

612.4

We can now drop the rightmost digit, and since its value is less than 5, no further number changes need be made. The final number, significant to three places is 612.

25

Mathematics

SQUARE VS. CIRCULAR AREA

It is sometimes necessary to convert from square to circular area, or from circular to square area. Square area is equal to circular area when the side of the square is equal to 0.88623 multiplied by the diameter of the circle. Conversely, the areas are equal when the diameter of the circle equals 1.12838 multiplied by the side of the square.

☐ **Example:**

A circle has a diameter of 2 inches. What is the length of any side of a square so that the area of the square is the same as the area of the circle?

The length of any side of the square is equal to the diameter of the circle multiplied by 0.88623. Therefore, 2×0.88623 equals 1.77246 inches. The area of the square is 1.77246×1.77246, or 3.1416 square inches. The area of the circle is πr^2. The radius equals one-half the diameter, $\frac{1}{2} \times 2$, or 1. The area of the circle is equal to πr^2 equals 3.1416×1^2 equals 3.1416. Thus, a circle with a diameter of 2 inches has the same area as a square, each of whose sides has a length of 1.77246 inches.

Table 25-1 supplies the circumference and area of circles having diameters ranging from 1/32 inch to 100 inches.

Table 25-1. Circumference and Area of Circles.

Diameter	Circumference	Area
1:32	0.09817	0.0007
1'16	0.19635	0.0030
3 32	0.29452	0.0069
3 16	0.58904	0.0276
7 '32	0.68722	0.0375
9 '32	0.88357	0.0621
11:32	1.07992	0.0928
13:32	1.27627	0.1296
9/16	1.76715	0.2485
19'32	1.86532	0.2768
21 32	2.06167	0.3382
11'16	2.15984	0.3712
23'32	2.25802	0.4057
25'32	2.45437	0.4793
27/32	2.65072	0.5591
29/32	2.84707	0.6450
1	3.142	0.7854
2	6.283	3.1416
3	9.425	7.0686
4	12.566	12.5664
5	15.708	19.6350
6	18.850	28.2743
7	21.991	38.4845
8	25.133	50.2655
9	28.274	63.6173
10	31.416	78.5398
11	34.558	95.0332
12	37.699	113.097
13	40.841	132.732
14	43.982	153.938
15	47.124	176.715
16	50.265	201.062
17	53.407	226.980
18	56.549	254.469
19	59.690	283.529
20	62.832	314.159

Table 25-1. Circumference and Area of Circles (cont'd).

Diameter	Circumference	Area
21	65.973	346.361
22	69.115	380.133
23	72.257	415.476
24	75.398	452.389
25	78.540	490.874
26	81.681	530.929
27	84.823	572.555
28	87.965	615.752
29	91.106	660.520
30	94.248	706.858
31	97.389	754.768
32	100.531	804.248
33	103.673	855.299
34	106.814	907.920
35	109.956	962.113
36	113.097	1,017.88
37	116.239	1,075.21
38	119.381	1,134.11
38	122.522	1,194.59
40	125.66	1,256.64
41	128.81	1,320.25
42	131.95	1,385.44
43	135.09	1,452.20
44	138.23	1,520.53
45	141.37	1,590.43
46	144.51	1,661.90
47	147.65	1,734.94
48	150.80	1,809.56
49	153.94	1,885.74
50	157.08	1,963.50
51	160.22	2,042.82
52	163.36	2,123.72
53	166.50	2,206.18
54	169.65	2,290.22
55	172.79	2,375.83

Table 25-1. Circumference and Area of Circles (cont'd).

Diameter	Circumference	Area
56	175.93	2,463.01
57	179.07	2,551.76
58	182.21	2,642.08
59	185.35	2,733.97
60	188.50	2,827.43
61	191.64	2,922.47
62	194.78	3,019.07
63	197.92	3,117.25
64	201.06	3,216.99
65	204.20	3,318.31
66	207.35	3,421.19
67	210.49	3,525.65
68	213.63	3,631.68
69	216.77	3,739.28
70	219.91	3,848.45
71	223.05	3,959.19
72	226.19	4,071.50
73	229.34	4,185.39
74	232.48	4,300.84
75	235.62	4,417.86
76	238.76	4,536.46
77	241.90	4,656.63
78	245.04	4,778.36
79	248.19	4,901.67
80	251.33	5,026.55
81	254.47	5,153.00
82	257.61	5,281.02
83	260.75	5,410.61
84	263.89	5,541.77
85	267.04	5,674.50
86	270.18	5,808.80
87	273.32	5,944.68
88	276.46	6,082.12
89	279.60	6,221.14
90	282.74	6,361.73

Table 25-1. Circumference and Area of Circles (cont'd).

Diameter	Circumference	Area
91	285.88	6,503.88
92	289.03	6.647.61
93	292.17	6.792.91
94	295.31	6,939.78
95	298.45	7,088.22
96	301.59	7,238.23
97	304.73	7,389.81
98	307.88	7,542.96
99	311.02	7,697.69
100	314.16	7,853.98

FUNCTIONS OF ANGLES

Table 25-2 supplies the trigonometric functions of the angle included between the base and the hypotenuse of a right-angle triangle. The sine (sin) of the angle is the ratio of the altitude to the hypotenuse; that is, the altitude divided by the length of the hypotenuse. The cosine (cos) is the base divided by the hypotenuse. The tangent (tan) is the altitude divided by the base while the cotangent (cot) is the base divided by the altitude. The sec (secant)

Table 25-2. Angles and Their Functions.

Angle A	sin A	cos A	tan A	cot A	sec A	csc A
$0°$	0	1	0	∞	1	∞
$30°$	$1/2$	$\sqrt{3}/2$	$\sqrt{3}/3$	$\sqrt{3}$	$2\sqrt{3}/3$	2
$45°$	$\sqrt{}/2$	$\sqrt{2}/2$	1	1	$\sqrt{2}$	$\sqrt{2}$
$60°$	$\sqrt{3}/2$	$1/2$	$\sqrt{3}$	$\sqrt{3}/3$	2	$2\sqrt{3}/3$
$90°$	1	0	∞	0	∞	1
$120°$	$\sqrt{3}/2$	$-1/2$	$\sqrt{3}$	$-\sqrt{3}/3$	-2	$2\sqrt{3}/3$
$180°$	0	-1	0	∞	-1	∞
$270°$	-1	0	∞	0	∞	-1
$360°$	0	1	0	∞	1	∞

Table 25-3. Polarity of Trigonometric Functions.

Quadrant	sin θ	cos θ	tan θ	cot θ	sec θ	csc θ
I	+	+	+	+	+	+
II	+	−	−	−	−	+
III	−	−	+	+	−	−
IV	−	+	−	−	+	−

is the ratio of the hypotenuse to the base while the csc (cosecant) is the ratio of the hypotenuse to the altitude.

POLARITY OF TRIGONOMETRIC FUNCTIONS

Whether a trigonometric function is plus or minus depends on the size of the angle. Table 25-3 supplies the polarity of the trigonometric functions in each of the four quadrants.

Table 25-4. The Ratio Tan X/R and Corresponding Values of Phase Angle, Θ.

Phase Angle (in degrees)	Ratio	Phase Angle (in degrees)	Ratio	Phase Angle (in degrees)	Ratio
0	0.0000	30	0.5774	60	1.7321
1	0.0175	31	0.6009	61	1.8040
2	0.0349	32	0.6249	62	1.8807
3	0.0524	33	0.6494	63	1.9626
4	0.0699	34	0.6745	64	2.0503
5	0.0875	35	0.7002	65	2.1445
6	0.1051	36	0.7265	66	2.2460
7	0.1228	37	0.7536	67	2.3559
8	0.1405	38	0.7813	68	2.4751
9	0.1584	39	0.8098	69	2.6051
10	0.1763	40	0.8391	70	2.7475
11	0.1944	41	0.8693	71	2.9042
12	0.2126	42	0.9004	72	3.0777
13	0.2309	43	0.9325	73	3.2709
14	0.2493	44	0.9657	74	3.4874
15	0.2679	45	1.0000	75	3.7321
16	0.2867	46	1.0355	76	4.0108
17	0.3057	47	1.0724	77	4.3315
18	0.3249	48	1.1106	78	4.7046
19	0.3443	49	1.1504	79	5.1446
20	0.3640	50	1.1918	80	5.6713
21	0.3839	51	1.2349	81	6.3138
22	0.4040	52	1.2799	82	7.1154
23	0.4245	53	1.3270	83	8.1443
24	0.4452	54	1.3764	84	9.5144
25	0.4663	55	1.4281	85	11.43
26	0.4877	56	1.4826	86	14.30
27	0.5095	57	1.5399	87	19.08
28	0.5317	58	1.6003	88	28.64
29	0.5543	59	1.6643	89	57.29

Table 25-5. Angular and Linear Dimensions of Oblique Triangles.

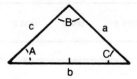

Parts Given	Parts to be Found	Formula
a b c	A	$\cos A = \dfrac{b^2 + c^2 - a^2}{2bc}$
a b A	B	$\sin B = \dfrac{b \times \sin A}{a}$
a b A	C	$C = 180° - (A + B)$
a A B	b	$b = \dfrac{a \times \sin B}{\sin A}$
a A B	c	$c = \dfrac{a \sin C}{\sin A} = \dfrac{a \sin (180° - A - B)}{\sin A}$
a b C	B	$B = 180° - (A + C)$

Table 25-6. Trigonometric Functions.

Degrees	Sine	Tangent	Cotangent	Cosine	
0	.0000	.0000	. . . . .	1.0000	90
1	.0175	.0175	57.29	.9998	89
2	.0349	.0349	28.636	.9994	88
3	.0523	.0524	19.081	.9986	87
4	.0698	.0699	14.301	.9976	86
5	.0872	.0875	11.430	.9962	85
6	.1045	.1051	9.5144	.9945	84
7	.1219	.1228	8.1443	.9925	83
8	.1392	.1405	7.1154	.9903	82
9	.1564	.1584	6.3138	.9877	81
10	.1736	.1763	5.6713	.9848	80
11	.1908	.1944	5.1446	.9816	79

Table 25-6. Trigonometric Functions (cont'd).

Degrees	Sine	Tangent	Cotangent	Cosine	
12	.2079	.2126	4.7046	.9781	78
13	.2250	.2309	4.3315	.9744	77
14	.2419	.2493	4.0108	.9703	76
15	.2588	.2679	3.7321	.9659	75
16	.2756	.2867	3.4874	.9613	74
17	.2924	.3057	3.2709	.9563	73
18	.3090	.3249	3.0777	.9511	72
19	.3256	.3443	2.9042	.9455	71
20	.3420	.3640	2.7475	.9397	70
21	.3584	.3839	2.6051	.9336	69
22	.3746	.4040	2.4751	.9272	68
23	.3907	.4245	2.3559	.9205	67
24	.4067	.4452	2.2460	.9135	66
25	.4226	.4663	2.1445	.9063	65
26	.4384	.4877	2.0503	.8988	64
27	.4540	.5095	1.9626	.8910	63
28	.4695	.5317	1.8807	.8829	62
29	.4848	.5543	1.8040	.8746	61
30	.5000	.5774	1.7321	.8660	60
31	.5150	.6009	1.6643	.8572	59
32	.5299	.6249	1.6003	.8480	58
33	.5446	.6494	1.5399	.8387	57
34	.5592	.6745	1.4826	.8290	56
35	.5736	.7002	1.4281	.8192	55
36	.5878	.7265	1.3764	.8090	54
37	.6018	.7536	1.3270	.7986	53
38	.6157	.7813	1.2799	.7880	52
39	.6293	.8098	1.2349	.7771	51
40	.6428	.8391	1.1918	.7660	50
41	.6561	.8693	1.1504	.7547	49
42	.6691	.9004	1.1106	.7431	48
43	.6820	.9325	1.0724	.7314	47
44	.6947	.9657	1.0355	.7193	46
45	.7071	1.0000	1.0000	.7071	45
	Cosine	Cotangent	Tangent	Sine	Degrees

PHASE ANGLE FOR THE TANGENT

The ratio of the altitude of a right-angle triangle to its base is called the tangent. This ratio and the corresponding phase angle are supplied in Table 25-4.

OBLIQUE TRIANGLES

The included angles or the dimensions of the sides of an oblique triangle can be determined with the formulas supplied in Table 25-5.

Table 25-7. Coefficients, Powers, Roots, and Logs of π.

$1\ \pi$ = 3.14159265		$6\ \pi$ = 18.84955592	
$2\ \pi$ = 6.28318531		$7\ \pi$ = 21.99114857	
$3\ \pi$ = 9.42477796		$8\ \pi$ = 25.13274122	
$4\ \pi$ = 12.56637061		$9\ \pi$ = 28.27433387	
$5\ \pi$ = 15.70796327		$10\ \pi$ = 31.41592652	

$1/\pi$ = 0.31830989		$4/\pi$ = 1.27323954
$1/2\pi$ = 0.15915494		π^2 = 9.86960440
$1/3\pi$ = 0.106103		$\sqrt{\pi}$ = 1.77245385
$\pi/2$ = 1.57079633		$1/\pi^2$ = 0.10132118
$\pi/3$ = 1.04719755		$1/\sqrt{\pi}$ = 0.56418958
$\pi/4$ = 0.78539816		$\sqrt{3/\pi}$ = 0.97720502
$\pi/5$ = 0.62831853		$\sqrt{4/\pi}$ = 1.12837917
$\pi/6$ = 0.52359878		$\sqrt[3]{\pi}$ = 1.46459189
$\pi/7$ = 0.44879895		$\log \pi$ = 0.4971499
$\pi/8$ = 0.39269908		$\log 2\pi$ = 0.7981799
$\pi/9$ = 0.34906585		$\log 4\pi$ = 1.0992099
$\pi/10$ = 0.314159265		$\log \pi^2$ = 0.9942997
$4\pi/3$ = 4.18879020		$\log\sqrt{\pi}$ = 0.248
$3/\pi$ = 0.954929666		

Table 25-8. Mathematical Symbols.

Symbol	Name or Meaning	Description
0, 1, 2, 3, 4, etc.	Whole numbers	Cardinal (Arabic) numbers used in mathematics. Arithmetic numbers.
No.; no.; #	Abbreviation	Symbol that may precede a number.
i; ii; iii; iv; v I; II; III; IV; V	Roman numerals	Have limited use—introductory pages of books; watches; building cornerstones, etc.
2-1/2; 3-1/8; 4-1/9	Mixed Numbers	Consist of a whole number and a proper fraction.
1/2; 3/4; 4/7	Proper fractions	Numerator is always smaller in value than the denominator.
5/2; 4/3; 9/5	Improper fractions	Numerator is always larger in value than the denominator.
+	Plus sign	Positive; plus; addition. May be used to indicate direction.
−	Minus sign	Negative; minus; subtraction. May be used to indicate direction.

376

Table 25-8. Mathematical Symbols (cont'd).

Symbol	Name or Meaning	Description
±	Plus-minus	Positive or negative; plus or minus; addition or subtraction.
∓	Minus-plus	Negative or positive; minus or plus; subtraction or addition.
a; x; y; z; etc.	Literal numbers	Letters used as a symbol for a number.
25/100; 4/10; 6589/1000	Decimal fractions	Fractions having 10, 100, 1000, etc. in the denominator.
1.657; 0.354 0.25	Decimal point	Point used in decimal fractions when denominator is omitted. Also known as the radix point. The number to the right of the decimal point is the decimal fraction. In some countries a comma is used in place of the decimal point. In the U.S. the decimal point is put at the bottom; in England at the center or higher. In other number systems, such as the binary, it is known as the binary point. In the hexadecimal system it is called the hexadecimal point. The term decimal point refers only to the use of the radix point in the decimal system.
∞	infinity	An indefinitely great number or amount.
×	Multiply by	Multiplication. A form of repeated addition.
x • y	Dot between letters means multiplication	x is multiplied by y.
(x) (y)	Adjacent parentheses means multiplication	x is multiplied by y.
x ÷ y	Straight line with dot above and below means division	x is divided by y.
x/y	Slant line indicates division	x is divided by y. Also, ratio of x to y.
$\frac{x}{y}$	Horizontal line means division	x is divided by y. Also, ratio of x to y.
x : y	Dots indicate division or ratio	x is divided by y; the ratio of x to y.
= or : :	Equals signs	Signs representing equality between two quantities; is equal to.
≡	Identity sign	Sign representing identity between quantities.

377

Table 25-8. Mathematical Symbols (cont'd).

Symbol	Name or Meaning	Description
$\cong$	Approximation sign	Quantities separated by this sign are approximately equal. Congruent.
$\neq$	inequality sign	Quantities separated by this sign are not equal.
$<$	inequality	Is less than.
$<<$	Inequality	Is much less than.
$>$	Inequality	Is greater than.
$>>$	Inequality	Is much greater than.
$\begin{array}{c}=\\>\end{array}$	Inequality or equality	Is equal to or is greater than.
$\begin{array}{c}<\\=\end{array}$	Inequality or equality	Is equal to or is less than.
$\therefore$	Therefore	Symbol used in geometry and logic.
$=$ or $\rightarrow$ or lim	Approaches as a limit	Variable approaches a constant, but never reaches it.
$\perp$	Perpendicular. Normal	Lines which form right angles.
$\parallel$	Parallel	Lines which are parallel to each other.
$\angle$	Angle. Also positive angle	The angle formed by two lines.
$\angle$s	Angles	Two or more angles.
$\overline{}$	Negative angle	An angle whose sine, tangent, cosecant and contangent are negative.
Δ	Change. Increase or decrease. Increment	Capital Greek letter delta. A similar symbol is used to represent triangles in trigonometry.
$\sqrt{}$	Square root	Quantity raised to the one-half power.
$n\sqrt{}$	nth root	Quantity raised to the nth power.
sin	Sine of included base angle	Ratio of altitude to hypotenuse in a right-angle triangle
cos	Cosine of included base angle	Ratio of base to altitude in a right-angle triangle.

Table 25-8. Mathematical Symbols (cont'd).

Symbol	Name or Meaning	Description
tan	Tangent of included base angle	Ratio of altitude to the base in a right-angle triangle.
cot	Cotangent of included base angle	Ratio of base to altitude in a right-angle triangle.
sec	Secant of included base angle	Ratio of hypotenuse to base in a right-angle triangle.
csc	Cosecant of included base angle	Ratio of hypotenuse to altitude in a right-angle triangle.

VALUES OF π

Various values of π appear regularly in calculations involving electronic problems. This constant is used for finding values of inductive reactance, capacitive reactance, frequency, and in problems involving resonance. Table 25-7 supplies coefficients, powers, roots, and logs of π.

MATHEMATICAL SYMBOLS

Electronics is a science which relies on mathematics for the solution of problems. Consequently, the symbols used in mathematics have become an integral part of textbooks having electronics as their subject matter. Table 25-8 lists those mathematical symbols most often found associated with electronics.

Index

381

Edited by Brint Rutherford